T0210455

Lecture Notes of the Institute for Computer Sciences, Social Informatics and Telecommunications Engineering 154

More information about this series at http://www.springer.com/series/8197

Prashant Pillai · Yim Fun Hu
Ifiok Otung · Giovanni Giambene (Eds.)

Wireless and Satellite Systems

7th International Conference, WiSATS 2015
Bradford, UK, July 6–7, 2015
Revised Selected Papers

 Springer

Editors
Prashant Pillai
Faculty of Engineering and Informatics
University of Bradford
Bradford
UK

Yim Fun Hu
Faculty of Engineering and Informatics
University of Bradford
Bradford
UK

Ifiok Otung
Faculty of Computing, Engineering
  and Science
University of South Wales
Cardiff
UK

Giovanni Giambene
University of Siena
Siena
Italy

ISSN 1867-8211   ISSN 1867-822X   (electronic)
Lecture Notes of the Institute for Computer Sciences, Social Informatics
and Telecommunications Engineering
ISBN 978-3-319-25478-4   ISBN 978-3-319-25479-1   (eBook)
DOI 10.1007/978-3-319-25479-1

Library of Congress Control Number: 2015951770

Springer Cham Heidelberg New York Dordrecht London

Printed on acid-free paper

Springer International Publishing AG Switzerland is part of Springer Science+Business Media
(www.springer.com)

Preface

The 7th International Conference on Wireless and Satellite Systems (WiSATS) was held at the Norcroft Centre at the University of Bradford, Bradford, UK, during July 6–7, 2015.

This conference was formerly known as Personal Satellite Services (PSATS) and catered mainly for research in the satellite domain. Recent technological advances in both wireless and satellite communications have made it possible to bring value-added services directly to the user by reducing the overall cost as well as addressing many technological challenges such as achieving seamless mobility, security, having miniaturized antennas and terminal sizes, and providing high data rate links. The synergy between satellite and terrestrial wireless networks provides immense opportunities for disseminating wideband multimedia services to a wide range of audiences over large numbers of geographically dispersed people. The services enabled by wireless and satellite systems not only cover the requirements of an ordinary citizen but also have applications in various areas such as defence, vehicular field, health, etc. Therefore the scope of the conference has been widened to cover wireless systems too. And WiSATS was born! The two-day event explored such techniques and provided a platform for discussion between academic and industrial researchers, practitioners, and students interested in future techniques relating to wireless and satellite communications, networking, technology, systems, and applications.

WiSATS 2015 boasted two outstanding keynote speakers. We were fortunate to have Prof. David Koilpillai, Dean (Planning) of the Indian Institute of Technology Madras, India, and Dr. Sastri Kota, President Sohum Consultants, USA, as our keynote speakers. The keynote speakers set the tone of the topic during the two days of the conference with one focusing on the next-generation 5G wireless communications and the other focusing on the emerging areas of satellite communications.

WiSATS 2015 included two workshops along with the main conference. The first workshop, Communication Applications in Smart Grid (CASG 2015), focused on the merging area of using communication technology within the electricity power grid for smart monitoring and control. It had six papers and two invited talks. The second workshop, Advanced Next-Generation Broadband Satellite Systems (BSS 2015), focused on the use of satellite systems for providing next-generation broadband services. This workshop had six accepted papers from eminent experts in the area.

WiSATS 2015 included three technical sessions consisting of 18 high-quality regular papers and one special session on network coding for satellite communications that had five papers. WiSATS 2015 also included a panel discussion session on "Future Maritime VHF Digital Exchange System (VDES): A Combined Terrestrial and Satcom System." This panel brought together experts involved in the VDES standardization activities worldwide for presentations and discussion on the current situation, the challenges, sharing options for the VDE terrestrial and satellite components, the user and regulative needs, and viewpoints on this new digital maritime communications.

The conference had close to 60 participants both from industrial and academic sectors from various parts of the world including India, France, Germany, Greece, UK, The Netherlands, and Italy. The conference provided tea and coffee breaks as well as lunch at the Norcroft Centre for all the participants throughout the conference. A gala dinner event was also organized on the first day of the conference at the National Media Museum in Bradford. The quality of the venue, services provided by the center's staff, and especially the quality of the food and drinks were all highly spoken about during the conference by the participants and the Organizing Committee.

Last but not least, we would like to thank the Organizing Committee members, session chairs, Technical Program Committee members, all the authors and speakers for their technical contributions, and the attendees for their participation. Also on behalf of the Organizing Committee and the Steering Committee of WiSATS, we would like to thank our sponsors, the University of Bradford, CREATE-NET, and EAI for their generous financial support and their extended support in making this event a successful one.

September 2015

Prashant Pillai
Yim Fun Hu
Ifiok Otung
Giovanni Giambene

Organization

WiSATS 2015 was organized by the Faculty of Engineering and Informatics of the University of Bradford in cooperation with the European Alliance for Innovation (EAI), the Institute for Computer Sciences, Social Informatics and Telecommunications Engineering (ICST), and CREATE-NET.

General Chairs

Prashant Pillai	University of Bradford, UK
Yim Fun Hu	University of Bradford, UK

Technical Program Chair

Ifiok Otung	University of South Wales, UK

Steering Committee

Imrich Chlamtac	Create-Net, Italy
Kandeepan Sithamparanathan	RMIT, Australia
Agnelli Stefano	ESOA/Eutelsat, France
Mario Marchese	University of Genoa, Italy

Publications Chair

Giovanni Giambene	University of Siena, Italy

Workshops Chair

Paul Mitchell	University of York, UK

Sponsorship and Exhibit Chair

Rajeev Shorey	TCS Innovations Labs, India
Arjuna Sathiaseelan	University of Cambridge, UK

Publicity Chair

Ram Krishnan	University of Texas at San Antonio, USA

Local Organizing Chairs

Muhammad Ali   University of Bradford, UK
Kai Xu   University of Bradford, UK

Conference Coordinator

Petra Jansen   EAI, Italy

Website Chair

Mohammed Amir   University of Bradford, UK

Special Session Organizers

Workshop on Communication Applications in Smart Grid (CASG)

Organized by:

Haile-Selassie Rajamani   University of Bradford, UK
K. Shanti Swarup   Indian Institute of Technology - Madras, India

Workshop on Advanced Next-Generation Broadband Satellite Systems (BSS)

Organized by:

Paul Thompson   University of Surrey, UK

Special Session on Network Coding for Satellite Communication Systems

Organized by:

Jason Cloud   Massachusetts Institute of Technology, USA

Sponsoring Institutions

Faculty of Engineering and Informatics of the University of Bradford
European Alliance for Innovation (EAI)
Institute for Computer Sciences, Social Informatics and Telecommunications
Engineering (ICST)
CREATE-NET

Contents

Workshop on Communication Applications in Smart Grid (CASG) 1

Residential Energy Consumption Scheduling Techniques Under Smart Grid Environment

J. Santosh Kumar and K. Shanti Swarup$^{(\boxtimes)}$

Department of Electrical Engineering,
Indian Institute of Technology Madras, Chennai 600036, India
santoshlots@gmail.com, swarup@ee.iitm.ac.in

Abstract. In recent years the load demand by residential consumers are rapidly increasing due to the usage of many electric appliances in daily needs. Load demand during peak hours is becoming increasingly larger than off-peak hours, which is the major reason for inefficiency in generation capacity. Introduction of smart grid technology in Demand Side Management programs provides an alternative to installation of new generation units. Consumers can play a major role in reducing their energy consumption by communicating with utilities so that they can minimize their energy costs and get incentives, which also helps utilities in many ways. Smart grid technologies provide opportunities to employ different pricing schemes which also help in increasing the efficiency of appliance scheduling techniques. Optimal energy consumption scheduling reduces the peak load demand in peak hour. Peak average ratio (PAR) also minimizes the energy consumption cost. In this paper, we observes different energy consumption scheduling techniques that schedule the house hold appliances in real-time to achieve minimum energy consumption cost and to reduce peak load demand in peak hours to shape the peak load demand. Formulation and Solution methodology of residential energy consumption scheduling is presented with simulation results illustrating the working of the model.

Keywords: Smart grid · Demand side management · Energy consumption scheduling · Peak load demand · Energy pricing

1 Introduction

In recent years energy demand by the residential consumers are rapidly increasing due to the usage of many electrical appliances, the utilities are unable to satisfy users requirement because of this, there is power shortage in every place. Especially in small towns in India there is 6-8 h power cut daily during peak hours, due to this the efficiency of power utilization is decreasing. Utilities facing many problems, so there is huge need for the installation of new power generation units which needs lot of investment though it cannot solve the problem which is to increase efficiency. Due to the advancement of smart grid technology in Demand side management programs, it has become an alternative for the installation of new power plants. Demand side management programs are implemented by utilities control energy consumption of the consumers, utilities motivate consumers to participate in DSM programs and they offer incentives [1]. DSM programs provides energy efficiency programs, demand response

© Institute for Computer Sciences, Social Informatics and Telecommunications Engineering 2015
P. Pillai et al. (Eds.): WiSATS 2015, LNICST 154, pp. 3–17, 2015.
DOI: 10.1007/978-3-319-25479-1_1

programs, fuel substitution programs, and Residential or commercial load management programs. Residential load management programs consist of mainly 2 objectives reducing consumption and shifting consumption. Utilities motivate people to reduce their power consumption by using smart homes. Energy production costs vary with time according to the generation capacity, but utilities charge average price to the consumers according to their calculation of profits. People consume energy without depending on the time and sometimes there will be huge overlap in the usage of loads and peak load increases drastically, it causes power shortage or increases energy consumption cost. DSM programs helps users to reduce their energy cost by changing their energy consumption patterns or they will get incentives from utilities for reducing or shifting their usage when system reliability is jeopardized.

2   Demand Side Management

Demand side management (DSM) refers to the programs implemented by the utilities to control the energy consumption of the consumers to increase the efficiency in usage of available power without installing new generation plants [1]. Residential DSM programs mainly concentrates on reducing or shifting of energy consumption, utilities motivates users to reduce energy consumption by using energy efficient buildings like smart homes, there is a need to shift the house hold appliances to off peak hours from peak hours to reduce PAR(Peak Average Ratio) which improves the reliability of power system network. Shift-able appliances are two types, namely, (i) Power shift-able loads and (ii) Time shift-able loads. For power shift-able appliances we can vary the energy supply within standby power limits according to the availability of generation, but for Time shift-able appliances we cannot vary the power supply since they will have their own power consumption patterns and we can only shift the time of usage according to the user's preference. Load shifting is becoming very important as the usage of loads with high power requirement is increasing and they will affect demand curve very much, these loads will double the PAR sometimes, so we need to schedule these loads carefully. Demand response is a part of DSM strategies which is defined as "changes in electricity usage by end-use customer's form their normal consumption pattern in response to changes in the price of electricity over time, or to incentive payments designed to induce lower electricity use at times of high wholesale market prices or when system reliability is jeopardized". Figure 1 shows the concept of demand response under smart grid environment. Demand response is not a new word to the world but due to the advancement in Smart grid technology these DR options are becoming very popular because of the participation of the people is increasing. There are different types of DSM strategies some of them are Direct Load Control [3], Real Time pricing [4], Time-of-use pricing, Critical peak pricing, Emergency Demand Response, Interruptible load etc. Since the goal of DSM programs coincides with target of smart grid which is to build a secure, Reliable, economical, clean and efficient power system. If these are combined simultaneously, it is bound to greatly promote the development of power industry.

In further sections we will discuss about energy consumption scheduling techniques for Residential Network with multiple users and one energy source [5, 6].

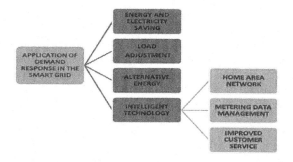

Fig. 1. Demand response under the smart grid environment

Consider a residential network having single energy source and multiple users which is connected to the electric grid. Each user contains ECS (Energy Consumption Scheduler) units in their smart meters which are connected to the power line coming from Energy source. These smart meters are also connected to LAN (Local Area Network) to communicate with utility and between users.

3 Energy Consumption Scheduling of a Residential Network

Consider a residential network having single energy source and multiple users which is connected to the electric grid.

Figure 2 shows a residential network having three appliances. Each user contains ECS (Energy Consumption Scheduler) units in their smart meters which are connected to the power line coming from Energy source. These smart meters are also connected to LAN (Local Area Network) to communicate with utility and between users.

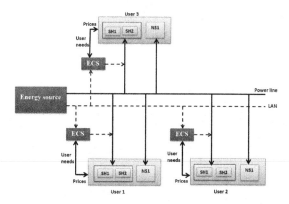

Fig. 2. Energy consumption scheduling model with 3 users each having 3 appliances

In our model we considered 3 users, each having set of appliances differentiated as shift-able and Non-shift-able loads. ECS units can alter only the power supplied to the shift-able loads according to the user needs. Users provide their Energy requirement, time preferences and usage of appliances to the utility through ECS and in return they will provide energy consumption cost. Each user n $\in$ N, (N is set of users and the number of users is N) have set of household appliances denoted with A_n. For each appliance $a \in A_n$ we will have an energy consumption scheduling vector

$$X_{n,a} = \left[x_{n,a}^1, x_{n,a}^2, \cdots x_{n,a}^H \right]$$

where H is total no of hours

The aim of ECS units is to get optimal solution for Energy Consumption scheduling vector according to the user's preferences and constraints.

$$PAR = \frac{DailyPeakLoad}{AveragePeakLoad} = \frac{L_{peak}}{L_{avg}}$$

$$PAR = \frac{\max_{h \in H}^{L_h}}{\frac{\sum_{h \in H}^{L_h}}{H}} = \frac{H \max_{h \in H} \left(\sum_{n \in N} \sum_{a \in A_n} x_{n,a}^h \right)}{\sum_{n \in N} \sum_{a \in A_n} E_{n,a}} \quad (1)$$

where
$x_{n,a}^h$ - Energy consumed by an appliance at an hour

$$l_n^h = \sum_{a \in A_n} x_{n,a}^h \quad (2)$$

$$l_h = \sum_{n \in N_n} l_n^h \quad (3)$$

$$L_{peak} = \max_{h \in H} L_h \quad (4)$$

$$L_{avg} = \frac{\sum_{h \in H} L_h}{H} \quad (5)$$

l_n^h - Total load at hour h
l_n^h - Total load at hour h
l_h - Daily load for user n (for all hours)
L_h - Total load of all users at each hour of the day
L_{peak} - Daily peak load
L_{avg} - Average load
$E_{n,a}$ - Predetermined total daily energy consumption

4 Energy Cost Model

In this model users can directly reduce their energy consumption cost so that more people participate and utilities can get benefits from it. Using concept of smart pricing utilities can offer different types of energy cost programs [7, 8]. We considered a simple quadratic cost function which represents actual energy cost as for thermal generators [9]. Let us denote the cost of energy consumption for an hour as $C_h(L_h)$. Energy cost varies to the energy consumed by the user. During peak hours (generally day time) the energy cost will be high compared to off-peak (generally night hours) hours.

Energy cost increases as energy consumption increases

$$C_h(L_h^1) < C_h(L_h^2) \text{ for all } L_h^1 < L_h^2$$

Energy cost function is

$$C_h(L_h) = a_h L_h^2 + b_h L_h + c_h \tag{6}$$

where a_h, b_h and $c_h \geq 0$
From (2) to (5)

$$C_h(L_h) + a_h \left(\sum_{n \in N} \sum_{a \in A_n} x_{n,a}^h \right)^2 + b_h \left(\sum_{n \in N} \sum_{a \in A_n} x_{n,a}^h \right) + c_h \tag{7}$$

5 Parameter Minimization

In parameter Optimization, the objective problem is formulated as follows

Objective Function: PAR Minimization

$$\min_{X_n \forall n \in N} \frac{H \max_{h \in H} \left(\sum_{n \in N} \sum_{a \in Ax} x_{n,a}^h \right)}{\sum_{n \in N} \sum_{a \in A} E_{n,a}} \tag{8}$$

Constraints:

$$a_{n,a} < \beta_{n,a} \tag{9}$$

For each appliance user needs to select the starting time $(\alpha_{n,a})$ and ending time $(\beta_{n,a})$ for the operation, Power supply should be between the time interval $(\alpha_{n,a})$ and $(\beta_{n,a})$, starting time of the supply should be less than ending time.

$$\sum_{h=a_{n,a}}^{B_{n,a}} x_h^a = E_{n,a} \tag{10}$$

For each user n and appliance a there will be a predetermined energy $(E_{n,a})$ which is given by user, the total daily energy consumption for an appliance between the time interval $(\alpha_{n,a})$ and $(\beta_{n,a})$ should be equal to $(E_{n,a})$.

$$x_{n,a}^h = 0 \tag{11}$$

Energy consumed by an appliance at an hour other than time interval between $(\alpha_{n,a})$ and $(\beta_{n,a})$ is zero.

$$\beta_{n,a} - a_{n,a} \geq 0 \tag{12}$$

Time interval needed to finish the operation

For any appliance the time interval chosen by the user should be greater than the time interval needed to finish the operation

$$\sum_{h \in H} L_h = \sum_{n \in N} \sum_{a \in A_n} E_{n,a} \tag{13}$$

The total energy consumed in all hours should be equal to the energy consumed by all users' appliances

$$\gamma_{n,a}^{min} \leq x_{n,a}^h \leq \gamma_{n,a}^{max} \tag{14}$$

Energy consumed by an appliance at any hour should be between minimum standby power and maximum standby power. Standby power is a power which is consumed by an appliance in switched off mode or standby mode. In our objective function (8) the denominator (predetermined energy $E_{n,a}$) is fixed for optimization problem, so we can neglect it and optimize the simplified one, so minimization of peak load satisfies the objective

$$\min_{X_n \forall n \in N} \left\{ \max_{h \in H} \left(\sum_{n \in N} \sum_{a \in A} x_{n,a}^h \right) \right\} \tag{15}$$

The above objective function (15) contains two functions (min and max) it is difficult to optimize so it can be further simplified as

$$\min_{X_n \forall n \in N} L$$

Subject to $L \geq \sum_{n \in N} \sum_{a \in A} x_{n,a}^h$

and the above constraints (9) to (14)

Since our objective function and constraints are linear implies it is linear program and it can be solved using simplex method or interior point method. For solving this problem MATLAB optimization tool box has been for linear programming method.

6   Cost Minimization

In cost Optimization, the problem is formulated as following

Objective Function: Energy consumption cost minimization

$$\min_{X_n \forall n \in N} \left\{ \sum_{h=1} C_h \left(\sum_{n \in N} \sum_{a \in A} x_{n,a}^h \right) \right\}$$

Constraints: Eqs. (9) to (14)

Since our objective is convex function it can be solved using convex programming techniques. For solving this problem we have chosen MATLAB optimization tool box [10, 11] which has Quadratic programming method in it.

Case Study -1: We considered 3 users each having 3 appliances model with 2 shift-able and 1 Non-shift-able loads. ECS cannot schedule Non-shift-able appliances; some of them are Heater, Hob, micro oven, Refrigerator, freezer, electric stove and lighting for some standard bulbs etc. Tables 1 and 2 Provide the data for energy consumption before scheduling and users preference accordingly. Shift-able appliances have soft energy consumption scheduling constraints some of them are Water boiler, PHEV, Washing machine, Dish washer, clothes dryer etc. Before participating in any program users consume their energy randomly without scheduling, we took most possible worst case as data.

Table 1. Data for Energy consumption before scheduling

Users	Appliance type	Appliance	Total energy consumption (kWh)	Energy usage at different times (kWh)
User 1	NS	Refrigerator	1.32	1 to 24 -0.055kWh for each hour
	SH	Washing Mc	1.49	9 to 12-0.5kWh for each hour
	SH	PHEV	9.9	10 to 13-3.3 for each hour
User 2	NS	Refrigerator	1.89	1 to 24- 0.079kWh for each hour
	SH	Dish Washer	1.44	10th and 20th hour -0.72kWh
	SH	PHEV	9.9	11 to 14- 3.3 for each hour
User 3	NS	Lighting	1	1 to 24-0.042for each hour
	SH	Washing Mc	1.49	16 to 19 -0.5kWh for each hour
	SH	PHEV	9.9	13 to 16- 3.3 for each hour

Every user should give their predetermined energy, time preference, standby power of appliance, type of appliance whether it is shift-able or Non-shift-able. Data is provided by the users to the utility at the beginning of the day for the schedule of that particular day. For all Non-shift-able appliances 24 h (whole day) given as the time preference, users want these appliances to be ON continuously, standby power is taken as $\left(\frac{\text{predetermined Energy}}{\text{time interval}}\right)$ i.e. user 1's Non-shift-able appliance standby power = $\left(\frac{1.32}{24}\right)$ = 0.055. Standby power for other appliances is taken according to their power usage Time preference and predetermined energy is taken according to the practical data available. In our quadratic cost function (7) we assumed a_h = Rs. 2, b_h = Rs. 0.02 during dynamic hours from 8:00 AM to 12:00. a_h = Rs. 1.8, b_h = Rs. 0.018 during night time from 12:00 to 8:00 AM. It can be observed that Peak load is very high because users consumed their energy randomly and there is huge overlap in the usage of heavy loads. Figures 3 and 4 shows the energy consumption of users

Table 2. Data for scheduling according to user's preference

Users	Appliance type	Appliance	Predetermined energy (kWh)	Time preference	Standby power (kw)
User 1	NS	Refrigerator	1.32	1 to 24	0.055
	SH	Washing Mc	1.49	8 to 12,18to24	0.0005
	SH	PHEV	9.9	1 to 13	
User 2	NS	Refrigerator	1.89	1 to 24	0.07875
	SH	Dish Washer	1.44	10 to 15	
	SH	PHEV	9.9	11 to 24	
User 3	NS	Lighting	1	1 to 24	0.04166
	SH	Washing Mc	1.49	6 to10,16to24	0.0005
	SH	PHEV	9.9	10 to 19	

before and after scheduling along with values of reduces values of peak, average and peak average ratios.

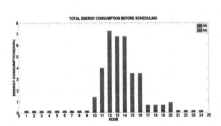

Fig. 3. Energy consumption of users before scheduling Peak = 7.27 Average = 1.59 PAR = 4.55

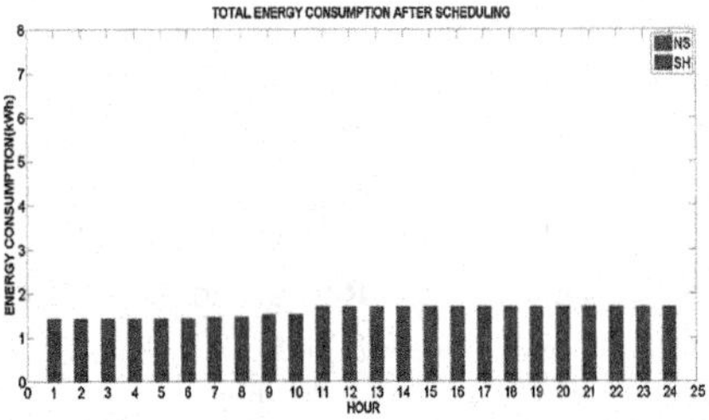

Fig. 4. Energy consumption of users after scheduling Peak = 1.69, Average = 1.59, PAR = 1.06

It can be observed that the energy consumption of users after scheduling, the peak load is reduced to 1.69 and PAR to 1.06, since the usage of heavy load appliances is scheduled without overlapping the peak load and PAR are highly improved. There is no change in the usage of Non shiftable appliances. In our case PHEV is the heavy load which shapes the load curve, it is distributed in the three zones of the day (considering user's preference) so that they won't overlap to decrease the peak load.

Figure 5 shows the Energy schedule consumption between users. In the Energy scheduling of individual users before and after ECS units, it can be observed that there is no change usage of Non shiftable appliances, for every user after scheduling PAR and Peak load is reduced Total energy consumption cost before and after scheduling. Cost is reduced from Rs. 221.59 to Rs. 53.17, we can observe that before scheduling most of the energy consumed in peak hours but after scheduling part of it is shifted to off peak hours.

Figures 6 and 7 shows the Total energy cost before and after scheduling. In the Total Energy consumption before scheduling with cost minimization, it can be observed that the PAR is reduced from 7.27 to 1.966.

Figure 8 shows the Energy consumption after scheduling with cost minimization, where a uniform energy consumption pattern can be observed.

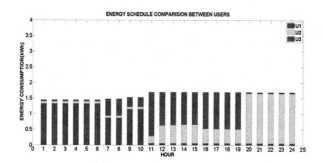

Fig. 5. Energy schedule comparison between users

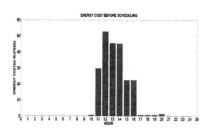

Fig. 6. Total energy cost before scheduling

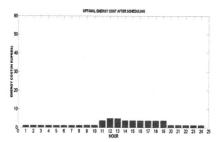

Fig. 7. Optimal energy cost after scheduling

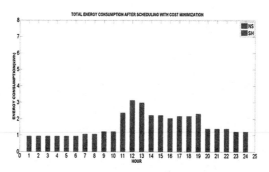

Fig. 8. Energy consumption after scheduling with cost minimization

Table 3 shows the Energy consumption cost before and after scheduling. It can be observed that there is significant reduction in the cost of each user after energy scheduling.

Table 3. Energy consumption cost before and after scheduling.

User	Energy consumption cost	
	Before (Rs)	After (Rs)
1	82.70	14.95
2	70.05	16.15
3	68.64	22.47
Total	**221.59**	**53.17**

Mixed Integer Programming Method was employed for energy consumption scheduling of Residential network. This technique is advancement of previous one which is not suitable for most appliances in practice, so we took single user to develop the model. Here we differentiated Shift-able loads as power shift-able and Time shift-able. For some appliances we cannot change its own power consumption pattern we can only shift its usage from one time to another according to our requirements. For example In previous technique we considered washing machine as power shift-able load and scheduled by reducing the power at different time slots, but in practice we cannot control the power we can only shift the usage of the appliance and after that it will run according to it power consumption pattern. For Time shift-able appliances smart meter (ECS) controls the switch and provide power according to its own consumption schedule during the available time slots considering user's preference. So we considered 3 types of loads in this model.

7 Single User Residential Network Model

Consider a single user connected to the power network for energy supply and also to LAN (Local Area Network) for information exchanges between other users and utility. In the above diagram power will be delivered from power line to the user, in the middle Energy Consumption Scheduler controls Time shift-able and power shift-able according to the user's preference. But cannot schedule or control non-shift-able appliances it only takes requirement from the user. In our proposed technique, the schedule of the energy consumption of user is done by taking Peak Average Ratio minimization as an objective (Fig. 9).

As explained earlier in our first technique we can simplify our objective function by neglecting Average load (since it is constant), so it is enough to minimize Peak load.

8 Problem Formulation for Peak Load Minimization

The problem formulation for peak load minimization cane be written as follows
Minimize of Peak load L

$$\min_{L=x_{a,h}h} L$$

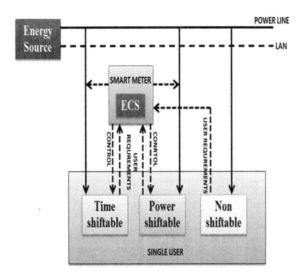

Fig. 9. Energy schedule model for single user in Residential Network

Subject to the Constraints

$$\sum_{a \in A} x_{a,h} \leq L, \forall h \in H \tag{16}$$

The total energy consumed by all appliances at an hour should be less than peak load.

$$x_{n,a}^{h} \geq 0 \tag{17}$$

The energy consumed by an appliance at an hour should be non-negative.

$$x_{n,a}^{h} \geq N_a, \forall a \in N, \forall h \in H \tag{18}$$

The energy consumed by every Non-shift-able appliance for an hour should be less than its fixed power requirement per hour (N_a). So the total predetermined energy for a Non-shift-able appliance is calculated as follows, EN,a = (no of working hours) x(N_a), N is set of Non-shift-able appliances.

$$\gamma_{n,a}^{min} \leq x_{n,a}^{h} \leq \gamma_{n,a}^{max} \forall a \in P \tag{19}$$

∀ th ∈ during its preferred working hours

Energy consumed by a power shift-able appliance at any hour should be between minimum standby power and maximum standby power. Standby power is a power which is consumed by an appliance in it's switched off mode or standby mode.

P is set of power shift-able appliances.

Table 4. Preferences given by the users for scheduling of their loads for 24 h

Appliance	Type	Hourly consumption (kWh)			Time preference			Energy requirement (kWh)	
		U1	U2	U3	U1	U2	U3	U1	U2
Fridge	NS	0.12	0.12	0.14	24 h	24 h	24 h	2.88	2.88
Heater	NS	1	1	1	9 pm to10 pm 3am to 5am	8 pm to 10 pm 3am to 6am	7 pm to 9 pm 4am to 7am	3	5
Water boiler	PS	0–1.5	0–1.5	0–1.5	24 h	8am to 8 pm	24 h	3	2
PHEV	PS	0.1–3	0.1–3	0.1–3	8pmto 8am	10 pm to8am	10 pm to 8am	5	4
Washing mc	TS	1kWh-1st hr 0.5kWh-2nd hr	1kWh -1st hr 0.7kWh-2nd hr	1.2kWh-1st hr 0.8kWh -2nd hr 0.6kWh -3rd hr	2 h	2 h	3 h	1.5	1.7
Dish washer	TS	0.8kWh-1 h	0.8kWh-1 h 0.6kWh-1 h	0.8kWh - 1 h	1 h	1 h	1 h	0.8	1.4

$$\sum\nolimits_{h=a}^{\beta}{}_{a \in A} x_{h,a} = E_a, \forall a \in P, N \tag{20}$$

For power shift-able and Non-shift-able appliances there will be a predetermined energy E_a which is given by user, the total daily energy consumption for an appliance should be in between time interval α and β should be equal to E_a. In previous technique above constraint is commonly taken for shift-able appliances, but it is not applicable for time shift-able appliances. We cannot supply flexible power between the time intervals provided by consumer. We can only shift the time of usage without changing its own power consumption pattern

$$s_a = \left[s_{a,1}, s_{a,2}, s_{a,3} \cdots s_{a,24}\right]^T; \quad s_a \in \{0,1\}^{24}.$$
$$0 \le s_a \le 1; \quad 1^T s_a = 1 \tag{21}$$
$$X = P_a^T s_a \; \forall a \in T$$

$P_a = \left[p_{a,1}, p_{a,2}, p_{a,3} \cdots p_{a,24}\right]^T$ is fixed power consumption pattern

The constraint mentioned above cannot be formulated using Linear Programming [11] method as in previous technique. Since our objective function and constraints consists of both integer and non-integer variables Mixed Integer Linear Programming method is suitable for solving this problem.

Table 4 shows the preferences by three users for 24 h load scheduling. Mixed Integer Linear Programming method has been chosen for solving this problem. We have 577 variables including objective function (peak load), 24 linear inequality constraints, 162 linear equality constraints with boundary conditions.

9 Conclusion

In this work, the authors have proposed and implemented Residential Energy consumption scheduling techniques, to reduce the energy consumption costs, minimize the Peak Average Ration (PAR) as well as Peak load by shifting the heavy loads form peak hours to off peak hours. All the techniques are implemented with case studies using MATLAB Optimization Toolbox with different programming methods. This work was extended to multiple energy sources, industrial load for commercial usage. Communication between users can be introduced through smart grid advanced technology so that users can involve more into these programs. Important conclusions of the techniques are given below

No	Technique	Objective	Optimization Method	Conclusion
1	Energy consumption scheduling of a Residential network	PAR	Linear Programming	Peak load is reduced from 7.27kWh to 1.69Kwh PAR is reduced from 4.55 to 1.69 Heavy load(PHEV) is scheduled in 3 different zones of the day to reduce peak load
2	Energy consumption scheduling of a Residential network	Energy Consumption cost	Quadratic Programming	Total Energy consumption cost is reduced from Rs 221.59 to Rs 53.57 PAR is also reduced from 4.55 to 1.96
3	Mixed Integer Linear Programming technique Energy consumption scheduling	Peak load	Mixed Integer Linear Programming	Hourly peak load is reduced to 3.68 and PAR to 1.73 More suitable for practical appliances Time shift-able and low energy required appliances are scheduled in off-peak hours

References

1. Masters, G.M.: Renewable and Efficient Electric Power Systems, Hoboken. Wiley, Hoboken (2004)
2. Assessment of Demand Response & Advanced Meter in Staff Report, FERC (2011)
3. Kondoh, J.: Direct load control for wind power integration. In: IEEE Power and Energy Society General Meeting, pp. 1–8 (2011)
4. Zhang, Q., Wang, X., Fu, M.: Optimal implementation strategies for critical peak pricing. In: 6th International Conference on the European Energy Market (2009)
5. Salinas, S., Li, M., Li, P.: Multi-objective optimal energy consumption scheduling in smart grids. Smart Grid. IEEE Trans. Smart Grid 4(1), 341–348 (2013)
6. Mohsenian-Rad, V., Wong, J., Jateskevich, R.: Schober and A. Leon-Garcia, "Autonomous Demand-Side Management Based on Game-Theoretic Energy Consumption Scheduling for the Future Smart Grid". IEEE transactions on Smart Grid 1(3), 320–331 (2010)
7. Wood, A.J., Wollenberg, B.F.: Power Generation, Operation, and Control. Wiley-Interscience, New York (1996)

8. Zhu, Z., Tang, J., Lambotharan, S., Chin, W.H., Fan, Z.: An integer linear programming and game theory based optimization for demand-side management in smart grid. In: Proceedings IEEE GLOBECOM Workshops (GC Workshop), Loughborough, UK (2011)
9. Strbac, G.: Demand side management: Benefits and challenges. Energy Policy **36**(12), 4419–4426 (2008)
10. Abebe Geletu, Dr.: Solving Optimization Problems using the Matlab Optimization Toolbox - a Tutorial (2007)
11. Venkataraman, P.: Applied Optimization With MATLAB Programming. Rochester Institute of Technology, New York (2005)

Residential Demand Response Algorithms: State-of-the-Art, Key Issues and Challenges

Rajasekhar Batchu and Naran M. Pindoriya[✉]

Electrical Engineering, Indian Institute of Technology Gandhinagar,
Ahmedabad 382424, Gujarat, India
{batchu.rajasekhar,naran}@iitgn.ac.in

Abstract. Demand Response (DR) in residential sector is considered to play a key role in the smart grid framework because of its disproportionate amount of peak energy use and massive integration of distributed local renewable energy generation in conjunction with battery storage devices. In this paper, first a quick overview about residential demand response and its optimization model at single home and multi-home level is presented. Then a description of state-of-the-art optimization methods addressing different aspects of residential DR algorithms such as optimization of schedules for local RE based generation dispatch, battery storage utilization and appliances consumption by considering both cost and comfort, parameters uncertainty modeling, physical based dynamic consumption modeling of various appliances power consumption at single home and aggregated homes/community level are presented. The key issues along with their challenges and opportunities for residential demand response implementation and further research directions are highlighted.

Keywords: Demand response · Distribution grid · Home energy management system · Price-based programs · Renewable generation · Battery storage · Load scheduling

1 Introduction

Deployment of smart grid technologies and integration of information and communication infrastructure in the existing electricity grid has brought immense automation, control and visualization in the grid. With the advent of ubiquitous data networks and advanced metering infrastructure (AMI) that enables bi-directional communication, the demand side energy management (DSM) has now attained an intelligent outlook in the smart grid framework. DSM is the cost-effective tool to intelligently control the customers' load demand; in general, it focuses on load shaping i.e. modifying the energy consumption pattern of users over time and at the same time improves service quality and customer satisfaction. Major thrust areas of DSM are: (1) demand response (DR) - an approach to reduce customers' consumption by shifting, shaving and shaping the electricity load in response to a peak energy signal from the power utility and (2) energy efficiency and energy conservation programs. Appropriate load-shifting is foreseen to be even more crucial with increasing penetration of distributed generation like roof-top solar photovoltaic (PV) in conjunction with/without battery energy storage

© Institute for Computer Sciences, Social Informatics and Telecommunications Engineering 2015
P. Pillai et al. (Eds.): WiSATS 2015, LNICST 154, pp. 18–32, 2015.
DOI: 10.1007/978-3-319-25479-1_2

system (BESS), plug-in hybrid electric vehicles (PHEV), power intensive HVAC loads and usage of intelligent appliances, making the customer load profile more stochastic.

Therefore, intelligent DSM algorithm to reduce peak load and manage the satisfactory quality of power has gained a lot of attention at the customer segment in the distribution network. Though, very limited papers [1, 2] that summarize the DSM algorithms and relevant technical challenges broadly in smart grid available in the literature, however, it is found that few recent residential DR algorithms covering state-of-the-art, key issues and research challenges are not properly highlighted. Therefore, this paper solely focuses on DR algorithms for residential customers at individual and multi-user levels (community segment) and highlights their benefits and challenges in effective design and implementation. The rest of the paper is organized as follows. Section 2 provides a background of the different scenarios of the problem. Section 3 provides a quick glance of optimization model used to describe the problem. Section 4 reviews important and recent DSM mechanisms individual and group of cooperative/competitive consumers. Finally, Sect. 6 provides identified challenges and opportunities for future research.

2 Residential Demand Response and Home Energy Management

A typical smart home consists of various types of power appliances, local renewable energy generation (such as rooftop solar PV, small wind turbine) with/without a battery energy storage, and an electric vehicle (EV) networked together to a home energy management system (HEMS) which is real enabler of residential demand management. HEMS is the key element comprises of a desktop or an embedded system that runs GUI monitoring software applications, as well as a communication technologies like ZigBee, Wi-Fi, etc. It is also to be noted that HEMS should have machine learning, pattern recognition, prediction capabilities and interface with the user [3] and demand response aggregator/community EMS. Loads in the residential sector are classified by EPRI's load database used by National Energy Modeling System (NEMS) into nine types viz. space cooling, space heating, water heating, cloth drying, cooking, refrigeration, freezer, lighting, others [4]. Based on their demand management potential, these are classified as critical loads - which might affect the day to day life of consumers when controlled. On the other hand, loads which do not have a major impact on consumer lifestyle are treated as controllable. Demand Response (DR) is effective mechanism which can provide residential consumers with an opportunity to reduce energy consumption costs and simultaneously help the utility to reduce the peak-to-average ratio (PAR) of power. Apart from peak load management, it provides various other applications and benefits like context-aware, power saving services, automation services, etc.

Depending on the price-scheme used, some DR programs may operate in real-time, whereas others may work on a day/hour ahead scheduling basis. The most general DR algorithm is typically formulated as an optimization problem that helps to minimize the cost of consumption of electricity on the customer side or maximize the profit on the utility side. Subject to a set of operating constraints like user comfort levels, priority and operating patterns of appliances, weather conditions, etc., and data uncertainty and user behavior considerations as shown in Fig. 1 The formulation of the algorithm and the load control strategy depends on the type of loads, typical usage patterns, working cycles, uncertainty considerations, behavior modeling, technical constraints and distributed renewable generation and storage facilities available, etc., Hence, demand response potential of various appliances needs to be assessed for designing a DR algorithm.

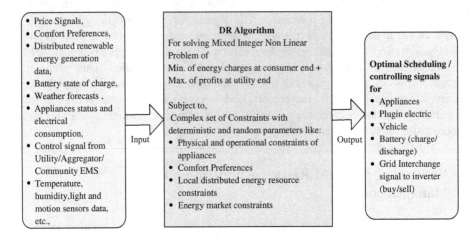

Fig. 1. Inputs and outputs for a typical smart home residential DR algorithm

However individual customer demand control mechanisms will have undesirable effects like peak rebound problem, blackout or brownouts, disturbing the load diversity if not properly coordinated. To avoid these side effects and further benefit the multiple consumers by utilizing the demand diversity and energy resources for grid oriented objectives and cost minimization. These multi customer demand management mechanisms mostly control/coordinate the task scheduling of appliances and dispatch of local distributed energy resources of customers community. This kind of scenario requires a distributed architecture with robust and generic model in order to handle the system complexity. Hence either coordinative multi-agent based optimization techniques or by competitive game-theoretic methods are used which are supervised by an aggregator/community EMS.

3 Demand Response Optimization Model: Mathematical Framework

A typical residential demand management is a mixed-integer non-linear programing problem with characteristics of stochastic, dynamic, multi-objective and multi-actor. Based on the types of loads, and pricing schemes and nature of decision variables involved, and the ability to include uncertainties, scalability, responsiveness, communication requirements, various mathematical formulations and optimization techniques have been suggested for DR management. In this section, a comprehensive description about the optimization model for individual home case and group of homes case related to a day ahead scheduling scenario is provided. Also a brief description about how they can be extended to real-time conditions.

3.1   DR Optimization Model: Single Home

Objectives: DR optimization problem is to manage the electrical consumption, generation and storage resources of the customer over a period of time (typically a day) divided into time slots of a few minutes to the extent of an hour. The objectives could be:

1. Minimize the total electricity cost, usage cycle cost of the battery storage
2. Minimize the inconvenience experienced by the users for delayed operation of time and power shiftable appliances and thermal discomfort level operation of HVAC loads out of their lower/upper limit of the user's comfort zone.
3. Maximization of local generation and storage resources by self-use (or) buying/selling from/to the grid.
4. Minimization of peak demand and/or peak-to-average ratio (PAR).

The optimization model for above objectives are considered from [2] is given below for ready reference.

$$\min. \sum_{t \in T} \left(c_t \cdot y_t - d_t \cdot z_t - \left(R_t^U \cdot a_t^U + R_t^L \cdot a_t^L \right) + \left(E_t^{B+} \cdot R_b + E_t^{B-} \cdot R_b \right) \right)$$

$$\min. \sum_{t \in T} \left(\alpha \cdot \sum_{s \in S} f_{st}^S + \beta \cdot \sum_{e \in CB} f_{et}^{CB} \right), \ \max. \frac{\sum_{t \in T} \left(E_t^{PV} - E_t^{net} \right)}{\mu \cdot E_t^{ToT}} \text{ and } \min. \frac{L_{peak}}{L_{avg}} \quad (1)$$

Where c_t, d_t are buying, selling cost of energy and y_t, z_t are energy brought and injected into the grid at time $t \in T$. R_t^U, R_t^L can be reward paid/penalty collected depending on demand request satisfaction of high and low limits at time $t \in T$ respectively. a_t^U, a_t^L are binary variables indicating the constraint satisfaction. R_b is battery storage system utilization cost per cycle, E_t^{B+}, E_t^{B-} are t is the time slot. α, β are weights and f_{st}^S, f^{CB} are discomfort associated with shiftable, comfort based appliances respectively. And $E_t^{RES}, E_t^{net}, E_t^{TOT}$ and μ are energy generated, net generation, total demand and energy conversion parameter respectively. And L_{avg} average load demand and L_{peak} is peak load demand.

Constraints: A residential customer has four major types of appliances based on controllability: fixed/critical, time-shiftable, power-shiftable, comfort based and local renewable energy generation and storage devices. The major constraints could be energy balance and equality and inequality constraints of loads. A brief description of the constraints are:

Fixed/Critical Loads: whose power consumption and usage cannot be controlled (refrigeration, lighting, TV etc.).

Time Shiftable: are that can only be shifted in time and operates on its own power consumption pattern (e.g., washing machine, dishwasher).

Power Shiftable: are the appliances which have a prescribed energy requirement depending upon the usage of customer (e.g., pool pump, EV). Their constraint modeling can be found in [5].

Comfort Based: The devices that are used to control a physical variable that influences the user's comfort (e.g., HVAC, water heater). Modeling details are presented in [6].

Local Energy Generation and Storage Systems: The local RE based generators such as PV, micro wind turbine, can be either used locally, stored in a battery or injected into the grid depending the buying and selling pricing [7].

This kind of day ahead model's tentative scheduling is extended to real time by a second stage short term stochastic programing problem for optimal scheduling and control; where the time interval is typical a few minutes for considering demand response signal and the uncertainties in price, load demand and local generation forecast etc., with receding time interval. A multi time scale model predictive approach for stochastic modeling is presented in [8].

3.2   DR Optimization Model: Community Level/Aggregated Customer Level

In this case, a community of residential user's will cooperate/negotiate with Aggregator/Community energy management system (CEMS) in managing the power exchange with the grid. The goal of a typical coordinated model is to minimize the global daily energy bill of the group of users by scheduling the users with in their allowable time limits.

$$\min \sum_{u \in U} \sum_{t \in T} \left(c_t . y_t^u - d_t z_t^u \right) \qquad (2)$$

The first term is the cost of energy purchase and the second one cost of energy selling. Where y_t^u, z_t^u energy brought and injected into the grid at time $t \in T$ by the user $u \in U$ is total number of users. The constraints could be total peak load of the users cannot exceed global peak power, network operational constraints, etc. the detailed modeling of the constraints can be found in [9].

4   Residential DR Algorithms: State-of-the-Art

A critical review of very recent DR algorithms for residential energy management is presented in this section. The focus is on the works related to single home and multi-home scenarios, their targets and solution techniques used. And how they tackled the challenges related to work like data uncertainty handling, user behavior modeling, customer involvement, pricing scenarios, cases with and without distributed RE based generation considerations. It was expected that the total number of publications would go up to 500 by the end of 2014 [1].

4.1 DR Methods at Individual Home Level

Single user optimization methods are defined to control load and energy resources of customer. Several classical and heuristic algorithms and techniques are proposed for scheduling and control of appliances along with/without distributed energy resources, under day ahead and real-time pricing environments. Genetic Algorithm (GA) is considered in [10] for load scheduling in an environment containing distributed RE based generation with an objective of peak load minimization and compared with mixed integer linear programing (MILP) approach. Along the same lines, an MILP problem formulation is proposed in [11] to minimize electricity cost subjecting to energy phase and operational constraints. In [12] proposed a convex programing (CP) optimization model for and demonstrated its computation complexity reduction capabilities. A multiple knapsack method is proposed (MKP) [13] for optimal load scheduling.

Real time pricing (RTP) is combined with inclining block rate (IBR) model in [14] to address the problem of possibility that most appliances may operate during the time with the lowest electricity price which may damage the entire electricity system due to the high PAR and solved using GA. Authors in [15] developed an Adaptive Neuro-Fuzzy Inference System (ANFIS) enabled Master Controller (MC) for HEM, where MC schedules the appliances as per user desires and communicates the same with the appliance nodes and ANFIS predicts the customer profiles and sends it to an aggregator.

To tackle the problem of uncertainty and randomness in the considered data many stochastic and robust algorithms that are used are: A typical automated optimization based residential load control scheme using RTP combined with IBR is proposed in [16] which predicts the price ahead time interval. An model predictive control (MPC) based appliance scheduling algorithm is proposed in [17] for buildings, for both thermal and non-thermal appliances. Where as in [8], a two timescale based MPC is proposed for DR considering stochastic optimization model. The authors in [18] proposed a control strategy for peak load reduction by adjusting set point temperature of HVAC loads with in customers preferred tolerance levels by comparing the retail price with threshold price level set by consumers. The uncertainty in behavior of consumer appliance usage is addressed by fuzzy-logic approach in [19]. A least square SVM mechanism for predicting load demand is presented in [20].

An integrated planning and controlling approach for optimal energy management in residential areas with RE based generation is proposed in [21]. The problem of renewable source and electricity price variations is addressed in [22, 24] by dynamically allocating different priorities to appliances according to their status and scheduling them according to the predicted output of renewable sources and the electricity market price forecasting. For customers who need to schedule their consumption, generation and storage, in [23] the authors have proposed an optimization algorithm for power scheduling using MILP. A simple and robust optimization technique is proposed in [22] considering uncertainties in price, renewable power generation prediction. A real time DR algorithm for limiting the load based on user comfort levels or priorities of appliances is given in [25] and its hardware demonstration in [26].

The challenge of consider multiple conflicting objectives with a meaningful balance between them, is partially addressed in [27, 27] by proposing a framework and indices for considering cost, user's convenience and comfort as a mixed objective function and demonstrated with a real data based simulation. Whereas [29] has a multi objective optimization along with uncertainties in the input data. Authors in [30] proposed and demonstrated a task scheduling cum energy management strategy for demand response management in a smart home. But these individual control techniques leads to large peaks during low cost periods and causes rebound peak, service interruptions etc., and to address these problems, the control strategies for community level energy management and DR algorithm has been proposed in literature.

4.2 DR Methods at Community Level/Aggregated Homes

Optimization methods for aggregated users are two types: (1) centralized scheduling (optimization approaches) – an extension of single customer methods to multiuser level and (2) distributed scheduling can be competitive or collaborative approaches (Game theoretic approaches) based on load diversity. The problem scenario and the distributed optimization schemes applied are as discussed below.

In [31], an incentive-based consumption scheduling scheme was proposed for multiple users connected to a single source and solved using coordinate ascent method. Game theoretic approaches are given in [32, 33]. Using an optimal stopping approach defined in [34], a real-time distributed scheduling scheme [35] considering randomness in pricing, appliance priority and power constraint to tackle peak load is presented. The optimization model that adapts the hourly load level in response to forecasted hourly electricity prices is presented in [36].

Authors of [9] have proposed two approaches for evaluating the real-time price-based demand response management for residential appliances namely, stochastic optimization and robust optimization. In [37] a scalable and robust Lagrange relaxation approach (LRA) has been proposed for minimizing energy cost and maximizing consumer satisfaction taking into consideration of variations in renewable generation and price uncertainties. An online algorithm, called Lyapunov-based cost minimization algorithm (LA), which jointly considers the energy management and demand response decisions is proposed in [38]. Vickrey–Clarke–Groves (VCG) auction based mechanism for maximizing social welfare of aggregated users is presented in [39].

Authors in [40–42] have proposed a distributed and coordinated control approaches respectively with focus on overcoming the peak rebound problem. On similar grounds in [43] general algebraic modeling (GAMS), in [44] Stackelberg non-cooperative game theory, agent-based model is developed in [45, 46] Q- learning an online reinforcement learning method is proposed for distributed control of time shiftable appliances. Reference [47] focuses on PHEV in which the authors have proposed a distributed algorithm whereas previous works solely dealt with centralized algorithms. In [48] authors proposed two different approaches for residential load scheduling combined with bi-directional energy trading using their EV's.

To tackle the demand uncertainties and randomness in real time environment in [49] DRSim, a physical simulator is proposed, which can be used to analyze algorithms performance at different case studies.

In [50] proposed an enhancement to the decentralized approach for deferrable and thermal loads of large group of customers. It showed that by adjusting to agent's behavior for market prices, by using adaptive Widrow-Hoff learning rule for deferrable load pattern and modeling their thermal load profile variations. Their strategy gives the emergent behavior of a centrally coordinated mechanism. A single approach will not suit for all the needs in [51] a comparison of various coordinated algorithms viz. balancing responsible, round robin, negotiation and centralized algorithms for a community with multiple house agents for appliance scheduling with respect to diversity, participation of community and amount of peak load reduction.

Table 1. DR optimization methods overview

	Individual home level	Aggregated homes level
Typical objectives	Min. of cost/Max. of comfort/Min. of peak load/Max. of self-consumption of local RE generation (or) combination of these.	Min. of cost, Min. of carbon emissions and Max. of social welfare
Typical constraints	Thermostatic and non-thermostatic controlled appliances, BESS, EV and their parameter limits, time & usage limits.	Non-thermostatic controlled appliances, BESS, EV constraints and price & load uncertainties, distribution network operational constraints etc.,
Optimization methods	GA [14, 18], ANFIS [15], CP [12], MPC [17], MILP [11], PSO [54], MKP [13]	GT [32, 33], MILP [33], Stochastic [9], LRA [37], LA [38], GAMS [43], Q-learning [46], Stackelberg GT [47], Heuristic [53], Agent based model [45]
Architecture and components	Single customer with Home energy manager that communicates with utility/aggregator.	Hierarchical architecture with distributed agents (i.e., home energy managers), centralized agent (aggregator or community energy manager)
Benefits and limitations	Self-use of renewable energy, cost and discomfort minimization, Privacy and Limitations are peak rebound, black out.	System wide perspectives, social network based sharing the useful information for mutual benefit. Difficulty in coordination between house agents, dependency on aggregator, privacy issues.
Popular simulation platforms used	PSCAD, MATLAB along with its optimization, fuzzy logic and ANFIS capabilities and solvers like GAMS.	JADE [55], GridLab-D [56], MATLAB environments. Most of the times a combination of these E.g. Agent-based architecture modelling of household devices in MATLAB and agents in JADE [57].

Although a large number of papers are available on DR algorithms for residential consumers, very few [52, 53] have focused on including power generation by addition of RE based generation at single home level and distributed generation (DG) at the community level respectively, considering together with their intermittent nature. This is one of the major challenges in going ahead with smart grid implementation. An overview of individual and multi user level approaches is presented in Table 1.

5   Residential DR: Key Issues, Opportunities and Challenges in Implementation

Based on the above survey, the derived important aspects, identified key issues and challenges for DR optimization methods implementation in residential sector and research directions are:

- Firstly an intelligent HEMS system [58] interconnected with local RE based energy sources, battery storage and loads is required at the consumer end for DR participation. The participation of the user depends on economic DR programs and efficient and secure information tools from the Aggregator or utility.
- Some works focused only on mathematical modeling and solution strategy for case specific simulation studies. Research works considering a realistic problem scenarios and possible problems, potential effects and cost benefit analysis is the further scope for research.
- In single home scenario, as the number of objectives increases, the tradeoffs are likely to become complex. Also the weightage to the objectives are likely to change with respect to customer's requirement. Hence the effectiveness of fuzzy stochastic multi-objective programing [59] approaches and evolutionary multi and many-objective optimization algorithms [60] needs to be explored to address the challenges in modeling the DSM optimization problem to real scenarios.
- The choice of solution strategy or algorithm depends on many factors user load diversity, amount of user participation, tariff structure etc. there is a need for a unified and robust solution which fits to most of the problem case scenarios.
- Popularly, two level time scales are used for demand management namely a day ahead scheduling typically one hour to 12 min interval and an intraday/online optimization with time interval of few minutes as tradeoff between problem computational complexity, uncertainty considerations and useful optimal solution. Similarly, two level control optimization have been used namely individual customer level and aggregated level. Developing effective techniques for coordination between these strategies remain open.
- Implementation of effective real time renewable generation and load demand forecasting methods and challenges associated still needs to be addressed. Algorithms based on machine learning, state space models, and ANN's can be further explored [61].
- As suggested in Sect. 4 of [62], exploring and studying the applications of various game theoretic methods for multi-home level demand response management needs to be looked at.

Table 2. A summary of Key Issues along with their challenges and opportunities

Key issues	Challenges	Opportunities
Load modelling	DR enabled physical based load modeling helps to know the consumption changes with respect to customer behavior and utility signals. The key is to develop a characteristic model with following qualities comprehensive, reasonably aggregated and DR enabled as in [64].	Development of a sophisticated load model based on historical data, physical parameters, occupant comfort and DR signals at the user operation. E.g. Weather based model helps for precool and/or arrival departure preparation, reduced comfort settings of cooling load for DR response participation.
Consumer behaviour modeling	Model complexities depends on various parameters	Exploring the use of machine learning, fuzzy logic and ANFIS systems
Seamless integration of hardware & software platform	Since multiple ICTs are used at different levels of communication with mesh networking between devices and EMS for control. The issue of communication protocol for integration of HEM interoperability with smart phone, Tablets etc.,	HEMS with support of multiple communication protocols and development of standards- based open platform for easy integration.
Computational & integration challenges at individual level	A coordination is required between costumer level HEMS and community level Aggregator for DR management.	Having an integrated and hierarchical multistage optimization strategy with time receding interval.
DR integration challenges aggregated-customer level	Peak rebound occurs multiple users adopt similar algorithms for load scheduling/may not cooperate with aggregator. Uncertainty in generation, interactions of multiple renewable resources with network.	Exploring the usage of stochastic game-theoretic approaches for interactive decisions. Self-use by storing in a battery during peak times for increased system flexibility. Buy/sell to grid or supply local load.
Coordination strategies	How to obtain coordination between HEMS and Hybrid Grid connected inverter with Rooftop solar PV and battery storage. Coordination between HEMS and CEMS/Aggregator.	Development of a hybrid grid connected inverter's having programmable discharge power, time and duration for ON/OFF control by a HEMS.

(Continued)

Table 2. (*Continued*)

Key issues	Challenges	Opportunities
Forecasting of local RE based power generation and load demand	Near real-time generation forecast models depends on local weather profile, time interval, site specific physical shading and clouding effects etc., Load demand depends on uses occupancy, behaviour, season, time of the day etc.,	Time-series and Neural network models needs to be explored for short term generation forecasting with consideration of uncertainty in weather parameters. Demand prediction and uncertainty modeling by machine learning together with model predictive approaches
Creating Awareness	DR programs are usually voluntary, resulting in self-selection, limitation in cost effectiveness and participation	Customer education and focusing on marketing and adaptation strategies
Pricing structure	Need for profitable and attractive dynamic pricing structure	Designing a simpler pricing scheme with dynamic dependency on power and time of use
Privacy and security	User's data gives critical information about a user life style which puts a user at risk.	Data encryption, safe cloud storage systems needs to be explored to ensure privacy

- Penetration of Plugin-electric vehicles and distributed energy resources brings an increased flexibility in load shaping and integration challenges [63], effective scheduling algorithms to reduce their impact needs to be further studied.
- Dynamic modeling and scheduling of appliances power consumption, prediction of price and uncertainties in renewable energy, making DR context-aware are issues needs to be addressed by collecting a large set of time-series data.

Finally, a summary of key issues and their challenges and opportunities are presented in Table 2.

6 Conclusion

This paper presents the background of smart home energy management functions and optimal DR models for residential users. And reviewed recent methods addressing different aspects of single and multi-user residential energy management and demand response. Based on this review, it is observed that different modelling approaches are explored for household devices, uncertainty in forecasted data and user behavior, and multiple conflicting objectives. Also many scheduling optimization techniques and methodologies are proposed; however, these methods should be further studied by applying on a similar problem scenarios for appreciating their relative merits, suitability, computational complexities, and integration challenges.

Coordination of day-ahead scheduling and real-time demand response in a home needs to be focused by considering time receding optimization strategies, for integration of RE based generation and loads under the scenario of real-time pricing with effective uncertainty consideration and moderate computational complexity. Development of case specific single home and aggregated home models with common set of time-series data such as device consumption pattern, occupancy patterns, and roof-top PV/Wind generation for over a period of time for future research and analysis.

At aggregated home levels, cooperative methods will use have more impact on efficient and economic operation of micro grid/distribution network environment. The future optimization tools for residential homes must offer intelligent ways for collective management of electric loads and resources of the multiple customers with effective coordination/negotiation strategies between HEMS and CEMS/aggregator for overall optimization.

References

1. Vardakas, J.S., Zorba, N., Verikoukis, C.V.: A survey on demand response programs in smart grids: pricing methods and optimization algorithms. IEEE Commun. Surv. Tutor. **17** (1), 152–178 (2014)
2. Barbato, A., Capone, A.: Optimization models and methods for demand-side management of residential users: a survey. Energies **7**, 5787–5824 (2014)
3. Hu, Q., Member, S., Li, F., Member, S.: Hardware design of smart home energy management system with dynamic price response. IEEE Trans. Smart Grid **4**, 1878–1887 (2013)
4. RELOAD Database Documentation and Evaluation and Use in NEMS. (2001)
5. Nair, A.G., Rajasekhar, B.: Demand response algorithm incorporating electricity market prices for residential energy management. In: Proceedings of 3rd International Workshop Software Engineering Challenges Smart Grid - SE4SG 2014, pp. 9–14 (2014)
6. Shao, S., Pipattanasomporn, M., Rahman, S.: Development of physical-based demand response-enabled residential load models. IEEE Trans. Power Syst. **28**, 607–614 (2013)
7. Hopkins, M.D., Pahwa, A., Easton, T.: Intelligent dispatch for distributed renewable resources. IEEE Trans. Smart Grid **3**, 1047–1054 (2012)
8. Yu, Z., Mclaughlin, L., Jia, L., Murphy-hoye, M.C., Pratt, A., Tong, L.: Modeling and stochastic control for home energy management. Power Energy Soc. Gen. Meet. **2012**, 1–9 (2012)
9. Chen, Z., Wu, L., Fu, Y.: Real-time price-based demand response management for residential appliances via stochastic optimization and robust optimization. IEEE Trans. Smart Grid **3**, 1822–1831 (2012)
10. Fernandes, F., Sousa, T., Silva, M., Morais, H., Vale, Z., Faria, P.: Genetic algorithm methodology applied to intelligent house control. In: 2011 IEEE Symposium on Computational Intelligence Applications In Smart Grid (CIASG), pp. 1–8. IEEE (2011)
11. Sou, K.C., Weimer, J., Sandberg, H., Johansson, K.H.: Scheduling smart home appliances using mixed integer linear programming. In: IEEE Conference on Decision and Control and European Control Conference, pp. 5144–5149. IEEE (2011)

12. Tsui, K.M., Chan, S.C.: Demand response optimization for smart home scheduling under real-time pricing. IEEE Trans. Smart Grid **3**, 1812–1821 (2012)
13. Kumaraguruparan, N., Sivaramakrishnan, H., Sapatnekar, S.S.: Residential task scheduling under dynamic pricing using the multiple knapsack method. In: 2012 IEEE PES Innovative Smart Grid Technologies (ISGT), pp. 1–6. IEEE (2012)
14. Corno, F., Razzak, F.: Intelligent energy optimization for user intelligible goals in smart home environments. IEEE Trans. Smart Grid **3**, 2128–2135 (2012)
15. Ozturk, Y., Senthilkumar, D., Kumar, S., Lee, G.: An intelligent home energy management system to improve demand response. IEEE Trans. Smart Grid **4**, 694–701 (2013)
16. Mohsenian-Rad, A.-H., Leon-Garcia, A.: Optimal residential load control with rrice prediction in real-time electricity pricing environments. IEEE Trans. Smart Grid **1**, 120–133 (2010)
17. Chen, C., Wang, J., Heo, Y., Kishore, S.: MPC-based appliance scheduling for residential building energy management controller. IEEE Trans. Smart Grid **4**, 1401–1410 (2013)
18. Yoon, J.H., Baldick, R., Novoselac, A.: Dynamic demand response controller based on real-time retail price for residential buildings. IEEE Trans. Smart Grid **5**, 121–129 (2014)
19. Zuniga, K.V., Castilla, I., Aguilar, R.M.: Using fuzzy logic to model the behavior of residential electrical utility customers. Appl. Energy **115**, 384–393 (2014)
20. Edwards, R.E., New, J., Parker, L.E.: Predicting future hourly residential electrical consumption: a machine learning case study. Energy Build. **49**, 591–603 (2012)
21. Zhao, Z., Lee, W.C., Shin, Y., Song, K.: An optimal power scheduling method for demand response in home energy management system. IEEE Trans. Smart Grid **4**, 1391–1400 (2013)
22. Boynuegri, A.R., Yagcitekin, B., Baysal, M., Karakas, A., Uzunoglu, M.: Energy management algorithm for smart home with renewable energy sources. In: 4th International Conference on Power Engineering, Energy and Electrical Drives, pp. 1753–1758. IEEE (2013)
23. Hubert, T., Grijalva, S.: Modeling for residential electricity optimization in dynamic pricing environments. IEEE Trans. Smart Grid **3**, 2224–2231 (2012)
24. Ivanescu, L., Maier, M.: Real-time household load priority scheduling algorithm based on prediction of renewable source availability. IEEE Trans. Consum. Electron. **58**, 318–326 (2012)
25. Pipattanasomporn, M., Kuzlu, M., Rahman, S.: An algorithm for intelligent home energy management and demand response analysis. IEEE Trans. Smart Grid **3**, 2166–2173 (2012)
26. Kuzlu, M., Pipattanasomporn, M., Rahman, S.: Hardware demonstration of a home energy management system for demand response applications. IEEE Trans. Smart Grid **3**, 1704–1711 (2012)
27. Anvari-Moghaddam, A., Monsef, H., Rahimi-Kian, A.: Optimal smart home energy management considering energy saving and a comfortable lifestyle. IEEE Trans. Smart Grid **6**, 324–332 (2015)
28. Anvari-Moghaddam, A., Monsef, H., Rahimi-Kian, A.: Cost-effective and comfort-aware residential energy management under different pricing schemes and weather conditions. Energy Build. **86**, 782–793 (2014)
29. Jacomino, M., Le, M.H.: Robust energy planning in buildings with energy and comfort costs. 4OR **10**, 81–103 (2011)
30. Zhou, S., Wu, Z., Li, J., Zhang, X.: Real-time energy control approach for smart home energy management system. Electr. Power Compon. Syst. **42**, 315–326 (2014)
31. Mohsenian-Rad, A.-H., Wong, V.W.S., Jatskevich, J., Schober, R.: Optimal and autonomous incentive-based energy consumption scheduling algorithm for smart grid. In: 2010 Innovative Smart Grid Technologies (ISGT), pp. 1–6. IEEE (2010)

32. Li, D., Jayaweera, S.K., Naseri, A.: Auctioning game based demand response scheduling in smart grid. In: 2011 IEEE Online Conference on Green Communications, pp. 58–63. IEEE (2011)

33. Zhu, Z., Tang, J., Lambotharan, S., Chin, W.H., Fan, Z.: An integer linear programming and game theory based optimization for demand-side management in smart grid. In: 2011 IEEE GLOBECOM Work. (GC Wkshps), pp. 1205–1210 (2011)

34. Zheng, D., Ge, W., Zhang, J.: Distributed opportunistic scheduling for ad hoc networks with random access: an optimal stopping approach. IEEE Trans. Inf. Theory 55, 205–222 (2009)

35. Conejo, A.J., Morales, J.M., Baringo, L.: Real-time demand response model. IEEE Trans. Smart Grid 1, 236–242 (2010)

36. Yi, P., Dong, X., Iwayemi, A., Zhou, C., Li, S.: Real-time opportunistic scheduling for residential demand response. IEEE Trans. Smart Grid 4, 227–234 (2013)

37. Giannakis, G.B.: Scalable and robust demand response with mixed-integer constraints. IEEE Trans. Smart Grid 4, 2089–2099 (2013)

38. Guo, Y., Pan, M., Fang, Y., Khargonekar, P.P.: Decentralized coordination of energy utilization for residential households in the smart grid. IEEE Trans. Smart Grid 4, 1341–1350 (2013)

39. Samadi, P., Schober, R., Wong, V.W.S.: Optimal energy consumption scheduling using mechanism design for the future smart grid. In: 2011 IEEE International Conference Smart Grid Communication, pp. 369–374 (2011)

40. Kishore, S., Snyder, L.V.: Control mechanisms for residential electricity demand in smartgrids. In: 2010 First IEEE International Conference Smart Grid Communication, pp. 443–448 (2010)

41. Pedrasa, M.A.A., Spooner, T.D., MacGill, I.F.: Coordinated scheduling of residential distributed energy resources to optimize smart home energy services. IEEE Trans. Smart Grid 1, 134–143 (2010)

42. Guo, Y., Pan, M., Fang, Y., Khargonekar, P.P.: Coordinated energy scheduling for residential households in the smart grid. In: 2012 IEEE Third International Conference Smart Grid Communication, pp. 121–126 (2012)

43. Safdarian, A., Member, S., Fotuhi-firuzabad, M.: A distributed algorithm for managing residential demand response in smart grids. IEEE Trans. Ind. Inform. 10, 2385–2393 (2014)

44. Tushar, W., Chai, B., Yuen, C., Smith, D., Wood, K., Yang, Z., Poor, V.: Three-party energy management with distributed energy resources in smart grid. IEEE Trans. Ind. Electron. 62, 2487–2498 (2014)

45. Saeedi, A.: Real time demand response using renewable resources and energy storage in smart consumers. In: 22nd International Conference on Electricity Distribution Stockholm, pp. 10–13 (2013)

46. O'Neill, D., Levorato, M., Goldsmith, A., Mitra, U.: residential demand response using reinforcement learning. In: 2010 First IEEE International Conference on Smart Grid Communications, pp. 409–414. IEEE (2010)

47. Fan, Z.: A distributed demand response algorithm and its application to PHEV charging in smart grids. IEEE Trans. Smart Grid 3, 1280–1290 (2012)

48. Kim, B.-G., Ren, S., van der Schaar, M., Lee, J.-W.: Bidirectional energy trading for residential load scheduling and electric vehicles. In: 2013 Proceedings of IEEE INFOCOM, pp. 595–599 (2013)

49. Wijaya, T.K., Banerjee, D., Ganu, T., Chakraborty, D., Battacharya, S., Papaioannou, T., Seetharam, D.P., Aberer, K.: DRSim: a cyber physical simulator for demand response systems. In: 2013 IEEE International Conference on Smart Grid Communications, SmartGridComm 2013, pp. 217–222 (2013)

50. Morais, H., Kádár, P., Faria, P., Vale, Z.A., Khodr, H.M.: Optimal scheduling of a renewable micro-grid in an isolated load area using mixed-integer linear programming. Renew. Energy **35**, 151–156 (2010)
51. Thevampalayam, A., Sathiakumar, S.: Peak demand management in a smart community using coordination algorithms. Int. J. Smart Home **7**, 371–390 (2013)
52. Aghaei, J., Alizadeh, M.-I.: Demand response in smart electricity grids equipped with renewable energy sources: a review. Renew. Sustain. Energy Rev. **18**, 64–72 (2013)
53. Barbato, A., Carpentieri, G.: Model and algorithms for the real time management of residential electricity demand. In: 2012 IEEE International Energy Conference and Exhibition (ENERGYCON), pp. 701–706. IEEE (2012)
54. Faria, P., Soares, J., Vale, Z., Morais, H., Sousa, T.: Modified particle swarm optimization applied to integrated demand response and DG resources scheduling. In: 2014 IEEE PES T&D Conference Exposition, p. 1 (2014)
55. Bellifemine, F., Caire, G., Greenwood, D.: Developing Multi-Agent Systems with JADE. John Wiley & Sons Ltd, Chichester (2007)
56. GridLab-D software. http://www.gridlabd.org/
57. Asare-Bediako, B., Kling, W.L., Ribeiro, P.F.: Integrated agent-based home energy management system for smart grids applications. IEEE PES ISGT Eur. **2013**, 1–5 (2013)
58. Khan, A.A., Razzaq, S., Khan, A., Khursheed, F.: HEMSs and enabled demand response in electricity market: an overview. Renew. Sustain. Energy Rev. **42**, 773–785 (2015)
59. Zhang, X., Huang, G.H., Chan, C.W., Liu, Z., Lin, Q.: A fuzzy-robust stochastic multiobjective programming approach for petroleum waste management planning. Appl. Math. Model. **34**, 2778–2788 (2010)
60. Deb, K., Jain, H.: An evolutionary many-objective optimization algorithm using reference-point based non-dominated sorting approach, part I: solving problems with box constraints. IEEExplore IEEE Org. **18**, 577–601 (2013)
61. Spiess, J., Joens, Y.T., Dragnea, R., Spencer, P.: using big data to improve customer experience and business performance. Bell Labs Tech. J. **18**, 3–17 (2014)
62. Saad, W., Han, Z., Poor, H.V., Başar, T.: Game theoretic methods for the smart grid. IEEE Signal Process. Mag. Spec. Issue Signal Process. Tech., Smart Grid (2012)
63. Zhao, J., Kucuksari, S., Mazhari, E., Son, Y.-J.: Integrated analysis of high-penetration PV and PHEV with energy storage and demand response. Appl. Energy **112**, 35–51 (2013)
64. Shao, S., Pipattanasomporn, M., Rahman, S.: Development of physical–based demand response–enabled residential load models. IEEE Trans. Power Syst. 28, 607—614 (2013)

A Roadmap for Domestic Load Modelling for Large-Scale Demand Management within Smart Grids

Alexandros Kleidaras[✉] and Aristides Kiprakis

Institute for Energy Systems, School of Engineering,
The University of Edinburgh, Edinburgh, EH9 3JF, UK
{A.Kleidaras,Aristides.Kiprakis}@ed.ac.uk

Abstract. This paper discusses the potential of the domestic sector to provide Demand Side Management (DSM) services. The inherent drawback of the domestic sector is its structure, consisting of numerous small loads, the high variety of sub-types, the deviation of consumption profiles between households but also the daily variation of each household's demand. In order for DSM to be coordinated and controlled effectively there is a need to create appropraite load clusters and categories. Moreover, there is a variety of domestic loads which can be considered controllable or 'smart'. These smart loads have different characteristics, constraints and thus suitability for DSM services. Hence, typical clustering of load profiles is not optimal and the problem needs to solved on a lower level. A promising method is proposed, some initial results are shown, and finally future work and possible imporvements are discussed.

Keywords: Demand side management · Demand response · Residential loads

1 Introduction

High penetration rates of Renewable Energy Sources (RES) in the distribution side of the power system introduces considerable fluctuations, making it difficult to maintain balance between power supply and demand in the grid. At the same time, the overall power consumption is increasing over the years; in particular, the peak electric demand is rapidly growing [1–3]. Energy storage systems (ESSs) have been proposed as an effective solution to this problem, but recent research is focusing on more feasible methods, namely Demand Side Management (DSM) and Demand Response (DR) using controllable loads [1–9]. The main concept behind Demand Response derives from the potential of some loads as controllable loads, thus making use of already existing components of the grid. Demand Side Management services can be procured by electricity system operators through monitoring, aggregation and control of loads and distributed generation to maintain reliability of electric power systems.

The control strategies for Demand Response can be divided into indirect (or decentralized), where users are prompted to alter their demand profile, through dynamic tariffs or other incentives [10, 11]; and direct control (or centralized), [12], a central or automated control, such as in [10]; for cases of faults, lost generation, RES fluctuation etc., where immediate response sometimes is needed [13].

© Institute for Computer Sciences, Social Informatics and Telecommunications Engineering 2015
P. Pillai et al. (Eds.): WiSATS 2015, LNICST 154, pp. 33–47, 2015.
DOI: 10.1007/978-3-319-25479-1_3

The main strategies for indirect control are time of use pricing (TOUP), real-time pricing (RTP), critical peak pricing (CPP) and peak time rebate (PTR) [11, 14–16]. These can be used for peak shifting/shaving, valley filling or RES following methods. A percentage of customers is expected to alter the starting time of their appliances (or automated systems [16]). Exact response (number of customers) to dynamic tariffs and effect on load shaping depends on prices themselves and human behavior.

Direct Load Control (DLC) services can provide various services, mainly for grid reliability [14, 17]. Some of these can be load shaping (RES integration or price following) [1, 2, 4–7], frequency control [3, 4, 8, 14] voltage control [4], overload relief (transmission and distribution) [4], grid reliability [5, 9, 10], peak load reduction [9, 12], reserve (in the form of positive or negative regulation) [9]. Based on the service provided, different loads or groups of loads are utilized. For instance, T. Masuta & A. Yokoyama [3] simulate frequency control with WHs (Water Heaters) and EVs (Electric Vehicles), which can be switched on/off for short intervals without affecting the quality of service. Hernando-Gil et al. [9] utilize wet loads for DSM, but in this case shifting the appliances' operating time to achieve peak demand reduction.

2   DSM in the Residential Sector

Industrial, commercial and residential sectors have combined consumption at 91.55 % of the total. Currently, the focus of DSM is primarily on the industrial sector due its inherently large loads, existing metering infrastructure (sensors and metering technologies available) and staff with expertise on power systems. Also, the commercial sector, though it tends to have a more distributed consumption (smaller loads), facilities (or groups of them) with enough flexibility have the ability to participate. Residential loads are gaining more attention, but have not been largely used since the loads are small, distributed, and not automated [2, 4, 6, 9–12].

The major challenge lies in the domestic sector, having the highest consumption of the three (35.76 % [18]). A lot of small consumption units need to be aggregated and controlled simultaneously to achieve same results as large commercial or industrial units [4]. In addition, problems arise from the deviation in load profiles, limited knowledge of load composition (how many flexible/deferrable loads operate and at which times/ conditions) and limited knowledge of their potential for DSM, including the end users' awareness and thus willingness to participate. Therefore, knowledge of the composition of the residential sector is essential. This effectively means analyzing the loads and their potential for DSM, their total volume (aggregated power), how much of it can be utilized, which times during the day, week, season and the major driving factors.

Controllable loads fall mainly into two categories: flexible and deferrable. The first type (flexible loads) are those that can provide balancing services, through altering or interrupting their cycle for a short amount of time without affecting the quality of service [12, 19–21]. For instance, electric vehicles or water heating (which usually operate for a few hours) can be switched off (or reduce their consumption) for a few minutes, as long as the battery gets fully charged or the water temperature is within the thermostat's limits respectively [3, 12, 22].

The second type, referred as deferrable loads (also found in literature as load shifting) can shift their operation in time [10, 12, 23, 24]. For instance, washing machines can be programmed to postpone (or advance) start times to favorable times (i.e. lower price due to excess RES generation or off-peak use) [11, 12, 23]. Deferrable loads are suitable for indirect load control (dynamic pricing) and is a form of decentralized DSM. Though, because of its nature, human behavior (even when assisted by automated systems [16]) plays a big part.

Table 1. DR potential of basic domestic appliance categories [12, 19–21, 25, 26]

Load type	Potential	Main factors
Cold Appliances	Flexible	Human behaviour
Electric space heating	Both	Weather
Electric water heating (excluding showers)	Both	Weather
Heating circulation pumps (Gas & Electric)	Both	Weather
Air conditioning	Both	Weather
Wet appliances	Both	Human behaviour Weather (dryers)
Cooking (ovens)	Deferrable	Human behaviour
Lighting	–	Time of day, weather
Consumer Electronics & Home Computing	–	Human behaviour

Note: Each type and each appliance has different constraints and potential. In some cases (i.e. ovens), this can be very limited.

3   Domestic Load Clustering

For the purposes of Load Clustering the main algorithms usually used are Hierarchical, Centroid and Distribution-base. Most commonly Hierarchical Agglomerative, K-means, Fuzzy C-means. The clustering algorithms prefer clusters of approximately similar size and coherent profile, as they will always assign the nearest object (distance based). This often may lead to incorrect clusters, since the main objective is to cluster similar profiles in shape and not necessarily size which have a "coherent" profile on a daily basis. The first step before clustering is thus the normalization of the load profiles.

Domestic loads aren't consistent, as the habitual patterns of users drive the main periods of loads usage, which is the main problem faced when trying to cluster them based on demand profiles. Some activities, such as laundry, don't take place on a daily basis. Thus the same households have different profiles from day to day, causing a high deviation in short term periods (weekly), unlike commercial and industrial units. Clustering domestic demand profiles on a similar manner to commercial/industrial ones, would give inaccurate results due to daily deviation.

Comparing Figs. 1 and 2, it is obvious that using average or aggregated domestic load profiles for studies on LV networks yields errors. Even though the demand profiles of individuals cannot be predicted and vary daily, on a larger scale, aggregated demand profiles do have consistency. This is due to consistency of habitual patterns, the probability of using specific appliances can be predicted on a large "homogenous" group based on historical data (and conditions such as working days, holidays, weather etc.).

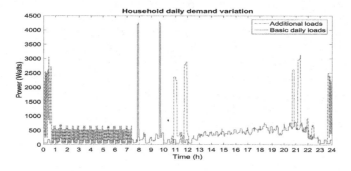

Fig. 1. Difference between two days with and without the use of Dishwasher, Washing machine, Tumble dryer, Water heater & Electric space heating

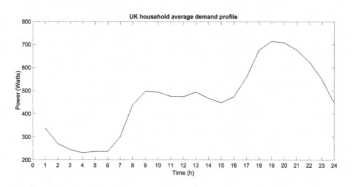

Fig. 2. Typical average household daily consumption in UK [25]

Another issue with demand response is the fact that whole load profiles do not give information about the availability of controllable loads (volume, time, etc.) but only the overall shape of profiles. Figure 2 compared to Fig. 1 has little to no information on the composition of the load profile, thus the controllable loads available. Thus, a large number of end users can be grouped in a few clusters based on their similarities, simplifying their management, supervision and forecasting. Moreover, it may allow unmonitored areas to be matched based on their characteristics to the closest template with a relatively low error (on an aggregated level) [27].

3.1 Modeling and Grouping

Knowledge of the availability of the controllable loads is essential. That means knowledge of quantity in the domestic sector. Table 2 shows ownership statistics, the model used for this paper was created for the case of UK. Things that need to be taken into consideration for the proper modelling are constrains of each type, driving factors, drawbacks and in case when loads are both deferrable and flexible how one affects the other. For example, electric heating as previously mentioned ([1–3, 5–8, 10]) can be used for DSM, but its availability depends on weather conditions (during cold weather mostly) and human behavior. If a low price signal caused the heating to operate at a t_i moment, it should be anticipated that there is extra load available for balancing services for that period. A known drawback is the rebound effect, the oscillation created when interrupted loads are switched on again, such as in [7]. Even though in this case it gets reduced over time because of random factors who affect thermal and cooling loads, initially it is still substantial, recreating similar fluctuations as the ones trying to correct.

Table 2. Appliances ownership statistics [25, 26]

Appliance	UK Ownership	EU Ownership
Fridge-freezer	69.4 %	106 %
Refrigerator	37.7 %	
Chest freezer	15.5 %	52 %
Upright freezer	31.4 %	
Electric oven	65.5 %	77 %
Electric hob	44.8 %	77 %
Microwave	93 %	–
Kettle	98 %	–
Washing machines	97 %*	95 %
Tumble dryers	56 %*	34.4 %
Dishwashers	42 %	42 %
Heating Circulation pumps	88.8 %**	70 %
Electric space heating (storage /direct)	6.13 % /0.74 %	–
Electric water heating [25]	4.8 %	–

*Includes washer dryers, ** DECC, based on number of dwellings with central heating/boilers [25], cooking, wet and heating loads are not operated every day, for example washing machines on average have 5 cycles/week and dishwasher 4.5 cycles/per week [30]

A bottom-up approach is needed, which takes into consideration the composition of the demand. One such approach has been proposed by A. Collin et al. in [28], which is driven by habitual patterns and user activities. Through the use of Markov chain Monte Carlo (MCMC) user activities profiles are generated. Then these are converted to electrical appliances use, which takes into consideration the operating cycle, power volume and other electrical characteristics of the appliances. Thus electrical load models are developed in ZIP form, which generates demand profiles.

The aggregation of those can give domestic load profiles while containing info about the composition of loads, as seen in Fig. 3. By utilizing the generated detailed demand profiles and the knowledge of the potential of smart appliances for DR, the mixture of flexible and deferrable loads during a day can be forecasted. VVPs (Virtual Power Plants) can be created, in the form of aggregated micro sources or batteries [29].

The next step is the creation of a suitable wide-scale network model. Low voltage networks differ based on many factors such as end users, number of customers, geographical location and more [9]. With the use of such LV network a larger realistic network can be created. Thus, simulations to test the potential of DSM strategies based on developed clustering of domestic loads can be evaluated. Basic DSM strategies include balancing services such as frequency control or fast reserve through central control and RES integration through decentralized control.

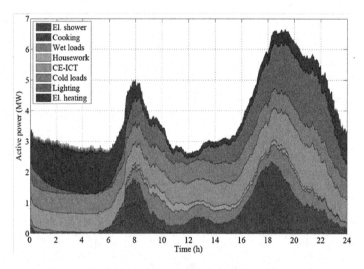

Fig. 3. Aggregated domestic demand [30] 10.000 households, bottom-up appliance specific model, urban areas

3.2   Overall Flowchart of the Methodology

In this paper, a method of combined classification and clustering is proposed, a flow chart of which can be seen in Fig. 4. Residential points with AMI (Advanced Metering Infrastructure) are assumed, with a 1 min interval of readings in the developed model for high definition, though less frequent sampling can also be used. Firstly, fixed data (classification rules) is utilized to examine characteristics that usually drive demand profiles. Then the classification rules used are estimated and analyzed for consistency. Based on the results, individuals are moved to other clusters or marked as anomalous data. The resulting clusters consist of "homogenous" individuals who have similar habitual pattern and thus aggregated demand profile.

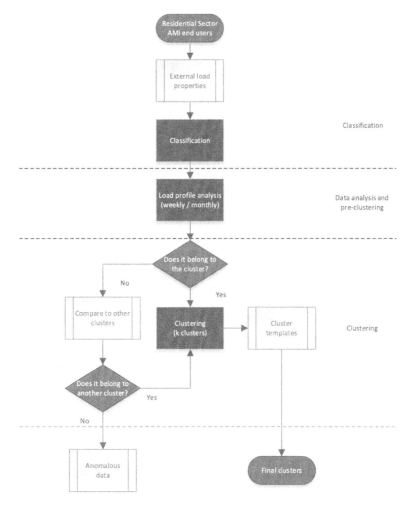

Fig. 4. Flow chart of the methodology

Step 1: Classification – The relationship and characteristics between households are examined. In this paper, occupation characteristics; total number and employment status, and demand characteristics; overall demand and time of use, are considered as main factors that drive the habitual behavior on a monthly basis.

Step 2: Data analysis – Similar households do not necessarily have similar habitual patterns and thus demand profiles. For instance, working occupants based from home or students (classified as non-working) being out of house during office hours, are some usual examples. Overall demand and especially demand in specific time frames of the day can be used to identify differences between similar households. In which case, historical data and analysis on monthly can be used to examine if individuals belong to their appointed cluster, another one or none (anomalous data).

Step3: Clustering - Finalizing clusters, which represent a homogenous group of individuals. On the individual level, demand profiles and thus smart appliances utilization cannot be predicted and is hard to monitor from day to day, but on an aggregated level, based on the habitual patterns of homogenous groups, the probability can be predicted. Thus, in a homogenous cluster of thousands of households, on a given day the overall use of smart appliances is known with high accuracy based on historic data and knowledge of driving factors (i.e. weather). Finally cluster templates are created, as a representative of the cluster, which can be seen as "one" micro source for VPPs, giving information on the available power for DR in specific times of the day. An important note is that households without AMI, can also be clustered based on some of the classification rules, such as occupancy characteristics and overall consumption with less accuracy.

3.3 Results

The classification used in this paper takes into consideration occupation characteristics; total number and employment status, demand characteristics; overall demand (consumption) and time of use (Table 3).

Table 3. Occupancy mixture of developed model

Number of Occupants	Working occupants				
	0	1	2	3	4
1	1210	2316	–	–	–
2	289	790	2290	–	–
3	105	395	1000	210	–
4	0	290	895	105	105

Generated through the Markov chain Monte Carlo (MCMC) method, based on UK population statistics [31]. Total number of households 10000, month January.

Combinations of household sizes of up to four occupants cover 95 % of the U.K. population [31], are thus suitable to represent the overall characteristics of the U.K. population. The correlation between occupation and consumption can be seen in Tables 4 and 5. Figures 5 and 6 are additionally presented to visualize some of these results.

Table 4. Consumption (kWh), random day January

Number of Occupants	Working occupants				
	0	1	2	3	4
1	8.416	6.553	–	–	–
2	11.134	9.850	8.962	–	–
3	12.599	11.063	10.658	10.085	–
4	–	12.588	11.761	11.387	11.617

Table 5. Relative standard deviation

Number of Occupants	Working occupants				
	0	1	2	3	4
1	32.9 %	36.7 %	–	–	–
2	30.4 %	32.1 %	36.2 %	–	–
3	53.3 %	31.6 %	34.9 %	29.9 %	–
4	–	27.1 %	28.7 %	28.2 %	28.4 %

Households of the same size, consisting of non-working occupants generally tend to have higher consumption, since the time spend in the house increases. Cases such as work based at home or students (classified as non-working) being out of the house on working days during office hours are just a few to name.

The approach suggested is a combination of occupancy characteristics and overall monthly consumption (historical data). For example, this allows case A to be placed in

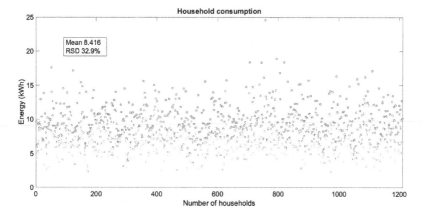

Fig. 5. Household consumption: 1 working occupant, random day in January

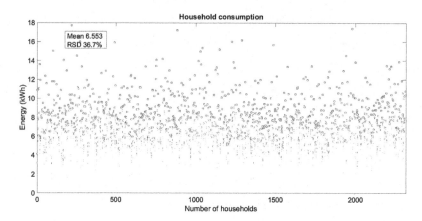

Fig. 6. Household demand: 1 non-working occupant, random day in January

a "non-working occupants" dominant group and case B in a "working occupants" dominant group, assuming similar consumption characteristics.

From Table 1, it can be seen that the households of 3 non-working, 3 working, 3 working & 1 non-working, 4 working occupants are small in number, thus their RSD values are examined with caution to avoid wrong conclusions (i.e. 3 non-working occupants has high RSD value compared to similar households). Nonetheless, we observe decrease in RSD values as the household size increases and as the number of working occupants decreases. The first one can be attributed to more consistent use of appliances, e.g. more frequent use of washing machine within a week for a bigger household thus less demand deviation. The second one can be attributed to occupants sharing more activities (habitual patterns) due to higher time flexibility opposed to working occupants, especially in cases where their working hours do not align.

The correlation between occupation and time of use (mainly controllable loads) can be seen in Figs. 7 and 8. The first one showing a typical office hours working occupant

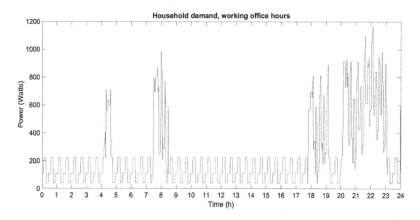

Fig. 7. Household demand: 1 working occupant

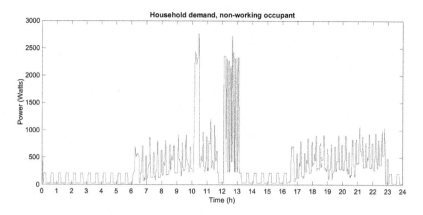

Fig. 8. Household demand: 1 working occupant

and the household demand which consists of cold loads and power electronics in a power down state (such as a TV which is plugged) between 8.30 am and 17.50 pm, the household can be considered in a "passive" consumption state. On the other hand, the second one has "active" consumption within the above hours.

A household with non-working occupants might match better a group consisting mainly of working occupants and vice versa as mentioned above. The approach suggested is a combination of occupancy characteristics and smart meter historical data. A comparison between households with different characteristics in an aggregated level can be seen in Fig. 9. In this case, two groups of the same household size but different occupancy characteristics. Night hours, from 20 pm till early morning hours, 8 am have small differences to almost none is certain hours, while the rest of the day has a substantial gap.

The low demand exhibited during office hours from the second group ("working occupant") is, mainly due to passive consumption. An important conclusion is that the

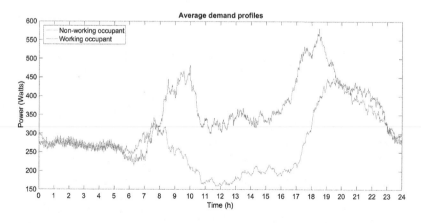

Fig. 9. Household demand differences

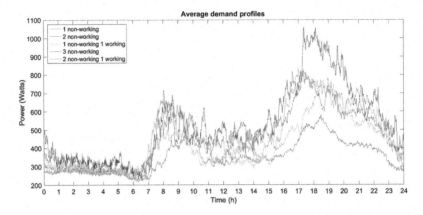

Fig. 10. Household demand similarities: non-working occupants Rest 4 combination follow the same pattern, not included for better visualization

second group ("working occupant") would have less interruption on cold loads (such as door opening or loading goods), thus higher availability than the first group ("non-working occupants"). In case a DSM service is needed, suitable to flexible loads such as cold loads, e.g. Frequency Control Demand Management (FCDM), sending signals to the second group would be the first choice.

On the other hand, heating loads, such as space heating (heating circulation pumps or electric heaters) and water heating loads are mainly expected to be operating within office working hours in the first group ("non-working occupants"). Additionally, for the second group, wet loads might potentially have a wider window to shift their operation, since ccupants will be absent during working hours.

Comparing clusters of different household sizes Figs. 10 and 11, consisting of similar occupants, the overall demand profile is similar with increased consumption. Oscillations increase is observed, an expected outcome for working occupants due to different

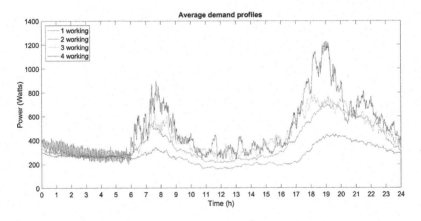

Fig. 11. Household demand similarities: working occupants

working hours, working from home or annual leave. Though the opposite would be expected from non-working occupants due to higher frequency of appliances usage such as wet loads and more shared activities.

4   Conclusions

This paper discusses the potential of domestic sectors to participate in DSM strategies in order to provide certain balancing services and load shaping. These are of great important for grid reliability, RES integration, reduction of cost and greenhouse gas emissions. Smart appliances are considered and their potential and suitability for DR and DSM strategies is discussed. Due to the wide variance of their ratings, electrical characteristics and use pattern, there is a need for a better modelling and then coordination and aggregation of domestic appliances for effective DSM. Due to the high deviation of the demand profile of single households on a daily basis, proposed clustering techniques in the literature are not always effective. However, it was observed that when aggregating similar types of domestic loads, the aggregated demand profile is more coherent, due to the habitual patterns of users which has a weekly /monthly frequency. The approach suggested in this paper is to cluster similar households based on external factors as well as demand profiles, creating a "homogenous" cluster, which on an aggregated level can be predicted and modelled using the probabilities of habitual patterns. The results show that classification based on occupancy and consumption is a good starting point but further analysis is needed. Additionally, demand during specific times of the day can be used to improve the homogeneity of the clusters. As such, a cluster can be used to identify available controllable loads (flexible and deferrable) with higher accuracy and thus based on the DSM service needed in specific times of the day, the proper cluster(s) can be selected to provide the service effectively.

Acknowledgments.   The research leading to these results has received funding from the European Community's Seventh Framework Programme (FP7-PEOPLE-2013-ITN) under grant agreement no 607774.

References

1. Kawachi, S., Hagiwara, H., Baba, J., Furukawa, K., Shimoda, E., Numata, S.: Modeling and simulation of heat pump air conditioning unit intending energy capacity reduction of energy storage system in microgrid. In: Proceedings of the 2011 14th European Conference on Power Electronics and Applications (2011)
2. Kajgaard, M., Mogensen, J., Wittendorff, A., Veress, A., Todor, A., Biegel, B.: Model predictive control of domestic heat pump. In: American Control Conference (2013)
3. Masuta, T., Yokoyama, A.: Supplementary load frequency control by use of a number of both electric vehicles and heat pump water heaters. IEEE Trans. Smart Grid **3**, 1253–1262 (2012)
4. Starke, M., Letto, D., Alkadi, N., George, R., Johnson, B., Dowling, K., Khan, S.: Demand-side response from industrial loads. In: 2013 NSTI Nanotechnology Conference and Expo, vol. 2 (2013)

5. Papaefthymiou, G., Hasche, B., Nabe, C.: Potential of heat pumps for demand side management and wind power integration in the German electricity market. IEEE Trans. Sustain. Energy **3**, 636–642 (2012)
6. Malík, O., Havel, P.: Active Demand-side management system to facilitate integration of res in low-voltage distribution networks. IEEE Trans. Sustain. Energy **5**(2), 673–681 (2014)
7. Lu, N., Chassin, D.: A state-queueing model of thermostatically controlled appliances. IEEE Trans. Power Syst. **19**, 834–841 (2004)
8. Kim, Y., Kirtley, J., Norford, L., Leslie, K.: Variable speed heat pump design for frequency regulation through direct load control. In: 2014 IEEE PES T&D Conference and Exposition (2014)
9. Hernando-Gil, I., Hayes, B., Collin, A., Djokic, S.: Distribution network equivalents for reliability analysis. Part 1: Aggregation methodology. In: 2013 4th IEEE/PES Innovative Smart Grid Technologies Europe (ISGT EUROPE), 6–9 October 2013
10. Chang, T., Alizadeh, M., Scaglione, A.: Real-time power balancing via decentralized coordinated home energy scheduling. IEEE Trans. Smart Grid **4**, 1490–1504 (2013)
11. Albadi, M.H., El-Saadany, E.F.: A summary of demand response in electricity markets. Electr. Power Syst. Res. **78**(11), 1989–1996 (2008)
12. Subramanian, A., Garcia, M., Callaway, D., Poola, K., Varaiya, P.: Real-time scheduling of distributed resources. IEEE Trans. Smart Grid **4**, 430–440 (2013)
13. National Grid: Frequency Control by Demand Management, 2015. http://www2.nationalgrid.com/uk/services/balancing-services/frequency-response/frequency-control-by-demand-management/
14. Strbac, G.: Demand side management: Benefits and challenges. Energy Policy **36**(12), 4419–4426 (2008)
15. Faruqui, A., Harris, D., Hledik, R.: Unlocking the €53 billion savings from smart meters in the EU: How increasing the adoption of dynamic tariffs could make or break the EU's smart grid investment. Energy Policy **38**(10), 6222–6231 (2010)
16. Du, P., Lu, N.: Appliance commitment for household load scheduling. IEEE Trans. Smart Grid **2**(2), 411–419 (2011)
17. National Grid: Short term Operating Reserve (STOR), 2015. http://www2.nationalgrid.com/UK/Services/Balancing-services/Reserve-services/Short-Term-Operating-Reserve/STOR-Runway/
18. Department of Energy and Climate Change (DECC) statistics, Electricity, 2013. https://www.gov.uk/government/uploads/system/uploads/attachment_data/file/337649/chapter_5.pdf
19. Shan-shan, Q., Jing, C., Xiao-hai: Interruptible load management in power market. In: 2014 China International Conference on Electricity Distribution (CICED 2014)
20. Tuan, L., Bhattacharya, K.: Interruptible load management within secondary reserve ancillary service market. In: 2001 IEEE Porto Power Tech Proceedings
21. Nistor, S., Wu, J., Sooriyabandara, M., Ekanayake, J.: Capability of smart appliances to provide reserve services. In: International Conference on Applied Energy (2013)
22. Pan, X., Li, Y., Wang, L.: Research on coordinative optimal dispatch of interruptible load on multi time scale. In: China International Conference on Electricity Distribution, CICED (2012)
23. Niyato, D., Dong, Q., Want, P., Hossain, E.: Optimizations of power consumption and supply in the smart grid: analysis of the impact of data communication reliability. IEEE Trans. Smart Grid **4**, 3–4 (2013)
24. Y, Z., Jia, L., Murphy-Hove, M., Pratt, A., Tuong, L.: Modeling and stochastic control for home energy management. IEEE Trans. Smart Grid **4**(4), 2244–2255 (2013)
25. Department of Energy & Climate Change, Energy consumption in the UK statistics. https://www.gov.uk/government/statistics/energy-consumption-in-the-uk

26. Stamminger, R., Broil, G., Pakula, C., Jungbecker, H., Braun, M., Rüdenauer, I., Wendker, C.: Synergy potential of smart appliances. In: D2.3 of WP 2 from the Smart-A project, November 2008
27. Li, R., Gu, C., Li, F., Shaddick, G., Dale, M.: Development of low voltage network templates — part I: substation clustering and classification. IEEE Trans. Power Syst. **30**(6), 3036–3044 (2014)
28. Collin, A., Tsagarakis, G., Kiprakis, A., McLaughlin, S.: Development of low-voltage load models for the residential load sector. IEEE Transa. Power Syst. **29**(5), 2180–2188 (2014)
29. Hao, H., Sanandaji, B., Poolaa, K., Vincent, T.: Aggregate flexibility of thermostatically controlled loads. IEEE Trans. Power Syst. **30**(1), 189–198 (2014)
30. Collin, A., Tsagarakis, G., Kiprakis, A., McLaughlin, S.: Modelling the electrical loads of UK residential energy users. IEEE Trans. Power Syst. **29**(5), 957–964 (2014)
31. 2011 census: Population and household estimates for the United Kingdom, Office for National Statistics, March 2013

Agent-Based Models for Electricity Markets Accounting for Smart Grid Participation

Sara Lupo$^{(\boxtimes)}$ and Aristides Kiprakis

The University of Edinburgh, Edinburgh EH3 5EU, UK
S.Lupo@ed.ac.uk

Abstract. A better understanding of the process of setting wholesale electricity prices does benet not only the generating companies but also the end users as it forces them to be responsible with their energy use in the time of peak electricity demand leading to smaller fluctuations in demand. Determining when a generator could maximise the prot based on demand fluctuations reduces risk and potential losses that could occur for generating companies. Based on this premise, this paper will outline the use of agent-based models (ABM) in future wholesale energy markets. By comparing agent-based modelling with methods currently employed by economists, this paper will show the impact ABM can have on developing a safer market structure. The results will propagate the idea of agent-based models influence on managing risks, controlling demand, and maximising prot in a time of smart grid technology. This paper is a proposal for work on smart grids in union with agent-based modelling being done in the future if suitable and useful.

Keywords: Agent-based modelling · Smart grids · Artificial intelligence · Equilibrium modelling

1 Introduction to Electricity Market Modelling

As the worldwide need for electricity grows, so does the necessary generating capacity and the associated costs. Using peaker plants to compensate for additional demand is expensive due to the plants only being operated during the time of peak electricity demand. To better understand electricity market dynamics and in order to offer a realistic visualisation of what might happen in the wholesale market in the future, economists use various mathematical tools, with one of the more common ones being equilibrium modelling. Equilibrium modelling follows the rational choice theory, the framework for modelling economic and social behaviour. It also gives consumers more credit about their knowledge of electricity markets and ability to make rational decisions than some research is inclined to agree with. These types of models are Nash Equilibrium models, which compute prices and quantities on the basis that all markets are in equilibrium (i.e., no single market participant has an incentive to choose a different price or quantity - doing so will only decrease its prots). This implicitly assumes

© Institute for Computer Sciences, Social Informatics and Telecommunications Engineering 2015
P. Pillai et al. (Eds.): WiSATS 2015, LNICST 154, pp. 48–57, 2015.
DOI: 10.1007/978-3-319-25479-1_4

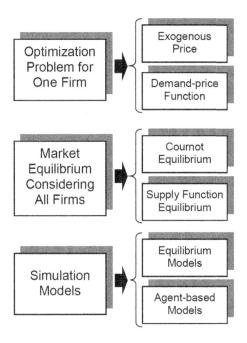

Fig. 1. Electricity market modelling trends, [2]

that everybody knows everything: particularly, the other market participants prot functions and constraints. If there are only a few market participants, this may be reasonable; if there are more, perhaps not. Equilibrium modelling is useful because basic assumptions that economists generally make about people (e.g., rationality) are not necessarily satisfied in agent-based models (if people were rational and prot maximising, they would not obey a simple set of rules but really try to gure out the equilibrium and adapt their behaviour accordingly), [1]. Furthermore, the structure of the wholesale energy market is unique and thus standard economic models cannot always be applied to better help us understand its dynamics. From electricity generators, energy travels to transmission system operators and then to distribution network operators to major suppliers, who then sell it to industrial and commercial customers or residential customers. The identities of these actors are incomparable and applying a broad general model to them would be naïve. Some of the main electricity market modelling trends are represented in Fig. 1.

This equilibrium approach lends itself to expression in equation form. And because equilibrium by definition is a pattern that does not change, in equation form it can be studied for its structure, its implications, and the conditions under which it obtains. Of course the simplicity that makes such analytical examination possible has a price. To ensure tractability we usually have to assume homogeneous (or identical) agents, or at most two or three classes of agents. It has to be assumed that human behaviour a notoriously complicated affair can be captured by simple mathematical functions. Agent behaviour that is intelligent but has no incentive to change has to be assumed; hence it must be assumed that agents and their peers deduce their way into exhausting all information they might find useful, so they have no incentive to change. Still, as a strategy of advancement of analysis, this equilibrium approach has been enormously successful.

2   ABM Applications in Electricity Markets

Actors in ABM provide us with useful tools we can use to design a hypothetical market, where each agent caters to our unique demands. Each agent can represent a distributor or a generator, or if we choose to, the end-user. The agents can also be used to portray, for example, a solely generator based market and we can use them to predict how a number of generators would act in a competitive market. The main difference between agent-based modelling and economists preferred manner of equilibrium-modelling, which is typically used in energy economics as in [6,7] is the fact that in ABM participants are not assumed to be omniscient and super smart, which is a more realistic approach. We have a large group of actors, who follow fairly simple rules and who do not know everything. Even if they did, they could not compute optimal strategies based on market equilibria because they are not super computers. Economists argue that using ABM takes away the rationality the participants have, however, applying these models to wholesale market transactions in which end users are not participating they are a lot more realistic as we are not dealing with the bounded rationality of human consumer behaviour. When additional consideration is given to the validity of the underlying model and the assumption and simplications that have been made, ABM allows us to control our own over-condense when interpreting the results from a simulation.

One thing noticeable about agent-based studies is that they are nearly always evolutionary in approach because agents are adaptive and heterogeneous. On first thought, this might seem to yield at most a trivial extension to standard homogeneous theory. If heterogeneous agents (or heterogeneous strategies or expectations) adjust continually to the overall situation they together create, then they adapt within an ecology they together create. And in so adapting, they change that ecology. Agent-based, non-steady-state economics is also a generalization of equilibrium economics. Out-of- equilibrium systems may converge to or display patterns that are consistent that call for no further adjustments. If so, standard equilibrium behaviour becomes a special case. It follows that out-of-equilibrium economics is not in competition with equilibrium theory. It is merely economics done in a more general, generative way, [3].

In this paper, we focus on an alternative modelling approach utilising artificial intelligence to replicate the behaviour of these actors by using agent-based modelling. Agent-Based Modelling (ABM) platforms are tools that allow the modelling of complex adaptive systems by using agents, providing a way to output the simulation results in a graphical manner according to several designed scenarios. The simulation results can be used to extract conclusions about the systems behaviour and consequently to rene the specification of the agent-based model. These tools provide an easy and powerful simulation capability which enables a fast testing and prototyping environment. ABM aims to recreate and predict the occurrence of complex phenomena. These platforms are being used to simulate agent-based models for different application domains, such as economics, chemical, social behaviour and logistics, [4]. Sophisticated ABM sometimes incorporates neural networks, evolutionary algorithms, or other learning techniques to allow realistic learning and adaptation. In agent-based modelling, a system is modelled as a collection of autonomous decision-making entities called agents. Each agent individually assesses its situation and makes decisions on the basis of a set of rules. Agents may execute various behaviours appropriate for the system they represent for example, producing, consuming, or selling, [5]. The advantage of agent-based modelling is that it allows mimicking each actor separately by using an individual agent. specie. This helps to obtain more realistic results, describing the behaviour and reactions of every energy market actor chosen to study. Comparisons and benefits of each type of modelling are shown in Table 1.

Table 1. Comparison of the available modelling methods for energy markets.

Features	Agent-Based Modelling	Equilibrium Modelling
Consumer Knowledge	Low (almost nothing)	High (assume knowledge of everything)
Rational Choice Theory	No	Yes
Nash Eq. all markets in eq	No	Yes
Incentive to choose another P or Q	To optimize the likelihood of achieving the end goal	Never
Individual demand satisfied	Yes - for each agent	No, patterns visible for a whole
Separate scenarios and agents	Yes	No
Level of modeling freedom	High	Low
Ability to alter individual agents during simulation	Yes	No
Easiness of implementation in artificial computation	High	Medium

3   Types of Agents

The basic agents are used to model entities related to elementary functions such as: the Consumer (C), the Generator (G), the Transmission Network (N), the Distributor (D), the Market Operator (M), the Wholesaler (W), the Retailer (R), and the Regulator (T), as we can see in Fig. 2. Each basic agent has well-dened roles and is characterized by a set of static attributes, a set of dynamic attributes and a set of capabilities, [7]. Consumers eventually use the electricity for any

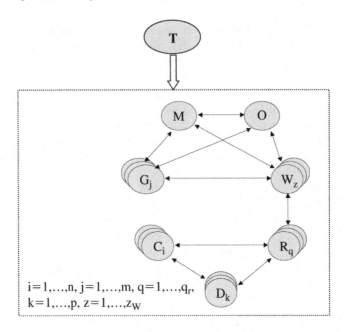

Fig. 2. Basic agents in an electricity market, [9]

purpose (from watching TV to heating to industrial production processes). There is a difference between small and large consumers, since the latter ones may be allowed to directly participate in the wholesale electricity markets. Generators own or lease one or multiple power plants, operate them and sell electricity to the spot and/or the multilateral market through wholesalers. Transmission networks companies own or lease one or several transmission networks.

4   Electricity Pricing

An increase in commodity demand or a decrease in its supply leads to a rise in the market price, which leads to additional investments and production capacity and a new equilibrium. There is a lack of elasticity when it comes to electricity demand so instead of witnessing gradual price increases on electricity markets we observe price spikes, which are very large increases in price over a short period of time, when demand begins to approach the total installed generation capacity. In times of tight but adequate demand to meet the load there is a sharp price rise during periods of peak demand as the market price is determined by the bids of generating units, which operate infrequently. However, under peak load conditions when all of generation capacity is in use price spikes get even higher. This could happen due to the current generation capacity not keeping up with the load growth, because generation capacity has been downsized or because of it is unavailable (like low or no wind leading to less wind supply). Under these conditions, the only factor that would limit the price increase is the elasticity of demand. Price spikes can thus be used as an indicator of insufficient capacity to meet the required demand and the extra revenue that they produce is essential to give generating companies the incentive that they need to invest in new generation capacity or keep older units available. Price spikes are very expensive for the consumers and should incentivise them to be more responsive to price signals. An increase in the price elasticity of the demand leads to a decrease in the magnitude of the spikes, even if the balance between peak load and generation capacity does not improve. Price spikes also give consumers a strong incentive to enter into contracts that encourage generators to invest in generation capacity. Based on this theory, an equilibrium should eventually be reached. At this equilibrium, the balance between investments in generation capacity and investments in load control equipment is optimal and the global welfare is maximum. Many issues, be it socio-behavioural or entirely pragmatic and their political consequences can stop the equilibrium from being reached. Currently, there is no technology to make demand responsive to short-term price signals. Until such technology becomes available, an implementation of quantity rationing instead of price rationing when demand exceeds supply may be needed. This means the system operator may have to disconnect loads to keep the system in balance during periods of peak demand. However, smart grids can help better meet those demands and decrease the risk of demands exceeding supply. Widespread load disconnections are widely unpopular and often have negative social consequences in addition to being economically inefficient. Their impact can be estimated using the value of lost load (VOLL), which is several orders of magnitude larger than the cost of the energy not supplied. Consumers are not used to such disruptions and their political representatives would not tolerate them for extended periods of time. Exposing consumers to spot prices and having them adjust their demand makes price spikes very unpopular. Why spiking occurs is not common knowledge among consumers so they often think they are being ripped off. Price spikes also force people with a lower income to cut back on essential electricity needs normally used for cooking and heating. For this reason electricity markets incorporate a price cap, which aims to prevent large price spikes. On the other hand, price caps decrease incentive for building or keeping generation capacity. An electricity market that relies on spikes in the price of electrical energy to encourage the development of generation capacity is not necessarily good for investors either. Price spikes may not materialise and the average price of electricity may be substantially lower if the weather is more temperate or if higher than average precipitations make hydroenergy more abundant. Basing investment decisions on such signals represents a significant risk for investors. This risk may deter them from committing to the construction of a new plant, [10].

5   Profit Maximization in Current Electricity Markets

Currently the way to maximize prot in an electricity market is by determining a cheaper way to generate electricity and continue selling it to the consumers for the same price, which is the socially desirable approach. However, generators can try and increase prots by gaining and exercising market power. This reduces efficiency, but market power is the ability to raise the market price, so it can be portable. Generally, there are three main ways to gain market power: gaining significant market share, generator collusion, and influencing the regulator to set higher prices. Gaining market share is easier on peak because you only need a small share, however, off peak you need huge market share. Colluding with other generators is generally illegal, however, it is extremely popular. The most popular way for generators is tacit collusion such as shadow pricing. Shadow pricing means generating companies implicitly agree on their prices. In game theory, this is a better long term decision, because if a generating company would try to undercut their competitors, there is nothing stopping the remaining companies from undercutting that generating company in the next game. Such approaches are dated and often illegal. This continues to show that there is a need for implementing a new technology into the electricity system that offers an unbiased way towards pricing and handling faults in the system such a peak demand and blackouts. Although there is a convincing reason for concern, staying put and not doing anything to change the way electricity is currently generated and transmitted benets no one in the long run especially generating companies, who themselves suffer from huge costs when dealing with demand not matching supply.

6   Objectives

In this paper we apply previously described assumptions to discuss a modern, efficient, and more representative way to model the future electricity markets of the United Kingdom, demonstrated in Fig. 3. Papers on modelling of electricity markets in Europe, [8], using agents with a focus on Central Europe, displayed an agent-based modelling framework, which uses the model predictive bidding algorithm to simulate the German electricity market under reference conditions as well as a higher wind energy contribution. However, this paper considers future smart grid technology participation in the energy market as well as large scale deployment of distributed energy resources, and how being modelled with agents benefits this incoming technology. There is a concern among generating companies about the consequences of introducing smart grid technology and demand side management and on the way they could influence the consumers. The general consensus is that such technology gives the user an opportunity to be more involved. It also allows the users to be more aware of the way their electricity costs arise. Because of such outcomes the generating companies are afraid of a potential loss in prot that would come along with a potentially more self-aware consumer.

It is important to help them realise how they can prevent such technology from hurting them, learn to adapt to it and see how in the long run it could even benet their business. Such technology could potentially lead to less operating power needed so the prot would remain proportional to the operating costs. However, the goal is to make these companies prot from the novel approaches,

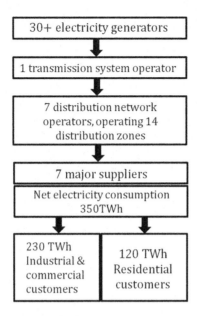

Fig. 3. Electricity market structure in the United Kingdom

as change is inevitable. Smart grid technology brings a new way to monitor users and helps generating companies decrease the need for peaker plant use. This prevents unnecessary costs that occur from starting and maintaining peaker plants and allows for the companies to be able to better fluctuations. Agent-based modelling is the approach that makes that easiest as it accounts for the individual needs of each agent and acts based on its memory, developed by the historical information it uses to function.

7   Conclusions

This paper merely aimed at describing the necessary adjustments in former and novel types of modelling to achieve an accurate representation of the models in the future. The hope for the future work is to apply both models in reality, compare them, and see how both can aid our predictions. Although equilibrium modelling has long been the go-to tool for depicting market dynamics in order to stay current and in touch with the incoming technology, there is a need for a new type of modelling that makes it easier to include novel energy technologies as they come along. Agent-based modelling allows for a modelling framework that will be able to process and essentially predict electricity market dynamics, accounting for the impact of smart grids on the prices and demand of electricity. Choosing to model a developing market with agents instead of somewhat dated economical modelling approaches helps pinpoint the issues in electricity markets when it comes to pricing and quantity determination and how to eliminate them without hurting end users. ABM illustrates the benets of smart grid technology and demand side management for generating companies as the general misconception is they can only be damaging for the prots of the generating companies. This novel technology agrees with the objective of not causing damage to the consumers yet it also helps generating companies with their demand predictions. If generating companies make a stride towards adopting to smart grid technology and invest in it first-handedly it will be much easier for them to regulate the transition and make sure there is no loss in profits. The longer they wait the less they will be able to influence and readjust the progress of smart grids the way it best suits them. If they become accepting of smart grids now, generating companies can have an influential and fundamental role in the path smart grids take. They can arrange the technology to benefit them as they can minimize unnecessary costs, especially those coming from persistent and unplanned peaker plant use. Allowing for consumers to have large control over their energy use and price could result in pointless and costly maintenance of larger generating capacity during times active consumers will learn to avoid. They face a decision of an opportunity cost. What is the best financial decision for them to make - working with smart grids or trying to reach an ultimatum and block them out? Not only would that result in a potential intervention of regulatory bodies it would also make them look unfavourable in the eye of the consumer, who do not want to have their generating companies taking advantage of them. Ultimately, it would lead to consumers choosing progressive generating companies accepting smart grids in their daily mechanisms, which could then lead to an increased market share for companies, which were astute enough to act first to aid their customers' needs. Taking advantage of both energy storage and smart grids would in my opinion lead to best financial results, even though the initial costs would be higher. It would teach the generating companies how to act responsibly with their generating capacity in terms of when to store energy for certain peak demand times.

Acknowledgements. The authors would like to thank Harry van der Weijde for his generous advice on writing this paper and his help explaining the economic aspects of electricity markets.

References

1. Besanko, D.A., Braeutigam, R.R.: Microeconomics: an integrated approach, pp. 534–536. Wiley, New York (2002)
2. Ventosa, M., Baíllo, Á., Ramos, A., Rivier, M.: Electricity market modeling trends. Energy Policy **33**(7), 897–913 (2005)
3. Arthur, W.: Out-of-equilibrium economics and agent-based modeling. Handb. Comput. Econ. **2**, 1551–1564 (2006)
4. Barbosa, J., Leitão, P.: Simulation of multi-agent manufacturing systems using agent-based modelling platforms. In: 9th IEEE International Conference on Industrial Informatics (INDIN), pp. 477–82 (2011)
5. Aletti, G., Naimzada, A.K., Naldi, G.: Mathematical Modeling of Collective Behavior in Socio-Economic and Life Sciences, pp. 203–204. Birkhauser, Basel (2010)

6. Gabriel, S.A., Kiet, S., Zhuang, J.: A mixed complementarity-based equilibrium model of natural gas markets. Oper. Res. **53**(5), 799–818 (2005)
7. García-Bertrand, R., Conejo, S.A., Gabriel, S.A.: Electricity market near-equilibrium under locational marginal pricing and minimum prot conditions. Eur. J. Oper. Res. **174**, 457–479 (2006)
8. Gnansounou, E., Pierre, S., Quintero, A., Dong, J., Lahlou, A.: Toward a multi-agent architecture for market oriented planning in electricity supply industry. Int. J. Power Energy Syst. **27**(1), 82–89 (2007)
9. Wehinger, L.A., Hug-Glanzmann, G., Galus, M.D., Andersson, G.: Modeling electricity wholesale markets with model predictive and prot maximizing agents. IEEE Trans. Power Syst. **28**(2), 868–876 (2013)
10. Kirschen, D.S., Strbac, G.: Fundamentals of Power System Economics. Wiley, Chichester (2004)

Workshop on Communication Applications in Smart Grid (CASG) 2

Securing the Information Infrastructure for EV Charging

Fabian van den Broek[1], Erik Poll[1](✉), and Bárbara Vieira[2]

[1] Radboud University, Nijmegen, The Netherlands
{f.vandenbroek,e.poll}@cs.ru.nl
[2] Software Improvement Group, Amsterdam, The Netherlands
b.vieira@sig.eu

Abstract. We consider the functional and security requirements for the information exchanges in the infrastructure for EV charging being trialled in the Netherlands, which includes support for congestion management using the smart charging protocol OSCP. We note that current solutions do not provide true end-to-end security, even if all communication links are secured (for instance with TLS), as some data is forwarded between multiple parties. We argue that securing the data itself rather than just securing the communication links is the best way to address security needs and provide end-to-end security.

Moreover, because of the number of parties involved and the fact that the precise roles of these parties are still evolving, we argue that more data-centric communication solutions, using pub/sub (publish/subscribe) middleware, may be better suited than using point-to-point communication links between all parties, given the flexibility and scalability provided by pub/sub middleware.

Keywords: EV charging · Congestion management · End-to-end security · Smart grids

1 Introduction

The introduction of electric vehicles (EVs) brings important new requirements on the information and control architecture of the electricity grid. Information needs to be exchanged for billing but possibly also for congestion management. Charging EVs consumes a lot of electricity, a single charge of an EV consumes roughly the same amount of energy as the daily demand of 2 to 3 houses, so controlling it is an important means to manage the grid's limited capacity. Furthermore, EV charging may become an important factor in balancing supply and demand. All this means that EV charging introduces a lot of information flows, adding a lot of complexity to the ICT infrastructure behind the electricity grid. Moreover, it involves many parties, and involves some data with high security requirements, especially for data which is used in actively managing the grid.

Looking at the solutions currently being used or trialled in the Netherlands, this paper considers the security requirements for data exchanged in the grid

© Institute for Computer Sciences, Social Informatics and Telecommunications Engineering 2015
P. Pillai et al. (Eds.): WiSATS 2015, LNICST 154, pp. 61–74, 2015.
DOI: 10.1007/978-3-319-25479-1_5

to support EV charging in Sect. 3. Here we note that there are many parties involved in exchanging or forwarding such data. This makes ensuring end-to-end security important, as otherwise the parties involved in EV charging have to put a great deal of trust in one another. The importance of end-to-end security is also stressed by standards such as IEC 62351 [15] and NIST guidelines for Smart Grid Cyber Security [14].

We then suggest possible directions to improve the situation, both when it comes to securing information and organising the new information flows. For securing information, Sect. 4.1 discusses the possibilities to introducing security measures at the level of the data being exchanged, and not the communications links. This seems the natural way to achieve end-to-end security in situations where data is exchanged and forwarded between multiple parties. For organising information exchanges, Sect. 4.2 discusses the use of pub/sub (publish/subscribe) middleware as a solution to exchange data between many parties that is more flexible and scalable than introducing direct communication links between all parties involved.

These two solutions form a natural combination. Indeed, the middleware solution developed in the C-DAX project (http://cdax.eu) combines them in a pub/sub solution that provides end-to-end security tailored to smart grid applications.

Concrete starting points for this paper are the solutions that are being rolled out and/or trialled in the Netherlands, which are described in detail in Sect. 2. These include the OCPP protocol for the communication between charge spots and operators, which is rolled out nationally, and the OSCP protocol for congestion management (using so-called smart charging), which is being trialled. The authors of this paper were involved in a security evaluation and resulting security design of smart charging, using OSCP. The security design focussed on achieving end-to-end integrity of data. This security design is currently being integrated within the EV charging system in the Netherlands. Anticipating on the roll-out of the security design, this paper aims to point out the more generic security problems at the heart of (smart) EV charging and present some generic solutions.

The use of EVs is still in its infancy: some solutions are still at the trial stage and, more importantly, the market models for EV charging, and the roles of the (many!) parties involved, are not yet clear and still evolving. However, this will not change the basic communication needs and associated security requirements. It is clear that EV charging will involve multiple parties, and some communication between these parties with high security requirements, as it involves information needed for billing, information that is privacy-sensitive, and information needed to actively control the grid. So money, privacy, and – most importantly – the stability of the grid are at stake. So even though we look at the concrete protocols currently being used or trialled in the Netherlands, we hope our conclusions will be relevant for any solution for EV charging.

Scope. This paper looks at EV charging from the grid perspective rather than the EV perspective. By this we mean that the focus is on the communication needs in the grid – between grid operators, charge spot operators, and energy suppliers – to manage EV charging, and we largely ignore the communication with the EV or its user.

Also, we will not consider the underlying physical networking infrastructure in the field, which may include PLC (Power Line Communication), cellular networks such as GPRS or LTE, or CDMA[1], or optic fibers for parts of the communication network. This underlying networking infrastructure may provide some security. For example, cellular networks will provide authentication and security at the transport level. Still, we believe that security solutions for the communication and information architecture needed to support EV charging should be designed to be independent of the underlying networking technologies. The infrastructure continues to evolve and change rapidly, and different grid operators are choosing different technologies. So, ideally solutions should not be tied to a particular networking technology, beyond imposing minimum bandwidth and latency requirements, or rely on security guarantees these technologies provide. Of course, such security guarantees are useful additional layers of defence (following the principle of 'defence in depth').

2   EV Charging

This section makes an inventory of the information and communication needs for managing EV charging, the various parties involved, and the associated security requirements. This includes communication for billing and for management of the grid, in particular for congestion management.

We consider the set-up and protocols that are being used or trialled in the Netherlands, where there are public EV-charge spots where customers with the right subscriptions can charge their EV. Still, the communication needs and security requirements are more general, and largely independent on the particular set-up and protocols used: Any solution for EV charging that involves billing and some form of congestion management will have similar requirements.

Figure 1 gives a schematic overview of the smart charging set-up in the Netherlands. The different parties or roles involved are described below.

– The *DSO (Distribution System Operator)* manages a regional electricity grid, and is responsible for a stable, reliable and well-functioning grid delivering electricity to consumers.
– The *EMSP (E-Mobility Service Provider)* (re)sells electricity to EV users for charging their car. So the EMSP will set up contracts with EV users and takes care of billing.
– The *CSO (Charge Spot Operator)* operates and maintains charge spots. CSOs play a important role in the EV market, as they interact with the DSO and the EMSPs.

---

[1] Alliander, one of the larger DSOs in the Netherlands, is rolling out its own CDMA cellular network, dedicated to communication with their equipment in the field and possibly other critical infrastructures.

– The *CSIO (Charge Spot Infrastructure Operator)* is typically a vendor of charge spots and will perform some maintenance, such as updating firmware, on behalf of the CSO. In some situations such maintenance is only performed through the CSO, i.e., updates are sent to the CSO and the CSO takes care of them, but in other cases it is done directly by the CSIO.

Figure 1 also includes the *Central Interoperability Register (CIR)*, which is an online customer database provided by the joint EMSPs, which can be queried by a charge spot (via the CSO) to see if a customer is allowed to charge his/her car.

The precise market models for EV charging are still in flux, and it is not yet clear which parties will play which role or roles. For example, some companies in the Netherlands play the role of both EMSP and CSO. One can also imagine that a DSO also plays the role of CSO.

One factor here are government goals of market liberalisation and fostering free competition in the energy sector: DSOs are natural monopolists in the region where they manage the grid, so there will be government regulations on what they are supposed to do and on what they are not allowed to do. However, such concerns may be in conflict with government aims to stimulate the use of EVs and roll-out of charge spots: as an important and resourceful party, DSOs may have to take the lead in some domains to encourage the use of EVs.

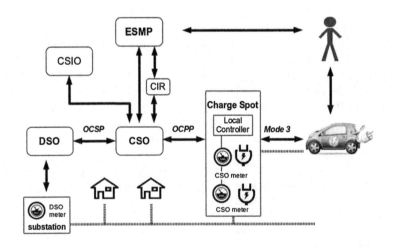

Fig. 1. Information flows for EV charging

We now turn to the physical infrastructure involved. The *charge spot (CS)* provides one or more sockets where EVs can be charged. A charge spot will include an electricity meter for each socket, which is owned by the DSO and controlled by the CSO. A charge spot or set of charge spots is managed by a *local controller* which has a communication link, for instance a GPRS connection, to the back-office of the CSO.

EV drivers with a subscription from an EMSP get an RFID card. A charge spot also contains an RFID reader, which is used to identify an EV-driver. When a charge starts, the charge cable is locked to the CS. The cable is only freed after identifying with the same RFID card that was used to start the charge.

Initially, charge spots in the Netherlands also contained a smart meter, similar to those placed in homes, which was under the direct control of the DSO, with its own GPRS connection. For cost reasons and size reasons (removing this additional meter allows for smaller charge spots) these smart meters are being phased out. New charge spots only have a traditional 'dumb' meter per socket, which communicates via a bus to the local controller. Even if the DSO no longer has a direct connection to a meter in a charge spot, it will have meters in the field, notably in the secondary sub-station that feeds a neighbourhood.

Each charging session is measured at the charge spot, and recorded at the CSO. The details (who charged how much, where and when) are then transmitted to the EMSP, who bills the customer. There are other billing chains, where energy providers bill the CSO for total consumption of its charge spots and the CSO bills the EMSP for charging of that EMSP's customers.

2.1 Protocols

EV charging in the Netherlands uses several internationally standardised protocols, incl. OCPP for communication with the charge spots by operators, and ISO 62196 Mode 3 for communication between EV and charge spot. A less standard solution being trialled in the Netherlands is the use of OSCP to dynamically control the capacity made available to charge spots for the purpose of congestion management. These protocols are discussed in more detail below.

Mode 3. ISO 62196 [9] standardises the charging of EVs, incl. the dimensions of different plugs and allowed current and voltage. It describes four possible modes for charging, of which the third describes EV charging at higher power stations. This specific connection is often referred to as Mode 3, and is supported by practically all currently available EVs.

There is a newer standard, ISO 15118 [10], that is essentially a successor of Mode 3. ISO 15118 still needs to see a wide roll-out, as hardly any EVs on the market support it. It includes several improvements, notably when it comes to security, as will be discussed later.

OCPP.[2] The charge spot communicates with the CSO through the Open Charge Point Protocol (OCPP). OCPP standardises the communication between the charge spot and the party that operates the charge spot (i.e., the CSO), thereby allowing CSO back-ends and charge spots of different vendors to communicate (preventing vendor lock-in). As part of that, OCPP also allows for remote maintenance of charge spots by the CSO or CSIO through monitoring and firmware updates. It also offers features needed for congestion management, notably limiting the maximum capacity that a charge spot can deliver to an EV in a certain time slot.

---

[2] http://www.ocppforum.net.

OCPP is a SOAP-based protocol[3] originally designed by the E-Laad foundation (http://www.e-laad.nl), a foundation set up by the joint Dutch DSOs, but currently used by most countries that offer public charge stations. The current release is version 1.5; version 2.0 is under development.

OSCP.[4] The large energy consumption of EVs poses a challenge for the electricity grid, given the limited capacity of the power lines at local level. OSCP (Open Smart Charging Protocol) allows a DSO to vary the capacity available to charge stations in time, given the varying predicted capacity needed for other consumers in an area.

This means that OSCP allows a DSO to do congestion management. Congestion management is about managing the limited capacity of the grid, given the physical infrastructure of transformers and cables, and sharing this capacity between charge spots, households, and commercial users in a neighbourhood.[5]

For congestion management, OSCP supports negotiation between a DSO and CSOs. The DSO creates a forecast, 24 hr in advance, for 15 min intervals, on the power usage for each cable, based on historic measurement data and weather forecasts. The DSO then divides the forecast power usage among CSOs, again using historic data and contracted capacity. Using OSCP, each CSO is informed of its allotted capacity and the remaining spare capacity. The CSO can negotiate for more or less capacity, again using OSCP. The CSO then creates a charge plan for the charge spots, specifying the limit of the power they can supply per time slot, and transmits this to the charge spots using OCPP.

There is an important trust assumption here on the part of the DSO, that the CSO will not consume more energy than it negotiated, as there is no way for the DSO to limit the energy flow, other than a tripping safety breaker on the cable, which would stop the electricity supply to all consumers on this cable.

2.2 Security Requirements

Any discussion of security is meaningless without considering the security requirements. A coarse classification in four overall security requirements can be made:

1. **Availability of Electricity.** Clearly availability of electricity is of paramount importance. Both the availability and the integrity of information could affect the electricity supply, namely if the absence or incorrectness of information could hamper operation of the grid.

---

[3] A JSON over websockets version (version 1.6) of OCPP is currently being developed.

[4] http://www.smartcharging.nl/smart-charging/open-smart-charging-protocol.

[5] Congestion management should not be confused which load balancing, which is about the more general issue of getting demand and supply in balance. The limited capacity of the grid is a (constant) factor here, but so is the variation in the supply of electricity – variation which will increase as there is more use of renewables (solar and wind power). So congestion management is always a local issue, and involves imposing limits on demand, whereas load balancing is also an issue on larger scale, and may involve influencing both demand and supply.

2. **Integrity and Non-Repudiation for Billing.** For billing integrity of the records of the charging is important. Some form of authentication of EVs or EV users will be needed for this. One may also want some form of non-repudiation, i.e. some evidence to settle disputes, say in case a customer of an EMSP disputes her bill. Non-repudiation is related to integrity, but, as we will see later, some measures to ensure integrity (notably the use of secure tunnels) do not provide a practical means to support non-repudiation.
3. **Privacy.** Confidentiality of information about an individual EV is important for the privacy of its user, as it for instance reveals the location where an EV was at a given time. Given that the user of an EV is typically a single person, such information will be personal information, and hence subject to legal requirements on the handling of personal information.
4. **Business Confidential Data.** Some of the companies involved may consider some of their data confidential for business reasons. For example, a CSO might not want its competitors to know how busy it's charge spots are, and an EMSP might not want its competitors to know customer information.

A more thorough evaluation of the security requirements, which would also involve the formulation of attacker models, is beyond the scope of the paper. Still, we do want to point out that EV charging introduces new players in the market, notably CSOs, that play an active part in congestion management and can affect the first security requirement above.

Here the introduction of smart EV charging seems to bring bigger risks than the introduction of smart metering. Smart meters also give new parties access to ICT infrastructure in the grid (for instance new service providers that read out metering data), but these are not meant to play an active part in managing the grid, as CSOs are expected to do in smart charging. Of course, this is not to say that smart meters are without risks to the availability of electricity, esp. if the smart meters allow consumers to be disconnected remotely [2]. A feature that was removed from smart meters in the Netherlands after a security evaluation.

Security risks can be mitigated at different levels: at the level of the ICT infrastructure, but also at the level of the application or service, as explained below.

1. At the level of the ICT infrastructure, risk to availability can be mitigated by redundancy, say by having back-up storage of critical data, or having a second communication link if a link fails. Risks to integrity and confidentially can be mitigated by various forms of access control, authentication, or the use of cryptographic checks, (i.e. digital signatures or message authentication codes (MACs) for integrity and encryption for confidentiality). Note that these are generic security measures, largely independent of the specific application. Of course, which measures and costs are reasonable will always depend of the specific application.
2. Independent of these more generic techniques at the level of the ICT infrastructure, it may also be possible to mitigate risk by more tailored measures at the level of the application. One such a measure is having fall-back

scenarios. For example, when the management of the grid uses smart charging as a way to do congestion management, there may be a fallback option on what to do if this system fails; a charge spot might have some default capacity that it will use in case it does not receive a dynamic capacity.

The security measures we discuss in the remainder of this paper will be of the former kind, but this does not mean one should overlook measures of the latter kind.

3 Security Shortcomings

It appears that security considerations have not played a very prominent role in rolling out the public charge spots in the Netherlands, or indeed in the design of the OCPP protocol. The OSCP protocol is still under development and has had a security evaluation on an initial functional design. While not exactly security-by-design, the early inclusion of a security evaluation is already a marked improvement on the design of OSCP over OCPP. This security evaluation yielded several security issues, not just in the OSCP link, but in the whole EV-charging chain, as discussed in more detail below.

3.1 Weak Authentication

At public charge spots drivers authenticate themselves using an RFID card. Surprisingly, only the static ID (the so-called UID) of the card is used for authentication here. In essence, this means every customer is identified through a password that is transmitted plaintext through the air. This makes copying the cards extremely simple: on legitimate RFID cards the UID is fixed and cannot be changed, but counterfeit cards with a configurable UID and equipment that can spoof the RFID communication are readily available.

The UID can be eavesdropped if one has access to the card by simply using a standard NFC-enabled phone. With electronic equipment it is also possible to eavesdrop on the UID when it is used at a charge spot. This is possible at a distance of several meters [6,7], but it would be simpler to stick eavesdropping equipment right on top of the RFID antenna of a charge spot. An attacker could also simply try out random UIDs until he finds one that the charge spot accept; by reading out the UID of a few legitimate cards it will be easy to determine the approximate range of UIDs used for EV charging.

That cloning cards is so easy does not necessarily mean there is a viable criminal business model. Blacklisting cloned cards can frustrate fraudulent use of cloned cards, at the expense of also creating hassle for innocent victims who had their card cloned. The real deterrent to fraud would probably be the risk that users of the cloned cards run of being caught red-handed. Especially since charging electric cars still takes a significant amount of time.

However, the weak authentication could be exploited to release the expensive charge cables, which in the Dutch setup are owned by the EV-driver.

3.2   Reliance on Secure Tunnels

As a security measure, the OCPP specification suggest the use of TLS to secure communication links. In practice, this suggestion may not be followed because of bandwidth restrictions (charge spots generate very small messages, where introducing TLS increases the overhead significantly) and cost: charge spots often communicate over cellular networks, and the use of this communication link will be charged per transmitted byte, making the overhead extra costly. This means that these OCPP links then rely on the security offered by the underlying cellular technology.

Note that even if TLS is used to protect both the OCPP and the OSCP links, this still has some security shortcomings: it would not always provide true end-to-end security, and it would not provide a practical means for non-repudiation, as explained below:

Lack of End-to-End Security. Some of the information for smart charging is forwarded across multiple links. For instance, measurement data generated at the charge spot meter should end up at the EMSP, so they can bill the customer accordingly. The CSO forwards the data received from the charge spot to the EMSP.

Even if both the communication links are protected by TLS, this does not provide end-to-end security between the charge spot and the EMSP. The TLS tunnels will prevent against tampering at intermediate points between the charge spot and the CSO, and at intermediate points between the CSO and the EMSP, but the CSO will have to be trusted not to change the data. The same goes for metering data that goes from the charge spot to the DSO, or, conversely, for the charge plans that go from the DSO to charge points.

To summarise: TLS does provides a secure tunnel, but only for one communication link, and not across multiple links.

Lack of Non-Repudiation. TLS ensures the integrity of the data sent between two parties: Message Authentication Codes (MACs) are added to any data sent and upon reception these are checked to rule out tampering with the data. As soon as data exits the TLS tunnel, all these integrity measures are stripped - what is left is the original data that was sent. This has the advantage of making the data protection completely transparent. But a downside is that there is no easy way for the receiver to later prove the integrity of the message to a third party. The only way to do this would be to provide a log of the entire TLS session, including the TLS handshake, which is hardly practical.

4   More Data-Centric Solutions

This section discusses directions to address the security shortcomings of secure tunnels above, and to provide more flexibility and scalability in handling the information flows between the many parties involved in EV charging. These directions are related in that they revolve around letting the data itself, rather than the communication links, play a central role.

4.1 Data-Centric Security

By data-centric security we mean providing security at the level of data messages, rather than at the level of the communication links. To illustrate the idea, we will first look at how integrity of meter readings in charging sessions is ensured in ISO 15118.

ISO 15118, the successor standard for Mode 3, provides built-in security measures that address some of the security concerns discussed in Sect. 3.2 above. In ISO 15118 metering data can be digitally signed by both the car and the charge spot. This means that the ultimate recipient of the data, say an EMSP, can verify that the data record comes from a particular customer and a particular charge spot. In case of any disputes, the digital signatures provide evidence that a particular EV was involved in charging. So this provides non-repudiation and end-to-end security, more specifically end-to-end integrity, between EMSP, charge spot and EV.

Note that these guarantees do not rely on any secure tunnels for the communication, and that the CSO does not have to be trusted not to change the data. The fundamental difference is that the security is added to the data messages themselves, and not to the communication channels over which the data is transferred.

More generally, similar to the way that ISO 15118 provides integrity checks on certain messages, data integrity and confidentiality of data messages can be handled at the level of individual messages using the same standard cryptographic mechanism: integrity of messages can be guaranteed using either digital signatures or MACs, and confidentially of messages can be handled by encryption.

These solutions overcome the limitations of generic secure tunnels discussed above in Sect. 3.2: they can provide end-to-end security, even for data forwarded between multiple parties, and provide non-repudiation, as messages come with their individual integrity checks.

4.2 More Flexible Architectures Using Pub/Sub Middleware

Adding security measures at the level of the data, as discussed above, rather than at the level of the communication links, opens up the possibility of using more flexible architectures to share data across multiple parties, as we will now discuss.

As shown in Fig. 1, the EV charging infrastructure requires a lot of communication links between various parties. In fact, the situation is more complex than Fig. 1 suggests: the figure only shows one DSO, CSO, EMSP, and one charge spot, whereas in reality there will be several DSOs, CSOs, and EMSPs, not to mention charge spots.

Organising communication links between all of these parties can be a challenge. One way to keep it manageable is to introduce some intermediaries or message brokers. Indeed, Fig. 1 already includes the CIR as a central intermediary acting on behalf of all EMSPs to allow a CSO to access client information irrespective of the EMSP.

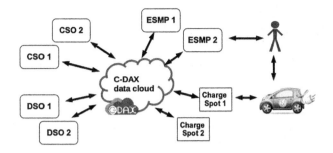

Fig. 2. C-DAX as pub/sub middleware for EV charging

A more structural way of organising information links between many parties is the use of a middleware solution such as pub/sub, short for publish/subscribe. This is a message-oriented middleware solution which provide a central 'data hub' that allows many parties to provide (aka publish) or receive (aka subscribe to) information.

Pub/sub middleware offers advantages of flexibility and scaling. It readily supports one-to-many communication as well as one-to-one communication. It does require a consistent data model to be shared between all parties, but in bilateral connections between individual parties data models have to be synchronised as well.

In the EU project C-DAX, a pub/sub information middleware solution [4] tailored to the smart grid has been developed. The solution has been inspired by and partially built on code of the earlier SeDAX system [11]. Although conceptually one can think of the C-DAX middleware as one central data cloud in which all the information is received and forwarded, as shown in Fig. 2, in reality this data cloud can be distributed over various geographical locations (for example to take into account bandwidth restrictions). Data may also be replicated across various locations to provide higher levels of resilience.

The C-DAX middleware also provides security mechanisms to provide confidentiality and/or integrity of messages, depending on the needs of the application, using either symmetric or asymmetric cryptography [8].

In using a pub/sub middleware solution to organise the information streams for EV charging there are still many configuration possibilities. For example, it may be useful to let EVs or their owners access data in the data cloud or provide data to the cloud. And instead of charge spots directly accessing the data cloud, one could also choose for a solution where they still only provide or obtain data via the responsible CSO.

5   Related Work

Whereas we look at the communication infrastructure and associated security requirements from the grid perspective, most of the literature on the communication infrastructure for EV charging, such as [3], takes the EV perspective, and for instance considers ways in which EVs – or their drivers – could communicate with the grid and what information would then be exchanged.

The idea to use pub/sub middleware solutions for smart grid application is not new. An overview of middleware solutions for smart grid applications, including pub/sub solutions, is given in Sect. 6 of [1]. One of the pub/sub solutions discussed there, the SeDAX system [11], provided the starting point for the pub/sub solution developed in the C-DAX project.

This overview in [1] does not consider the specific scenario of EV charging. Rivera et al. do explore the use and advantages of pub/sub middleware for EV charging, for a more specific goal of optimising distributed EV charging [13].

6 Future Work

The Dutch ElaadNL foundation is working on implementing a specific security design for smart charging, which focusses on end-to-end integrity of meter readings and stronger authentication of EV drivers.

We have not considered interaction with the user and/or the EV yet. This could be in the form of communication between the EV and the charge spot, to communicate wishes for charging, e.g. using ISO 15118 [10], but it could also involve communication between the user, e.g. using a smartphone app, and the EMSP.

One important aspect that we have not considered in this paper is privacy. How to take privacy into account in designing the overall information and communication infrastructure is an important issue. One interesting option to investigate is the use of privacy-friendly aggregation techniques as have been successfully applied for smart metering in homes [5,12], where certain parties can then only see aggregate usage.

7 Conclusions

There were two observations that motivated us to write this paper. Firstly, we observed that the solutions that are being rolled out or trialled for EV charging do not provide true end-to-end security across the whole communication chain of the various parties involved. Given the small scale of EV charging, security may not be much of an issue yet, as the stakes involved are relatively small. But a danger is that retro-fitting security afterwards when these initiatives do grow to larger scales will be difficult. Indeed, it is widely recognised that it is best to practice Security by Design, and take security into account from the earliest stages of any design.

Secondly, we noted that when security is being considered, the security solutions largely relies on the use of secure communication tunnels. This is the case for both OCPP and OSCP. While using secure communication tunnels is a good step, and using standard solutions such as TLS is then the wise thing to do, it is important to realise that this may not take care of all security needs: as we argue in Sect. 3.2, TLS can secure a communication link between two parties, but it will not provide end-to-end security if data is forwarded between parties, and even when it is used between two parties it does not provide practical support for non-repudiation.

We considered two compatible directions for organising and securing the communication needs associated with EV charging: (i) adding security measures at the level of the data, and not just at the level of the communication links, discussed in Sect. 4.1, and (ii) using middleware solutions such as pub/sub that provide a more flexible way of connecting the many parties involved in EV charging, discussed in Sect. 4.2.

W.r.t. (i), given that the management of EV charging involves forwarding communication between multiple parties – including DSOs, CSOs and EMSPs – the right way to tackle security seems to be secure the data being exchanged, and not (just) secure the communication channels over which the data is communicated (e.g. using TLS). We were pleased to note that the newer ISO 15118 standard does provide security guarantees in this way, by having charging records digitally signed by both the charge spot and EV.

W.r.t. (ii), given the number of parties involved, and the fact their roles and business models are still evolving, solutions where all these parties have to bilaterally exchange data may not be practical. Having some central party to collect data and/or coordinate the exchange of data may be a more scalable approach. An interesting analogy here is the exchange of metering data. For this, the DSOs in the Netherlands have set up a joint organisation, called EDSN, to provide a central intermediary for exchanging metering data between DSOs and energy suppliers. EDSN was set up well before the introduction of smart meters, to facilitate billing by energy suppliers who have customers served by different DSOs. One can envisage a similar solution for the exchange of EV charging data. One way to realise this is through the use of pub/sub as a middleware solution. The pub/sub middleware solution developed in the EU FP7 project C-DAX demonstrates that such middleware solutions are feasible even in high-volume, low latency applications and with high guarantees for resilience [4].

Irrespective of whether the solutions we propose are ultimately the best or even feasible, a broader aim of this paper is to raise awareness and encourage debate about ICT and security issues surrounding EV charging. We have only discussed the way EV charging is organised in the Netherlands. Presumably there will be similar initiatives in other countries. By sharing information on how this is organised, there may be much that initiatives in different countries can learn from each other here.

Acknowledgements. The research of Fabian van den Broek and Bárbara Vieira has been supported by the European Community's Seventh Framework Programme FP7-ICT-2011-8 under grant agreement no. 318708 for the C-DAX project (http://cdax.eu). The authors alone are responsible for the content of this paper.

References

1. Ancillotti, E., Bruno, R., Conti, M.: The role of communication systems in smart grids: architectures, technical solutions and research challenges. Comput. Commun. **36**(17), 1665–1697 (2013)
2. Anderson, R., Fuloria, S.: Who controls the off switch. In: International Conference on Smart Grid Communications (SmartGridComm). IEEE (2010)
3. Bayram, I.S., Papapanagiotou, I.: A survey on communication technologies and requirements for internet of electric vehicles. EURASIP J. Wirel. Commun. Netw. **2014**(1), 1–18 (2014)
4. Chai, W.K., Wang, N., Katsaros, K.V., Kamel, G., Melis, S., Hoefling, M., Vieira, B., Romano, P., Sarri, S., Tsegay, T., Yang, B., Heimgaertner, F., Pignati, M., Paolone, M., Develder, C., Menth, M., Pavlou, G., Poll, E., Mampaey, M., Bontius, H.: An information-centric communication infrastructure for real-time state estimation of active distribution networks. IEEE Trans. Smart Grid (2015, to appear)
5. Defend, B., Kursawe, K.: Implementation of privacy-friendly aggregation for the smart grid. In: ACM Workshop on Smart Energy Grid Security. ACM (2013)
6. Engelhardt, M., Pfeiffer, F., Finkenzeller, K., Biebl, E.: Extending ISO/IEC 14443 Type A eavesdropping range using higher harmonics. In: SmartSysTech 2013, pp. 1–8. IEEE (2013)
7. Habraken, R., Dolron, P., Poll, E., De Ruiter, J.: An RFID skimming gate using higher harmonics. In: RFIDsec 2015. Springer, Heidelberg (2015)
8. Heimgaertner, F., Hoefling, M., Vieira, B., Poll, E., Menth, M.M.: A security architecture for the publish/subscribe C-DAX middleware. In: IoT/CPS-Security (IEEE ICC 2015). IEEE (2015, to appear)
9. IEC 62196: Plugs, socket-outlets, vehicle couplers and vehicle inlets - conductive charging of electric vehicles (2003)
10. ISO/TC 22/SC 31 (Road Vehicles/Data communication): ISO 15118: Road vehicles - vehicle to grid communication interface. Technical report, ISO (2013)
11. Kim, Y.-J., Lee, J., Atkinson, G., Kim, H., Thottan, M.: SeDAX: a scalable, resilient, and secure platform for smart grid communications. IEEE J. Sel. Areas Commun. **30**(6), 1119–1136 (2012)
12. Kursawe, K., Danezis, G., Kohlweiss, M.: Privacy-friendly aggregation for the smart-grid. In: Fischer-Hübner, S., Hopper, N. (eds.) PETS 2011. LNCS, vol. 6794, pp. 175–191. Springer, Heidelberg (2011)
13. Rivera, J., Jergler, M., Stoimenov, A., Goebel, C., Jacobsen, H.-A.: Using publish/subscribe middleware for distributed EV charging optimization. Comput. Sci. Res. Dev. **21**, 1–8 (2014)
14. The Smart Grid Interoperability Panel Cyber Security Working Group: Introduction to NISTIR 7628 Guidelines for Smart Grid Cyber Security, September 2010
15. WG15 of IEC TC57: IEC 62351: Power systems management and associated information exchange data and communications security (2007)

Workshop on Advanced Next Generation Broadband Satellite Systems (BSS) 1

Capacity Enhancing Techniques for High Throughput Satellite Communications

Svilen Dimitrov[1]([✉]), Stefan Erl[1], Benjamin Barth[1], Rosalba Suffritti[2],
Niccoló Privitera[2], Gabriele Boccolini[3], Adegbenga B. Awoseyila[4],
Argyrios Kyrgiazos[4], Barry G. Evans[4], Stephan Jaeckel[5], Belén Sánchez[6],
Ana Yun Garcia[6], and Oriol Vidal[7]

[1] Satellite Networks Department, German Aerospace Center (DLR),
82234 Wessling, Germany
{svilen.dimitrov,stefan.erl,benjamin.barth}@dlr.de
[2] Mavigex S.r.l., ICT Solutions, 40125 Bologna, Italy
{rsuffritti,niccolo.privitera}@mavigex.com
[3] Galician R&D Center in Advanced Telecommunications (GRADIANT),
36310 Vigo, Spain
gboccolini@gradiant.org
[4] Centre for Communication Systems Research, University of Surrey,
Guildford GU27XH, UK
{a.awoseyila,a.kyrgiazos,b.evans}@surrey.ac.uk
[5] Fraunhofer Heinrich Hertz Institute, 10587 Berlin, Germany
stephan.jaeckel@hhi.fraunhofer.de
[6] Thales Alenia Space Espana, Madrid, Spain
{belen.sanchez,ana.yungarcia}@thalesaleniaspace.com
[7] Telecom Systems Department, Airbus Defence and Space, Toulouse, France
oriol.vidal@astrium.eads.net

Abstract. In this paper, we present physical layer and system level techniques that can increase the capacity of a multi-beam high throughput satellite (HTS) communication system. These include advanced predistortion and equalization techniques for the forward link, synchronization and non-linear distortion minimization techniques for the return link waveforms. Interference management techniques such as precoding, multi-user detection (MUD), interference cancellation and coordination are evaluated under realistic co-channel interference (CCI) conditions. Interference-aware radio resource management (RRM) algorithms are presented for the forward and return links with full frequency reuse. These include satellite-switched smart gateway diversity, interference-aware scheduling (IAS) for the forward link and scheduling based on multi-partite graph matching for the return link. The capacity enhancing techniques are evaluated in the Broadband Access via Integrated Terrestrial and Satellite Systems (BATS) framework.

Keywords: Satellite communications · Interference management · Radio resource management · Spectral and power efficiencies

© Institute for Computer Sciences, Social Informatics and Telecommunications Engineering 2015
P. Pillai et al. (Eds.): WiSATS 2015, LNICST 154, pp. 77–91, 2015.
DOI: 10.1007/978-3-319-25479-1_6

1   Introduction

In order to serve the increasing demand for satellite broadband [1], a multi-beam HTS system needs to employ a very high resource reuse over the coverage area, combined with intelligent interference management and resource management mechanisms to alleviate the resulting CCI. In addition, the utilization of higher pieces of bandwidth, for example in the Ka band, requires the application of suitable signal processing techniques and waveforms, in order to efficiently manage the interference in the physical layer.

In this paper, non-linear pre-distortion and equalization techniques are evaluated in a multi-carrier scenario in the forward satellite link. In addition, fast-Fourier-transform-(FFT)-based waveforms, such as single-carrier frequency division multiple access (SC-FDMA), orthogonal FDMA (OFDMA) and filter-bank multi-carrier (FBMC) are compared in a return link setup in terms of power and spectral efficiencies, and suitable synchronization algorithms for the satellite channel are proposed.

In addition, interference management techniques are evaluated under realistic CCI conditions. Techniques such as precoding, multi-user detection, interference cancellation and interference coordination allow the use of more aggressive frequency reuse patterns as compared to conventional frequency reuse 4. Significant gains in terms of available system bandwidth and total throughput are presented, in particular when multi-gateway cooperation strategies are considered.

The fixed superframe length introduced with the DVB-S2X standard allows for burst mode transmission, as is the case with the return link DVB-RCS2 standard. While the return link inherently benefits from RRM and scheduling to mitigate the CCI, the forward link can only benefit, if a number of the interfering beams are switched off at a given time step or if the system is not under full load. In this paper, the performance of several RRM solutions is evaluated in the forward and return satellite links of a practical HTS system.

The rest of the paper is organized as follows. Section 2 presents the study of physical layer techniques. Section 3 elaborates on interference management techniques, while Sect. 4 presents the radio resource management algorithms. Finally, Sect. 5 concludes the paper.

2   Waveform

In the forward link DVB-S2X standard, static predistortion is applied at the gateway to the time division multiplexing (TDM) waveform to compensate the constellation warping due to the nonlinearities in the channel. In addition, a linear equalizer is employed at the user terminal to counter both linear and non-linear distortion. However, the evaluation of more potent compensation techniques, such as memory polynomials predistortion and equalization [2], for the newly introduced roll-off factors of 10 % and 5 %, is still an open issue.

In this paper, we also study the application of FFT-based waveforms in the return link and compare their power and spectral efficiencies with time division multiple-access (TDMA) with 5 % and 20 % roll-offs. The FFT-based waveforms depend strongly on frame timing and carrier frequency synchronization of all users in order to mitigate inter-symbol interference (ISI), inter-carrier interference (ICI) and multiple access interference (MAI).

2.1 Non-linear Compensation by Predistortion and Equalization in Forward Link

In a multi-carrier satellite forward link, predistortion and equalization algorithms are applied jointly to more than one carrier with the goal of exploiting mutual correlation between the carriers that share the same satellite transponder. The mean-squared error (MSE) performance for 32-APSK is presented in Fig. 1(a). The optimum combination of non-linear predistortion and equalization has been selected on the basis on the MSE results w.r.t the baseline scenario without predistortion and equalization. The analysis has been performed for roll-off values in the set of $[0.25, 0.10, 0.05]$. The MSE improvements at the modulation-dependent working input back-off (IBO) have been computed, and the most promising techniques are summarized in Fig. 1(b).

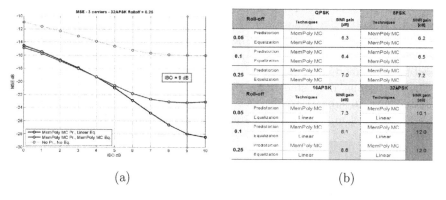

(a)   (b)

Fig. 1. (a) MSE vs. IBO for advanced predistortion/equalization, 32-APSK, 3 carriers, 0.25 roll-off. (b) Non-linear compensation techniques and MSE gains.

2.2 Synchronization Acquisition in Return Link

Synchronization acquisition focuses mainly on frame timing at log-on to correct for differential propagation delays in the presence of residual frequency offsets and amplifier nonlinearity. Timing alignment of the uplink transmissions is achieved by applying a timing advance at each user terminal [3]. However, the propagation delays in the satellite channel with round-trip time (RTT) of 500 ms and differential propagation delays within a spot beam in the order of ms, as compared to an order of μs in a terrestrial link, are considerably larger. Given a typical symbol duration of an FFT-based waveform in the order of μs, a guard time greater than the satellite RTT or even the maximum differential propagation delay is not affordable. In addition, the cyclic prefix (CP) cannot be dimensioned for using cyclic shifts of a Zadoff-Chu (ZC) sequence.

In order to solve these problems, we propose the use of global-positioning-system-(GPS)-based timing pre-compensation. A GPS device at each return channel via satellite terminal (RCST) can track its location. This can be combined with existing satellite and network control center (NCC) data to calculate the two-way propagation delay and pre-compensate it before transmitting the random access (RA) preamble. In this way, the timing misalignment between RCSTs can be reduced from ms to ns, resulting in a very low CP overhead.

2.3   Synchronization Tracking and Estimation in Return Link

After synchronization is acquired, estimation of residual timing error (TE) and carrier frequency offset (CFO) is used for fine synchronization and tracking. The Mengali and Morelli (M&M) algorithm [4] is chosen for tracking timing offset since it: (1) does not show a definite threshold effect in terms of signal-to-noise ratio (SNR); (2) requires low complexity due to the adoption of two implementation strategies, one for small number of points and the other for larger sets; (3) has an estimation range independent from the number of points in the observed sequence. The Moose algorithm [5] is selected for tracking frequency offsets due to its closeness to the Cramer-Rao bound. The remaining static phase offset can be recovered using phase locked loop (PLL) filters before the signal is passed to the demodulator [6,7].

The algorithms are tested in the SC-FDMA setup with 128 subcarriers in the frame with 16 subcarriers per user. A pilot period of 10 SC-FDMA symbols and a CP of 16 samples are considered. Figure 2 shows the MSE performance for different values of CFO, TE and SNR. A normalized CFO of 0.1 means that

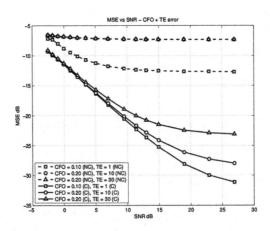

Fig. 2. MSE vs. SNR, CFO and TE compensation. NC stands for "Not Compensated", while C denotes "Compensated".

terminal and gateway oscillators are misaligned by $0.1 * \Delta f$, where Δf is the subcarrier spacing. TE of 1 means that the acquisition stage outputs a signal with a relative drift of 1 sample w.r.t. the nominal start of frame. The M&M and Moose algorithms present sufficiently low MSE at low SNR below 0 dB, and the performance is degraded only when the TE approaches the size of the CP.

2.4   Power Efficiency Optimization in Return Link

The power efficiency of the transmission schemes can be maximized by optimization of output back-off (OBO) and minimization of total degradation (TD) in the non-linear channel. In this study, the SNR requirement and spectral efficiency (SE) are compared for SC-FDMA, OFDMA and state-of-the-art TDMA (with 5 % and 20 % roll-off) with QPSK, 8-PSK and 16-QAM for a practical Ka-band amplifier model. In OFDMA and SC-FDMA, 2048 digital subcarriers per physical carrier are allocated to 32 users with 64 subcarriers per user. In addition, a 5 % guard band is included between the physical carriers [3]. An IMI cancelling receiver is studied, where an estimate of the intermodulation interference (IMI) is calculated and subtracted from the received signal in an iterative fashion [8].

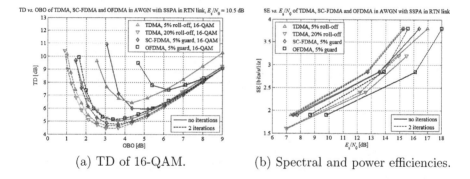

(a) TD of 16-QAM.   (b) Spectral and power efficiencies.

Fig. 3. Power and spectral efficiencies of TDMA, OFDMA and SC-FDMA in return link.

In order to quantify the power efficiency penalty induced by the amplifier nonlinearity, the TD metric for a target bit-error ratio (BER) is defined as follows:

$$TD[dB] = OBO + SNR_{\substack{req.non-linear \\ channel}} - SNR_{\substack{req.linear \\ channel}} . \qquad (1)$$

Results for the TD of 16-QAM and the SE of the DVB-RCS2 waveforms are presented in Fig. 3. The chosen reference SNR targets in the linear channel are 3.7 dB, 8.9 dB and 10.5 dB, respectively, for a BER target of 10^{-3}. The gains in power efficiency for the three modulation orders with 2 iterations of IMI cancellation can be summarized as follows: $0.1 - 0.7$ dB for TDMA with 20 % roll-off, $0.2 - 1.8$ dB for TDMA with 5 % roll-off, $0.3 - 1.1$ dB for SC-FDMA and

1.1 − 2.5 dB for OFDMA. SC-FDMA demonstrates an overall improvement of the power efficiency of more than 2 dB as compared to state-of-the-art TDMA with 20 % roll-off.

2.5 Spectral Efficiency Improvement in Return Link

In the return link, the received signal is the sum of all user contributions, each one with its own synchronisation offsets and non-linear distortion. These impairments can be mitigated with the insertion of guard band (GB) between users. The SE can be improved by the use of FBMC with staggered multitone (SMT). In OFDMA, 5 % of GB between physical carriers is included. In FBMC, 2 % of GB is considered to take into account non-linear distortions. Table 1 shows the efficiency in terms of percentage of subcarriers used for data over the total number of 2048 subcarriers. With 1 resource block (RB) of 12 subcarriers assigned to each user, the best efficiency is achieved by OFDMA without GB. However, as soon as the number of GB subcarriers is increased to reduce the effect of synchronization errors, SMT turns out to be more efficient than OFDMA. SMT efficiency increases, when a larger number of RBs is allocated to each user as reported in the table for 10 RBs.

Table 1. Comparing efficiencies of FBMC (SMT) and OFDM.

	GB Carriers	GB Users	N_u 1RB	Eff 1RB	N_u 10RB	Eff 10RB
OFDMA	5 %	0 SC	162	95 %	16	95 %
OFDMA	5 %	1 SC	149	87 %	16	94 %
OFDMA	5 %	2 SC	139	81 %	15	93 %
SMT	2 %	1 SC	154	91 %	16	97 %

3 Interference Management

A typical multi-beam satellite system relies on reusing the available frequency bands multiple times across the antenna beams. In such a scenario, the interference introduced by nonorthogonality between beam patterns represents an important issue to be solved in order to improve the capacity of the satellite network. The considered system model is based on a set of realistic data on a multi-beam satellite network covering the European region, composed of 302 spot beams of 0.21 degrees in beamwidth, and operating in Ka-band. Frequency reuse 4 (FR-4) represents the benchmark scenario for the forward and return links of the considered system that foresees two options for the feeder link: the Q/V feeder link and the optical feeder link, where in both cases 1.5 GHz of spectrum is allocated in the forward link and 500 MHz in the return link. In addition to FR-4, also FR-2 and FR-1 schemes have been considered. The beams have been grouped in clusters and every cluster is served by one gateway (GW). GWs addressing different clusters can also cooperate to apply jointly the interference management techniques to the entire set of considered beams, such as in an optical feeder link scenario.

3.1 Forward Link

The interference management schemes in the forward link can be applied at the transmitting GW. The signals referring to the same cluster can be combined using, for example, precoding algorithms before transmission over the uplink to minimize the distortions of every user-received signal resulting from nonorthogonality between the antenna beams. In addition, fractional frequency reuse (FFR) can be also taken into account as a smart resource reuse strategy. In the following, more details on each technique are provided along with the main outcomes of the performance assessment.

Precoding: Linear and non-linear precoding algorithms have been analysed and compared in the considered scenario [9]. The most promising results come from non-linear techniques such as Tomlinson-Harashima Precoding (THP) which uses modulo arithmetic in order to jointly achieve pre-equalization and power limitation of the transmitted signal [10]. The results show that increasing the collaboration between gateways up to the entire set of gateways (optical feeder link scenario) leads to complete cancellation of the interference [9]. The gains in terms of SE and s/N0 are presented in Fig. 4(a) for the case of 4 gateway cooperation.

Fractional Frequency Reuse: Two FFR schemes, A and B, have been evaluated [11] and compared with FR-4 which reuses 50 % of the total bandwidth in each beam. This is increased to 75 % in FFR scheme A, where 50 % are reused in

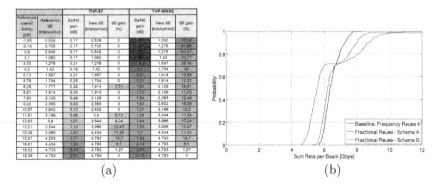

(a) (b)

Fig. 4. (a) Es/N0 and SE improvement with THP precoding, FR-4, 4 cooperating GWs. (b) Achievable sum rate for different reuse schemes.

the beam centres and 25 % are exclusively used for the beam edge. In the FFR scheme B, the full bandwidth is reused: 50 % in the beam centres and 50 % in the beam edges. The CDF of the sum rates is presented in Fig. 4(b). FFR strongly increases the maximum rate. However, the results also indicate a strong impact of the utilized beam patterns on the overall system performance, since the beam patterns have been optimized for FR-4. In order to show the anticipated gain of FFR in the satellite system, the power difference between beam center and beam edge (currently 3 dB) needs to be increased. Since the shape of the beam pattern is determined by the antenna, a reorganization of the beam layout is advisable in order to maximize the throughput of FFR.

3.2 Return Link

In a multi-user system, a suboptimal approach is to detect each user independently by means of a bank of matched filters, considering the interfering users as random noise. However, this approach does not take into account the information content in the interfering signals. A multi-user detector (MUD) exploits this information in order to achieve a higher system capacity. Within the MUD family, interference cancellation schemes have an important role since they achieve a good performance with a lower complexity. In addition, another class of techniques applicable in return link can be classified as static, semi-static, and dynamic inter-cell interference coordination. In the following, these techniques are elaborated, and the main outcomes of the performance assessment are reported.

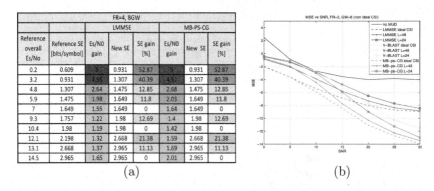

| FR=4, 8GW | | | | | | | |
| Reference overall E_s/N_0 | Reference SE [bits/symbol] | LMMSE | | | MB-PS-CG | | |
		Es/N0 gain	New SE	SE gain [%]	Es/N0 gain	New SE	SE gain [%]
0.2	0.609	5.	0.931	52.87	5	0.931	52.87
3.2	0.931	1.55	1.307	40.39	4.51	1.307	40.39
4.8	1.307	2.64	1.475	12.85	2.68	1.475	12.85
5.9	1.475	1.98	1.649	11.8	2.03	1.649	11.8
7	1.649	1.55	1.649	0	1.64	1.649	0
9.3	1.757	1.22	1.98	12.69	1.4	1.98	12.69
10.4	1.98	1.19	1.98	0	1.42	1.98	0
12.1	2.198	1.32	2.668	21.38	1.59	2.668	21.38
13.1	2.668	1.37	2.965	11.13	1.69	2.965	11.13
14.5	2.965	1.65	2.965	0	2.01	2.965	0

(a)   (b)

Fig. 5. (a) E_s/N_0 and SE improvement with MUD, FR-4, 8 GWs. (b) MSE vs. SNR, FR-2, 8 GWs.

Multi-user Detection: The following MUD techniques are studied: linear techniques, such as zero-forcing and linear MMSE filtering, successive interference cancellation (SIC), and a multi-branch decision feedback (MB) technique. MB exploits different ordering patterns to improve the performance of a SIC algorithm, and a maximum-likelihood detector makes the final decision over the

Reference overall Es/No [dB]	Reference SE [bits/symbol]	MMSE-SIC			MMSE-CPIC		
		Es/NO gain [dB]	New SE [bits/symbol]	SE gain [%]	Es/NO gain [dB]	New SE [bits/symbol]	SE gain [%]
0,2	0,609	5.57	0,931	52,87	5.37	0,931	52,87
3,2	0,931	5.61	1,307	40,39	5.42	1,307	40,39
4,8	1,307	2.78	1,475	12,85	2,85	1,475	12,85
5,9	1,475	2.12	1,649	11,8	2,23	1,649	11,8
7	1,649	1,67	1,649	0	1,81	1,649	0
9,3	1,757	1,46	1,98	12,69	1,61	1,98	12,69
10,4	1,98	1,44	1,98	0	1,62	2,198	11,01
12,1	2,198	1,56	2,668	21,38	1,74	2,965	34,9
13,1	2,668	1,85	2,965	11,13	1,8	2,965	11,13
14,5	2,965	1,97	2,965	0	2,12	2,965	0

Scenario (0.7 IUR)	Capacity Δ % wrt to unscheduled scenario
Unscheduled	0
Scheduled - Min	43.25
Scheduled - Average	109.5
Scheduled - Max	147.26

(a)   (b)

Fig. 6. (a) E_s/N_0 and SE improvement with IC, FR-4, 8 GWs. (b) Return link capacity with and without scheduling, FFR scenario.

list of candidates generated by the multiple branches. The ordering patterns are obtained from permutation of the first branch ordering which can be based on complex channel gain ordering (MB-CG) or best output signal-to-noise-and-interference ratio ordering (MB-SNIR). MB-CG achieves similar results as SIC with optimum ordering (SIC-SNIR), while requiring a significantly lower complexity, whereas MB-SNIR can achieve a gain of $3 - 5$ dB in interference-limited scenarios such as in the case of FR-1, but it does not improve significantly the performance of SIC-SNIR in case of FR-4 and FR-2. Nevertheless, at low SNR (less than 10 dB), all the considered techniques show similar performance, therefore LMMSE becomes a good low-complexity candidate for noise-limited scenarios as shown in Fig. 5(a), where spectral efficiency improvements achieved with LMMSE and MB-CG are reported.

MUD techniques are also sensitive to channel estimation errors but the diversity achieved through the MB approach allows to counteract the non-ideal channel state information (CSI) better than the other techniques [12]. As reported in Fig. 5(b) for FR-2, MB with training sequence L= 24 has the same MSE as V-BLAST with training sequence L= 48 for SNR greater than 15 dB.

Interference Cancellation: Interference cancellation (IC) techniques have been studied in the considered scenario extending the analysis also to lower frequency reuse factors and a variable number of collaborating gateways. In particular, one of the selected techniques is the MMSE successive interference cancellation (MMSE-SIC), where each iteration is composed of a spatial MMSE filtering followed by the demodulation and decoding of the user terminal having the highest value of SNIR [13]. In addition, MMSE filters can be also applied to the convention parallel interference cancellation (CPIC) technique [14].

The gains in terms of spectral efficiency and required E_s/N_0 in case of 8 cooperating gateways are shown in Fig. 6(a). MMSE-SIC and MMSE-CPIC have confirmed the results achieved in terms of MSE [9] also in terms of required SNR and spectral efficiency improvements.

Interference Coordination. Interference coordination (ICIC) schemes [15], such as FFR with scheduling to coordinate the users transmissions in the frequency and time domain, can also improve the SNIR performance and spectral efficiency. An average carrier-to-interference ratio (C/I) improvement of 3 dB is observed for the users in the outer beam, while the improvement is greater than 3 dB for the users inside the inner beam [16]. The resulting capacities for the FFR scenario with an inner beam user ratio (IUR) equal to 0.7 (with 50 % more bandwidth per beam as compared to FR-4), reported in Fig. 6(b), show an average improvement in the return link capacity of 109.5 %.

4   Radio Resource Management

Interference-aware RRM is considered as the key to the realization of very high resource reuse HTS systems. In the forward link, if the system is under full load, i.e. all the beams are active on all the resources, the CCI experienced by an intended user is constant and cannot be alleviated through scheduling. However, if there are free resources, users with low carrier-to-noise ratio (C/N) can be scheduled on resources which are lightly used across the beams. Alternatively, a number of interfering beams can be switched off at a given time step. In the return link, based on the association of users in the co-channel beams to share a system resource (even at full load), different CCI is experienced by the users. Through optimization of the user associations and scheduling over the available resources, the CCI can be minimized and the user throughput maximized.

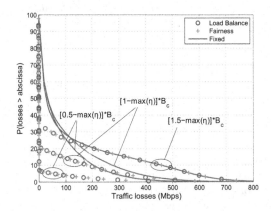

Fig. 7. CCDF of traffic losses for a fixed scheduling and the proposed algorithms for 4 gateways in a cluster.

4.1   Satellite-Switched Smart Gateway Diversity for Feeder Link

Considering a time switched payload, more flexibility in terms of interconnections between feeder and user links is introduced into the system. The time that each feeder link is connected with a user link may vary over time. A framework which takes into account the atmospheric conditions of each gateway plus the traffic demands of the beams and decides the number of time slots that each gateway is connected to a user beam is proposed allowing a match of the instantaneous user demands with the offered capacity of the gateways. The dwelling time of a gateway to a user beam can be tuned such as to minimize the capacity losses, keeping as low as possible the number of gateways in the system whilst providing the required availability. Two objective functions have been considered herein: (1) one that maximises the product of the satisfaction ratios (offered traffic/traffic demands) for the sake of load balance; and (2) another that maximises the minimum satisfaction ratio for the sake of fairness. Figure 7 shows the complementary CDF of the traffic losses for three traffic scenarios, where the users traffic demands are drawn from a uniform distribution $[\{0.5,1,1.5\}\text{-max}(\eta)]^*B_c$, considering 4 gateways in a cooperation cluster. The maximum achievable spectral efficiency, $\text{max}(\eta)$, depends on the link budget, detailed for the BATS system in [17] for $B_c = 447.3$ MHz. An important observation is the reduction of traffic losses due to the better match of feeder transmissions rates with the user beam demands. The effectiveness of the proposed methods is improved as the number of gateways in the cluster or the traffic variance increase.

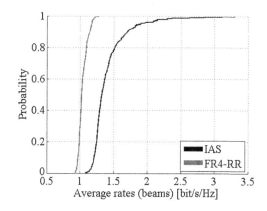

Fig. 8. Average rate per beam in the BATS scenario.

4.2   Interference-Aware Scheduling for Forward User Link

The IAS algorithm [18] for full reuse in the forward link utilizes partial channel state information (CSI) to reduce the number of interfering beams on the served users. Neighbouring beams are switched off, if a served user cannot handle the interference caused by these beams. As a result, it is best suitable for burst mode beam hopping applications. The performance of the IAS is compared to a FR4 Round Robin scheme. A noise power of -82.4 dBm and a transmit power of 47.4 dBm per beam are assumed. The clear-sky received power of the users' strongest beam varies between -74 dBm and -64.7 dBm due to path loss and antenna gains of transmit and receive antenna. The user rates R are calculated by the Shannon function. The system rate is calculated by averaging the rates of the beams normalized to the full system bandwidth of 2.9 GHz in the Ka band forward user link. In Fig. 8, the CDF of the average rate of the beams is presented. The IAS is able to increase the minimum, maximum and mean data rate of the all users. Since TDM is employed, disabling of beams does not relate to an outage, but rather a delay in serving the traffic request. However, no noteworthy latency is observed compared to the FR4-RR. Both the IAS and the FR4-RR presented no SNIR values below the outage threshold of -1.85 dB defined in the BATS system, therefore meeting the availability requirement of 99.7 %. The IAS is able to achieve an average system rate of 1.415 bit/s/Hz, while the FR4-RR only achieves an average system rate of 1.043 bit/s/Hz, which represents a gain of 35.7 %.

4.3 Scheduling Based on Multi-partite Graph Matching for Return User Link

Scheduling algorithms for full reuse based on multi-partite graph matching that maximize the minimum user rates have been presented in [19]. To reduce the computational complexity, users are split into clusters. The assumption is made that users in second tier beams contribute only negligibly to the CCI. Therefore, a virtual 4-color scheme is applied, and the users in a color are scheduled randomly. These random cluster schedules are merged together by applying a minimum deletion algorithm and graph matching techniques to find the optimum user association. The low-complexity distinct merging is a greedy approach that concatenates the clusters sequentially. The globally optimum common approach merges all clusters in one step. A variation of both techniques is the free slot assignment (FSA) method, where one slot per schedule is left unallocated to increase the flexibility in the scheduler. In this study, the scheduling performance is evaluated with and without the application of SIC at the gateway. A bandwidth of 1.05 GHz in Ka band is considered in 302 beams over Europe, and the DVB-RCS2 MODCODs are applied. The results for the capacity and availability improvements are summarized in Table 2. It has to be noted that either scheduling or SIC need to be employed with full reuse so that the system meets the availability requirement of 99.7 %. In terms of system capacity, the common merging gives the best results, i.e. the average system capacity is increased by 2.7 % in the case with SIC and by 3.5 % in the case without SIC as compared to the random scheduling method. The distinct method offers a good compromise between performance and computational complexity. In the case with SIC, the FSA algorithm has a performance comparable to the random scheduling, while it is slightly better in the case without SIC. However, it should be noted that the full frequency reuse system with scheduling and SIC is able to double the capacity as compared to the state-of-the-art 4-color scheme with random scheduling, while it still offers more than 50 % increase, when SIC is not employed.

Table 2. System capacity and availability improvements.

FR	SIC	Scheduling	Capacity [Gbps]		Availability
			Value	Increase	
4	No	Random	384.6	–	99.99 %
1	No	Random	575.7	49.7 %	98.92 %
		Distinct	585.2	52.2 %	99.97 %
		Distinct FSA	578.4	50.4 %	99.97 %
		Common	596	55 %	99.97 %
		Common FSA	586.7	52.5 %	99.97 %
	Yes	Random	746.2	94 %	99.99 %
		Distinct	762	98.1 %	99.99 %
		Distinct FSA	742.9	93.2 %	99.99 %
		Common	766	99.2 %	99.99 %
		Common FSA	746.8	94.2 %	99.99 %

5   Conclusion

In this paper, several capacity enhancing techniques have been evaluated for application in HTS communication systems. In the physical layer, non-linear predistortion and equalization techniques are essential to preserve the spectral efficiency gains of higher order modulations when high baudrates and very low roll-off factors are employed. Performance gains in terms of SIR improvement in the range from 7.3 dB to 12 dB have been reported. The application of FFT-based waveforms in the return link has shown the highest spectral efficiency in the non-linear channel, especially when an IMI cancelling receiver is employed, whereby SC-FDMA demonstrated an overall power efficiency gain of more than 2 dB as compared to state-of-the-art TDMA. GPS-based delay pre-compensation has been shown to enable synchronization acquisition with negligible CP overhead. Timing and frequency offsets are best estimated and tracked by means of the M&M algorithm and the Moose algorithm, respectively, showing significant MSE improvement even at low SNR. The adoption of interference management techniques in multi-beam satellite systems allows the use of more aggressive frequency reuse patterns and provides promising improvements in terms of available system bandwidth and total throughput. CCI-aware RRM is shown to be an enabling factor for realization of full resource reuse HTS communication systems. Satellite-switched smart gateway diversity with scheduling helps to reduce the total number of gateways in the system, while ensuring the necessary feeder link availability in rain fading. The IAS approach for full frequency reuse in the forward link and the scheduling based on multi-partite graph matching in the return link respectively offer 35.7 % and 99.2 % increase of the system capacity as compared to state-of-the-art 4-color reuse with random scheduling.

Acknowledgement. This work has been supported by the BATS research project funded by the EU-FP7 under contract n317533.

References

1. Broadband Access via Integrated Terrestrial and Satellite Systems (BATS), ICT-2011.1.1 BATS D4.1: Satellite Network Mission Requirements, European Project, Technical report (2012)
2. Ding, L., Zhou, G.T., Morgan, D.R., Zhengxiang, M., Kenney, J.S., Jaehyeong, K., Giardina, C.R.: A robust digital baseband predistorter constructed using memory polynomials. IEEE Trans. Commun. **52**(1), 159–165 (2004)
3. Technical Specifcation Group Radio Access Network; Evolved Universal Terrestrial Radio Access (EUTRA); Physical Layer Procedures (Rel. 11), 3rd Generation Partnership Project Std. 3GPP TS 36.213 (2013)
4. Mengali, U., Morelli, M.: Data-aided frequency estimation for burst digital transmission. IEEE Trans. Commun. **45**(1), 23–25 (1997)
5. Moose, P.H.: A technique for orthogonal frequency division multiplexing frequency offset correction. IEEE Trans. Commun. **42**(10), 2908–2914 (1994)
6. Benfatto, D., Privitera, N., Suffritti, R., Awoseyila, A., Evans, B., Dimitrov, S.: On acquisition and tracking methods for SC-FDMA over satellite. In: Proceedings of 7th Advanced Satellite Multimedia Systems Conference (ASMS 2014), Livorno, Italy, 8–10 September 2014
7. Dimitrov, S., Boccolini, G., Jaeckel, S., Benfatto, D., Privitera, N., Suffritti, R., Awoseyila, A., Evans, B.: FFT-based waveforms for high through satellite communications: opportunities and challenges. In: Proceedings of 20th Ka and Broadband Communications Conference, Vietri sul Mare/Salerno, Italy, 1–3 October 2014
8. Dimitrov, S., Privitera, N., Suffritti, R., Boccolini, G., Awoseyila, A., Evans, B.: Spectrally efficient waveforms for the return link in satellite communication systems. In: Proceedings of European Conference on Networks and Communications 2015 (EUCNC 2015), Paris, France, 29 June–2 July 2015
9. Suffritti, R., Privitera, N., Dimitrov, S., Katona, Z., Boccolini, G., Jaeckel, S., Raschkowski, L., Kyrgiazos, A., Evans, B., Rodriguez, J. M., Yun, A., Vidal, O., Inigo, P.: On interference management techniques in broadband satellite systems. In: Proceedings of Ka Conference (2014)
10. Liu, J., Krzymien, W.: A novel nonlinear precoding algorithm for the downlink of multiple antenna multi-user systems. Proc. IEEE Veh. Technol. Conf. **2**, 887–891 (2005)
11. Chang, R.Y., Tao, Z., Zhang, J., Kuo, C.-C.J.: Dynamic fractional frequency reuse (D-FFR) for multicell OFDMA networks using a graph framework. Wirel. Commun. Mob. Comput. **13**(1), 12–27 (2011)
12. Arnau, J., Mosquera, C.: Multiuser detection performance in multibeam satellites links under imperfect CSI. In: Proceedings of Asilomar (2012)
13. Gallinaro, G., Debbah, M., Muller, R., Rinaldo, R., Vernucci, A.: Interference mitigation for the reverse-link of interactive satellite networks. In: Proceedings of 9th International Workshop on Signal Processing for Space Communications (2006)
14. Chan, A., Wornell, G. W.: A class of asymptotically optimum iterated-decision multiuser detectors, In: Proceedings of IEEE International Conference on Acoustics, Speech, and Signal Processing (ICASSP 2001), vol. 4, pp. 2265–2268 (2001)

15. Pateromichelakis, E., Shariat, M., Quddus, A., Tafazolli, R.: On the evolution of multi-cell scheduling in 3GPP LTE / LTE-A. IEEE Commun. Surv. Tutorials **15**(2), 701–717 (2012)
16. Ng, U.Y., Kyrgiazos, A., Evans, B.: Interference coordination for the return link of a multibeam satellite system. In: Proceedings of 7th Advanced Satellite Multimedia Systems Conference (2014)
17. Dimitrov, S., Erl, S., Jaeckel, S., Rodriguez, J. M., Yun, A., Kyrgiazos, A., Evans, B., Vidal, O., Inigo, P.: Radio resource management for forward and return links in high throughput satellite systems. In: Proceedings of 20th Ka and Broadband Communications Conference, Vietri sul Mare/Salerno, Italy, 1–3 October 2014
18. Dimitrov, S., Erl, S., Barth, B., Jaeckel, S., Kyrgiazos, A., Evans, B.: Radio resource management techniques for high throughput satellite communication systems. In: Proceedings of European Conference on Networks and Communications 2015 (EUCNC 2015), Paris, France, 29 June–2 July 2015
19. Boussemart, V., Berioli, M., Rossetto, F.: User scheduling for large multi-beam satellite MIMO systems. In: Proceedings of the IEEE Asilomar Conference on Signals, Systems and Computers 2011 (ASILOMAR 2011), pp. 1800–1804 (2011)

Intelligent Gateways Enabling Broadband Access via Integrated Terrestrial and Satellite Systems

Luc Ottavj[1(✉)], Emmanuel Duros[1], Jacques Webert[1], Fabrice Gamberini[1], Christian Niephaus[2], Javier Perez-Trufero[3], and Simon Watts[3]

[1] OneAccess Networks, 2455 Route Des Dolines – BP355,
06906 Sophia Antipolis, France
{luc.ottavj,emmanuel.duros,jacques.webert,
fabrice.gamberini}@oneaccess-net.com
[2] Schloss Birlinghoven, Fraunhofer-Fokus, 53757 Sant Augustin, Germany
Christian.niephaus@zv.fraunhofer.de
[3] Avanti Communications Ltd, 20 Black Friars Lane, London, EC4V 6EB, UK
{javier.pereztrufero,simon.watts}@avantiplc.com

Abstract. Satellite broadband systems will play a key role in reducing the Digital Divide by complementing terrestrial networks in the delivery of next generation broadband to users in remote and rural locations. We describe an integrated broadband delivery system to fixed users that makes simultaneous use of heterogeneous access networks in order to optimize the end-user QoE. The design of the overall network architecture and the key building blocks of the routing entities at both ends of the integrated system are presented. The paper argues that MPTCP is an appropriate mechanism to offer hybrid communications over heterogeneous links and details a specific implementation of the intelligent routing gateways. Results from in-factory validation tests of the prototyped platforms are presented and discussed.

Keywords: Satellite broadband · Integrated terrestrial and satellite systems · MPTCP · Intelligent routing · Digital divide

1 Introduction

The research project BATS (Broadband Access via integrated Terrestrial & Satellite systems), co-funded by the European Commission (EC) under the FP7 programme, addresses the delivery of Broadband (BB) future services in Europe according to the EC Digital Agenda [1] objective to reliably deliver > 30 Mbps to 100 % of European households by 2020. Next generation geostationary (GEO) broadband satellite systems will play a key role in achieving such objectives as the accelerated deployment of terrestrial broadband technology will not be able to satisfy this requirement in the most difficult-to-serve locations. Market studies indicate that in a significant number of regions of Europe, more than 50 % of premises, will lack access to superfast Broadband [2, 3], either due to a lack of coverage in areas where the revenue potential for terrestrial service providers is too low

© Institute for Computer Sciences, Social Informatics and Telecommunications Engineering 2015
P. Pillai et al. (Eds.): WiSATS 2015, LNICST 154, pp. 92–102, 2015.
DOI: 10.1007/978-3-319-25479-1_7

(unserved areas) or due to technological limitations which diminish the available end-user data rate in some suburban and many rural environments (underserved areas).

The BATS project [4] aims to bridge the potentially widening Broadband divide between urban and rural areas and fulfil the Digital Agenda targets in the underserved areas via an integrated network that combines the flexibility, large coverage and high capacity of future multi-spot beam satellites, the low latency of fixed DSL lines, and the pervasiveness of mobile-wireless access. The integrated broadband service will be delivered to the end-user via an *Intelligent User Gateway (IUG)* and *Intelligent Network Gateway (ING)*, dynamically routing traffic flows according to their service needs through the most appropriate broadband access network, with the goal of optimizing the user's Quality of Experience (QoE). Due to the heterogeneity of the various technologies (e.g. satellite broadband offers high bandwidth but higher latency as opposed to narrow-band xDSL low latency links) randomly distributing traffic among the different connections despite their different characteristics could in turn affect negatively the service quality. More sophisticated methods are required which allow for a simultaneous use of the different links in a seamless manner fully exploiting their particular benefits.

In this paper we present the IUG and ING design approach followed in the BATS project and argue that MPTCP [8] is an optimal mechanism to enable Quality-of Service (QoS)-aware Link Selection, key requirement for the integration of satellite and terrestrial networks.

The remainder of this article is organized as follows. Section 2 describes the overall integrated broadband architecture and focuses on the functional design of the key enabling modules in the IUG and ING. Section 3 provides results from the in-factory tests of the prototype IUG and ING to be used in the laboratory and field trials of the project. Finally, Sect. 4 concludes the article and provides some insight into our future activities.

2   Network Architecture and Key Enabling Components

2.1   The Overall Integrated Network Architecture

As illustrated in Fig. 1, the overall network architecture comprises the three broadband access segments, namely xDSL, cellular and satellite, whose connections are terminated at the IUG on the end-user side and at the ING on the operator side. The IUG is the routing entity located at the end-user premises serving as the focal point for the integration of the terrestrial and satellite connections. As the counterpart of the IUG on the network side, the ING has the functionalities of both managing a set of associated IUGs and acting as a single connection interface to the public internet. For the downstream traffic, an ING has equivalent building blocks and routing functionalities to the IUG. The main functionality of IUG and ING is to route the outgoing traffic towards the most suitable access network segment considering the QoE requirements of each particular traffic flow and the real-time status of each of the links. Based on this, the ING shall be located closer to the Point of Presence (PoP) of the terrestrial operators involved in the integrated system, in order not to increase the latency of the services routed terrestrially (which are meant to be the most delay sensitive).

Due to TCP performance degradation over satellite links, Performance Enhancement Proxies (PEPs) [6] are currently the most commonly adopted solution to achieve good transport performance (in terms of link utilization and user experience) whatever the available TCP stack at both ends (clients and servers). The location of PEP terminations in the BATS architecture has important impacts on the overall network design. Trade-off analysis carried out in the project concluded that the best compromise between performance, impact and complexity is to locate a high capacity PEP in a central point of the network near to the PoP or the ING, alleviating the internal re-routing and synchronization issues compared to a case where the PEP is located at the nominal satellite gateway and traffic needs to be re-routed to a redundant gateway to avoid service interruption due to fading or failure at the nominal one.

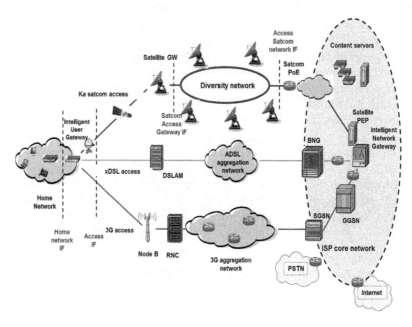

Fig. 1. Overall network architecture

2.2 The Intelligent User and Network Gateways

Due to the heterogeneity of the various access segment technologies, distributing traffic arbitrarily among them despite their different characteristics, e.g. satellite systems offer large bandwidth but introduce additional latency compared to typical terrestrial networks, has proved to be sub-optimal. A multipath routing entity such as the IUG and ING, which can seamlessly select the right access technology for the different types of network traffic, requires the ability to know the QoS requirements (e.g. bandwidth, latency and packet loss) of each particular traffic flow and to compare them with the real-time status of each of the available access links. Figure 2 illustrates the key building modules of the gateways. The heart of the system is the *Link Selection* algorithm itself, which operates on certain attributes or parameters and selects the most optimal link for a particular traffic unit. In order to do that, a module called *Link Status Estimation* evaluates the RTT, maximum bandwidth and packet losses of each of the paths between IUG and ING. In addition, the *Traffic Classification* module classifies the traffic as required to be able to take the link selection decisions and identifies the Class of Service (CoS) for QoS purposes. The IUG and ING have been conceived with a modular approach and well-defined connecting interfaces between the different components which allow for specific improvements on individual modules without modifying the entire system. In the following sub-sections details are provided on the approach BATS has followed for these different modules.

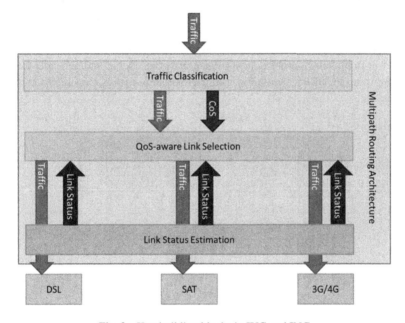

Fig. 2. Key building blocks in IUG and ING

Multipath Routing Architecture

Since TCP (and HTTP) are now the main protocols dominating the Internet traffic [7], they become a de-facto convergence layer. This trend is also driven by the Internet and content service provider infrastructures that are largely based on this protocol set. A perfect example is the evolution of the streaming services that are evolving from RTP/UDP to adaptive streaming (such as MPEG-DASH) over HTTP/TCP networks. For this reason, in BATS, MPTCP [8] has been identified as an appropriate mechanism for the core of the multipath routing architecture, as it aims at using multiple paths between a source and a destination host while providing the same interface to both the application and the network layer as in conventional TCP. Several TCP sessions, called MPTCP sub-flows, can be established on the different access links and used simultaneously whilst preserving the ordering of the packets at the far end. As with regular TCP, MPTCP is also designed to be an end-to-end transport layer protocol. Hence, in order to exploit the multipath features of MPTCP only between intermediate routers, MPTCP proxies need to be used at the IUG and ING, as shown in Fig. 3. In the BATS design, the implementation of MPTCP proxies at both IUG and ING allows the definition of a completely integrated architecture based on a common set of mechanisms independent from the heterogeneity of the access networks. Furthermore, the need for TCP acceleration in the satellite access segment perfectly fits with MPTCP proxies, as TCP PEPs can be co-located at the IUGs and INGs on both ends of the network.

In the BATS implementation, the negotiation between IUG and ING is done using TCP options to avoid losing time during payload handshakes. These TCP options need to be protected from layer 4 filtering equipment that may be in-between the IUG and ING. As an example, many firewalls and core network appliances modify the TCP options they do not understand, which would lead to a breakage of the negotiation between the BATS gateways. Therefore, the communication between IUG and ING happens within tunnels to protect encapsulated protocols including the header information. In this case, a UDP encapsulation is added in order to ease processing through NAT equipment.

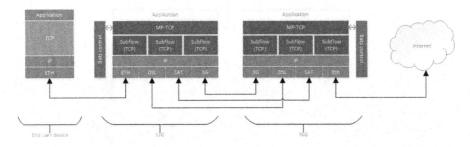

Fig. 3. Multipath routing architecture based on MPTCP proxies

Link Status Estimation Module

The link estimation module analyses each path between the IUG and the ING and provides the real-time status of different path characteristics, e.g. latency, bandwidth, loss rate, in order to assist the decision on link selection. Benefiting from the MPTCP architecture in which the IUG and ING become the end-points for the MPTCP subflows, in BATS the link estimation module is implemented by measuring information from the TCP connections, updating the statistics each time a TCP segment is sent or received.

The bandwidth estimation evaluates the traffic bandwidth towards a peer through each tunnel, estimating the maximum TCP bandwidth that a TCP connection could achieve. In order to evaluate the maximum TCP bandwidth internal variables of the TCP control block are evaluated. These include the amount of unacknowledged data in the TCP sending buffer and the number of consecutive duplicate ACKs (DUPACKs) that have been received. Upon reception of 3 DUPACKs, the TCP sender deduces that the link is congested and reduces the sending window. At this moment, the link estimation module provides the maximum bandwidth a TCP connection can achieve. In parallel, the currently used bandwidth per connection is estimated every RTT by low-pass filtering the rate of returning acknowledge packets from a peer. When an ACK is received by the source, it conveys the information that a certain amount of data corresponding to a specific transmitted packet was delivered to the destination. If the transmission process is not affected by losses, simply averaging the delivered data count over time yields a fair estimation of the bandwidth currently used by the source. When DUPACKs, indicating an out-of-sequence reception, reach the source, they should also count toward the bandwidth estimate, and a new estimate should be computed right after their reception. In addition the amount of UDP being sent needs to be accounted for. Each subflow over each connection is estimated individually and the sum of all estimations provides accurate current bandwidth estimation for each tunnel.

The latency and packet loss estimations are computed from the RTT and segment retransmission counters in the TCP kernel control blocks of each of the MPTCP subflows. This passive method allows evaluation of any path between an IUG and an ING at IP/transport layer in a technology and vendor agnostic manner without having to make assumptions on the underlying media. Another strategy could be to benefit from the implementation of the Dynamic Link Exchange Protocol (DLEP) [9] on the modems, which will then feed real time link parameters to the IUG and ING. However this protocol is currently not complemented nor widely adopted by modem manufacturers.

QoS-aware Link Selection Module

As the reference MPTCP implementation [10] always routes the traffic first towards the sub-flow with lowest RTT value, in BATS we have defined a novel QoS-aware Link Selection algorithm named "Path Selection based on Object Length (PSBOL)" that takes account of traffic requirements and the real-time status of each of the links. In the initial implementation, the link selection is performed by analyzing the so-called TCP object size, where an object is typically an Application Protocol Data Unit (APDU), e.g. an HTML request. The link selection algorithm is based on the assumption that for long objects (e.g. greater than a certain threshold, which can be selected dynamically based on the link characteristics and the class of traffic) the priority is to benefit from high bandwidth links to reduce the total transmission time, whereas for short objects is preferable to benefit from low latency and faster delivery times. Thus, for a particular TCP connection, long objects are routed via the highest bandwidth link (e.g. satellite) and short-objects are routed towards the lowest latency link (e.g. terrestrial). The discovery of objects boundaries is based on an algorithm that computes packets Inter-Arrival Times (IAT) and differentiates between the reception of packets from current or new objects. Based on this process, the size of the objects is updated on the fly with the reception of each TCP segment; hence, segments are initially considered as part of a small object and transmitted over the lowest RTT link. Once the object size exceeds the threshold, segments start to be transmitted over the highest bandwidth link.

Due to the IUG/ING architecture, optimized link selection decisions are performed independently for the inbound and outbound, which can benefit the routing of asymmetric traffic. Obviously, this approach is based on MPTCP and thus only optimized for TCP traffic. Link selection decisions for UDP traffic are based on policy based routing prioritizing first the links offering lowest latencies.

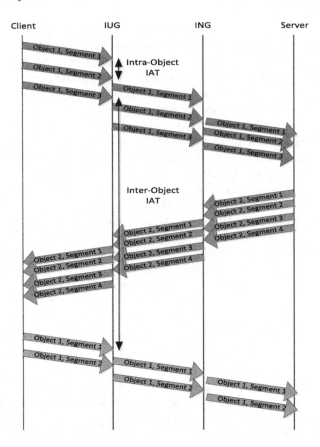

Fig. 4. TCP object boundary detection based on segment inter-arrival time

Traffic Classification Module

The proposed BATS solution identifies and sorts TCP payloads based on object charac-teristics. As aforementioned, the link selection decision for each packet is taken based on the size of the TCP object it belongs. It is therefore necessary to be able to compute the size of each particular object. In our implementation, objects are classified based on the IAT of the different TCP segments. The proposed process takes place in both IUG and ING and analyses the IAT of consecutive TCP segments belonging to a TCP connection. Hence, on the ING the process analyses the IAT for TCP segments received from Internet, and on the IUG the process analyses the IAT for TCP segments received from end-user devices. The algorithm compares the IAT of a segment to a threshold value to decide whether the segment belongs to the current object (Intra-object IAT being below threshold) or it is the first segment of a new object (Inter-object IAT above threshold), concept illustrated in Fig. 4. This threshold is computed from parameters that are locally available on the machine (IUG or ING) in order not to require protocols to exchange information between the two units. This method is applicable to any TCP-based applica-tion protocol and works on unencrypted as well as encrypted TCP connections.

3   Experimental Results

For the purpose of the lab and field trials of the BATS project, the IUG and ING have been prototyped on real routing platforms [11, 12]. As part of the in-factory functional, performance and stability validation tests, the software implementation of the different key building blocks has been evaluated first on a controlled environment with the IUG, ING, and network simulators for the different satellite and terrestrial paths implemented on Linux machines. Focusing on the Link Status Estimation module, Fig. 5 illustrates the performance of the bandwidth estimation when the satellite link is used and 20 parallel TCP connections are established between the two end-points. Note how during the test the bandwidth configured on the satellite simulator is reduced by half and the estimation curve is able to follow this variation. Similarly, Fig. 6 shows the output of the RTT estimation for a test during which the RTT value of the simulators is increased during a specific period of time. As we are measuring the RTT at the end-points, this accounts not only for the delay on the transmission link but also processing and buffering delays at the gateways.

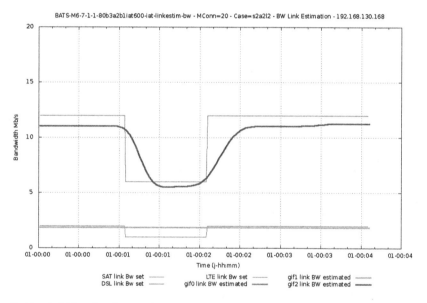

Fig. 5. Bandwidth estimation results on satellite interface with 20 parallel TCP connections *(thin red curve being real value and thick red curve being estimated value).*

In order to test the implementation in conditions as near as possible as a live experiment, the prototyped network appliances were connected to real networks (i.e., 7 Mbps satellite link, 500 kbps LTE link, and 1 Mbps DSL link) as illustrated in Fig. 7. Different configurations were tested: ADSL only, Satellite only, Cellular only, BATS implementation with Weighted Round Robin (WRR) as Link Selection algorithm with weights equivalent to each link's bandwidth, and finally BATS implementation with PsBOL as described earlier in this paper. Note that in these tests the threshold for classifying short

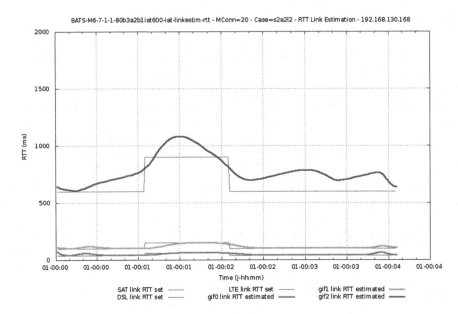

Fig. 6. RTT estimation results with 20 parallel TCP connections *(for the satellite link, thin red curve being real value and thick red curve being estimated value).*

and long objects was set to a default static value of 30 KB. In order to distinguish the BATS MPTCP implementation from the implementation in [10], it is referred in the figures as MCTCP. Figure 8 shows the result of a performance test consisting of loading 35 different websites. In each repetition of the test every website is loaded three times, results being the average of 9 repetitions. Figure 8 illustrates the average loading time and the standard deviation for each of the configurations. In this test, both configurations with the BATS architecture outperform the single link cases. However, no clear benefit is brought by PsBOL in comparison with WRR. This is due to the fact that the in WRR works well when weights are pre-defined in alignment with the available bandwidth in each connection. PsBOL shows a good performance when the link characteristics are not known a-priori or are varying over time.

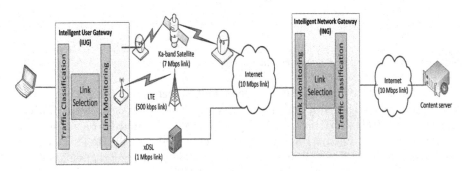

Fig. 7. Lab setup for in-factory tests with real networks

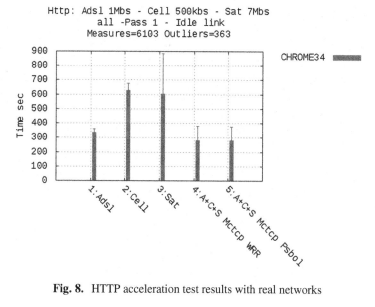

Fig. 8. HTTP acceleration test results with real networks

In the same setup, the performance of the system with SSH encrypted sessions has also been tested. Three different scenarios have been considered:

- Short Objects or Interactive: 60 commands generating small objects (e.g., ls, cd, etc.) are sent for a total amount of data equal to 48 KB.
- Long objects: 15 commands generating long objects (e.g., cat file, etc.) are sent with a total amount of data equal to 5.4 MB
- Mixed: the combination of both previous scenarios.

Table 1 details the time in seconds, averaged over 12 passes, that has been required to complete all the commands in the different scenarios and for the different configurations. As expected, for short objects BATS PsBOL matches the performance of the lowest RTT link (i.e., DSL) as it is the only one used for most of the time. For the scenarios with Long and Mixed-sized objects, PsBOL outperforms all other options benefiting from the intelligent use of the heterogeneous links.

Table 1. SSH acceleration test results with real networks (time is in seconds)

SSH Tests : ADSI 1Mbs - Cell 500kbs – Satellite 7Mbs				
Case	Description	Interactive	Long Objects	Mixed
1	Adsl Only	5.5	50	57
2	Cell Only	8	99	106
3	Sat Only	33	31	43
4	MCTCP WRR	24	20	34
5	MCTCP + PSBOL	6	15	16

4   Conclusions

We have presented a novel broadband delivery system designed in the frame of the ongoing FP7 project BATS which integrates satellite with terrestrial access communication networks. We have described the design of integrated gateways in both user and operator sides of the system intelligently using all available access technologies to provide high speed broadband and improved QoE to users in under-served areas. We have discussed a specific implementation of the routing entities and provided results from the in-factory tests of the prototyped platforms.

Our future work will focus on running extensive Laboratory and Field trials with the prototype IUG and ING involving real end-users in both controlled and real-world environments to be able to assess the benefit of the designed system in terms of QoE.

Acknowledgments. This work has been supported by the BATS research project which is funded by the European Commission Seventh Framework Programme under contract number 317533. The authors of this paper would like to acknowledge the European Commission Seventh Framework Programme (FP7) and all the members of the BATS Consortium.

References

1. European Commission, A digital Agenda for Europe, COM(2010) 245, 26, August 2010
2. Point Topic, Mapping is vital to broadband investment, ITU Telecom World (2011)
3. BATS Project, Deliverable D5.2 "Cost Benefit Analysis", April 2015. www.batsproject.eu
4. Pérez-Trufero, J., Peters, G., Watts, S., Evans, B., Dervin, M., Fesquet, T.: Broadband Access via integrated Terrestrial and Satellite systems (BATS), Ka and Broadband Communications, Navigation and Earth Observation Conference 2013, October 2013
5. Corbel, E., Peters, G., Sperber, R., Bosch, E.: TERASAT: high-throughput satellite system by 2020, Ka and Broadband Communications, Navigation and Earth Observation Conference 2013, October 2013
6. Peng, F., Cardona, A.S., Shafiee, K.: V.C.M Leung, TCP Performance Evaluation over GEO and LEO Satellite Links between Performance Enhancement Proxies, Vehicular Technology Conference (VTC) 2012, September 2012
7. Lee, D., Carpenter, B., Brownlee, N.: Observations of udp to tcp ratio and port numbers, in Internet Monitoring and Protection (ICIMP) 2010, pp. 99–104, May 2010
8. Ford, C., Raiciu, Handley, M., Bonaventure, O.: TCP Extensions for Multipath Operation with Multiple Addresses, RFC 6824 (Experimental), Internet Engineering Task Force, January 2013
9. Ratliff, S., Berry, B., Harrision, G., Jury, S., Satterwhite, D.: Dynamic Link Exchange Protocol (DLEP), Internet Draft, Internet Engineering Task Force, February 2014
10. Paasch, C., Barre, S., et al.: MultiPath TCP – Linux Kernel implementation. http://www.multipath-tcp.org/
11. OneAccess Networks, One540: Cloud Services Gateway & Multi-Service Access Router, Datasheet.   http://www.oneaccess-net.com/products/multi-service-data-routers-and-eads/medium-business-regional-offices/item/69-one540
12. OneAccess Networks, WXD2450: Wan Optimization Platform, Datasheet. http://www.oneaccess-net.com/products/wan-optimization-platforms/item/131-wxd2450

Modeling the Lifecycle Greenhouse Gas Emissions of a Hybrid Satellite System

David Faulkner[1(✉)], Keith Dickerson[1], Nigel Wall[1],
and Simon Watts[2]

[1] Climate Associates Ltd, 1 Westland Martlesham Heath,
Ipswich IP5 3SU, UK
davewfaulkner@gmail.com,
keith.dickerson@climate-associates.com,
nigel.wall@shadow-creek.co.uk
[2] Avanti Communications Ltd,
Cobham House, 20 Black Friars Lane,
London EC4V 6EB, UK
Simon.Watts@avantiplc.com

Abstract. The aim of this paper is to present the approach used to model the greenhouse gas emissions of a hybrid broadband terrestrial/satellite system over its lifecycle. The lifecycle analysis showed that the electricity used by the customer premises equipment was responsible for the majority of the GHG emissions, assuming that the power plants continue to use fossil fuels. Emissions from manufacture, transport and waste treatment represented only 0.00453 % of the total emissions. Under a 1 % cut-off rule only the in-use emissions from on grid electricity would need to be considered. Manufacture, transport, and waste treatment can be safely ignored. This includes emissions from the manufacture of the satellite launch vehicle and the transport of the satellite into geostationary orbit.

Keywords: Hybrid satellite systems · Broadband Access · Energy-efficiency · Environmental assessment · Greenhouse gas emissions · Lifecycle

1 Introduction

Hybrid satellite networks combine different types of communications paths to provide a service. The EU 7th Framework Programme 'BATS' project (Broadband Access via integrated Terrestrial and Satellite systems) proposes such a hybrid system aimed at proving coverage to Europe at 30 Mbit/s with emphasis on rural areas which are unlikely to be well-served by terrestrial-only solutions.

Other publications concerning the environmental impact of this system focus on perspectives such as a comparison of the BATS solution with possible alternative terrestrial-only systems over the lifecycle [1]. In this paper we focus on how the inventory was produced, what emissions factors were used and what assumptions were made. The methodology used to carry out the environmental assessment was based on life cycle analysis (LCA) standardised by ETSI [2] with the ITU methodology [3] used

© Institute for Computer Sciences, Social Informatics and Telecommunication Engineering 2015
P. Pillai et al. (Eds.): WiSATS 2015, LNICST 154, pp. 103–115, 2015.
DOI: 10.1007/978-3-319-25479-1_8

to provide more detailed guidance where needed. The model is described in some detail so that the parameters which have most impact on the overall emissions can be identified and so that steps can be taken to minimize emissions. At the outset of this project we had no idea of the relative importance of the different lifecycle stages which include: raw material extraction, manufacturing, transport, use-phase and waste treatment. The life-cycle analysis (LCA) includes the launch vehicle which was chosen to be Ariane 5 as an example. The use-phase was considered to be important as it operates over the full 16 year life of the BATS system using on-grid electricity, which in most European countries is dominated by the burning of fossil fuels. These release greenhouse gases (GHGs) such as carbon dioxide (CO_2) into the atmosphere. The aim of this report is to explain how an LCA was carried out with focus on the hybrid satellite network elements and to show the relative importance of the life cycle stages.

2   System Boundary and Identification of Network Elements

The system boundary for carbon accounting is shown in Fig. 1. Note that the components of the hybrid satellite system are depicted in red. The terrestrial system, for example ADSL, is assumed to be already in place when the satellite system is added. The emissions associated with the construction of the terrestrial system are therefore excluded from the model, although the user modems (both satellite and ADSL) energy are taken into account.

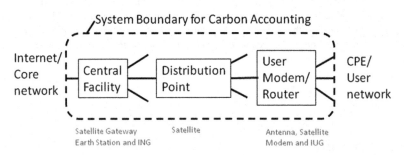

Fig. 1. Identification of network elements and system boundary

The network elements include: Satellites (GEO), Central Facilities added for BATS, Intelligent Network Gateway (ING), Earth stations including antenna, User satellite router including upstream power amplifier and power supply, Intelligent User Gateway (IUG), a home gateway which includes the terrestrial modem(s), and power supply. A system Block diagram is shown in Fig. 2. Note that the ING is shown as Integrated Core Network.

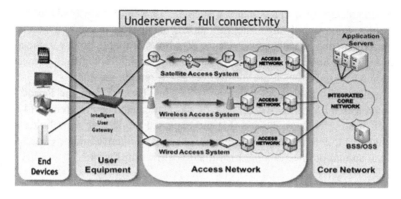

Fig. 2. Identification of network elements

3 Addressable Market

A key parameter is the number of terminals added and in operation each year which is equal to:

Number of households x take rate x market share for BATS x market annual profile.

The addressable market for rural services in Europe used in this analysis is shown in Fig. 3 where the blue bars represent households and the red bars the number of small businesses.

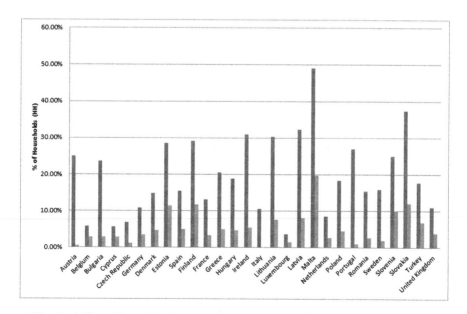

Fig. 3. Addressable market for 30 Mbit/s services in year 2020 (Color figure online)

It was calculated from Fig. 3 that, on average, 14.4 % of households in the EU27 countries plus Turkey will have satellite (shown in blue) as the only available technology for contracting broadband services at 30 Mbit/s or more. However, the average percentage of total households which will take up a satellite broadband connection is 3.72 % (shown in red), which are mostly located in remote areas.

The number of households in Europe (E28 plus Turkey) over the Period 2020–2035, the projected life of the BATS system, was estimated using references [4–7]. An analysis of these references showed that the expected number of households in year 2020 would be 245.5 M and a growth rate of 0.41 % per annum could be expected. The take-up rate annual profile (timeline) assumed is shown in Fig. 4.

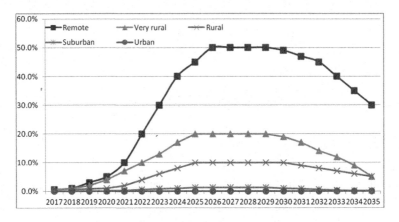

Fig. 4. Take-up annual profile for BATS services

It is assumed that the take up rate for BATS services versus other technologies such as satellite-only is 50 %. In the GHG emissions model the peak number of households taking BATS services was calculated to be 6.4 M households in year 2029.

4   Emission Factors

Emissions factors are parameters are normally provided by tables which when multiplied by a value attributable to a network element such as energy in kWh, weight in kg or distance travelled in km produce a value of carbon dioxide equivalent (CO_2e) emitted in kg. In some cases no emission factor could be found in tables or other references, such as the manufacture and combustion of butadiene which is used as a binder in the solid fuel booster. In that case estimates were made from first principles.

4.1   Electricity

A key emission factor is that of electricity. The GHG emission factor of electricity supply each year for the EU will vary over the period 2020–2035. The report "EU

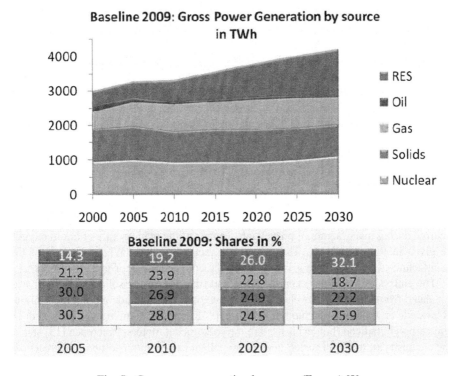

Fig. 5. Gross power generation by source (Europe) [8]

Energy Trends to 2030" provides a perspective of the energy source mix [8] over the life cycle of the BATS deployment.

From Fig. 5 the annualized decrease in fossil fuel is 0.75 % between 2020–2030. This trend is assumed to continue until 2035. The average GHG conversion factor for the EU in 2013 is given as 0.34723 kg CO_2 per kWh [9]. This was treated as the base year from which subsequent conversion factors are derived. In 2020 Fig. 5 shows that the proportion of fossil fuel is predicted to have fallen from 51.8 % to 49.5 %. The emission factor will then have fallen pro-rata to $0.34723 \times 49.5/51.8 = 0.33181$. Linear interpolation was used to derive the emission factors for each year over the period 2020–2035.

5   Life Cycle Stages of the Network Elements

5.1   Satellite Payload and Launch Vehicle

The manufacture of the satellite system includes all the physical components including the launch vehicle its payload. In addition the emissions arising from combustion of the fuel used to take the satellite into geostationary orbit must be accounted for. This should be accounted for under the Transport phase of the life cycle.

To estimate the emissions due to manufacturing, tables [9] were used to obtain a carbon factor. In this case the weight (mass) was multiplied by the emission factor to arrive at a CO_2e value in kg.

The Ariane 5 launch vehicle has a Gross Take Off launch capacity of 10,000 kg (22,000 lb.) for dual payloads or 10,500 kg (23,100 lb.) for a single payload [10]. The emissions of the launch vehicle were therefore scaled according to the actual payload (6400 kg) assuming a second load is carried.

The launch vehicle mass includes the main stage the upper stage, the solid fuel boosters, the payloads and the propellants.

Hydrogen and oxygen are the two chemicals used in the cryogenic main engine. The masses are 133 and 26 tonnes respectively. These react to produce water which is not classified as a GHG. No account is taken of the impact of water emission into the stratosphere. Note that this reaction would be in the transport phase of the life cycle as the satellite is transported to its geostationary orbit. Account is taken of the emissions incurred during manufacture of propellants. Details of the manufacture of liquid oxygen are given in Reference [11]. The conversion factor is 0.310 kWh/kg. Details of the manufacture of hydrogen are given in [12]. The conversion factor is 8.5 CO_2e kg/kg.

The emission due to the manufacturing and transport of the solid fuel booster was calculated from the mix of its chemical constituents which include a propellant mix of 68 percent ammonium perchlorate (oxidizer), 18 percent aluminium (fuel), and 14 percent polybutadiene (binder) is used in the solid rocket motors. Reference [13] states: The chemistry of the solid rocket booster propellant can be summed up in this reaction:

$$10 \; Al + 6 \; NH_4ClO_4 \rightarrow 4 \; Al_2O3 + 2 \; AlCl_3 + 12 \; H_2O + 3 \; N_2$$

The CO_2e emission arising from the manufacturing solid fuel boosters was estimated by consideration of the detailed manufacturing processes of each chemical multiplied by their molecular weights. The chemicals included: aluminium (both as a fuel and in the tank casings), ammonium perchlorate, hydroxyl-terminated polybutadiene (binder). Boosters are fabricated from 62 tonnes steel.

The CO_2e emissions arising from the launch of Ariane 5 may be considered as the transport phase of the life cycle. The reactant products are not classified as greenhouse gases by the IPCC. They are emitted as a white powder. In the stratosphere they are likely to reflect direct sunlight and so may have an overall cooling effect. However they may trap infrared radiation and have a warming effect. More research is being carried out to measure the overall impact on surface temperature.

The second reaction arising from the solid fuel booster may be estimated from the combustion of 1-3 Butadiene (C_4H_6) binder. It is assumed this short chain dominates but longer chains exist in the binder.

$$C_4H_6 + 11O_2 \rightarrow 4CO_2 + 3H_2O$$

The mass of polybutadiene was calculated to be 66.6 tonnes. From consideration of the molecular weights the mass of CO_2 ejected is 217 tonnes per launch assuming all the binder is consumed. No account was taken of the waste treatment as no parts of the Ariane 5 are recovered. The environmental impact of steel in seawater was assumed to be negligible.

5.2   Ground Segment (Earth Station) Life Cycle Parameters

The ground segment includes: Antenna, high power amplifier (HPA), frequency convertors, electronics for control and management and air-conditioning.

Table 1. Breakdown of parameters for satellite gateway

ID	Component	Wt (kg)	Pwr (W)	QTY	Tot wt	Tot pwr	Source
1	Antenna	2,500	15	1	2,500	15	http://comsatsystems.co.in/admin/?page_id=25
2	HPA	25	821	3	75	2,463	http://www.cpii.com/docs/datasheets/256/M KT-1511.pdf
3	Frequency convertors	7	60	21	147	1,260	GD Satcom, 10 uc 10 dc tc
4	Other BATS electronics	50	500	1	50	500	Estimate
5	Air con	150	8,000	0	0	381	In existing room with other services. 30% duty cycle assumed at PUE=1.3
	Total				2,772	4,605	

The breakdown of parameters for a single satellite gateway is given in Table 1. 38 of these are required throughout Europe.

It was assumed that the buildings already exist and that there is sufficient existing HVAC, UPS and standby generator capacity is sufficient.

The Earth Station manufacturing emissions were estimated from the incremental weight of 2772 kg multiplied by a carbon factor of 0.537 kg CO_2e per kg for WEEE – large [9]. The 2014 figure was used. It was assumed that the equipment and antenna are made mostly from aluminium.

Incremental installation activities include: site survey, installation of equipment and commissioning. Transport was assumed to be approximately 400 km. This would be via 10 visits per gateway in a light commercial van with a 40 km round trip. The emission factor of a light commercial van was estimated to be 0.164 kg CO_2e in year 2020 falling to 0.12 kg CO_2e in year 2035 [14].

The manufacturing emissions were estimated from the incremental weight of 10 kg (including modem cable and antenna) in kg multiplied by a carbon factor of 0.537 kg CO_2e per kg for WEEE – large [9]. The 2014 figure was used as the latest available.

The delivery vehicle for installation is assumed to be a light commercial van. 40 km was assumed initially (as for VDSL and fibre examples). However 15 % users are assumed adopt self-install. Therefore a nominal 40 km reduces to 34 km plus 2 km round trip for mail van delivery of the modem and IUG (final proportion of trip).

The waste treatment was calculated from the mass of the components multiplied by the conversion factor for WEEE [9]. The 2014 figure was used as the latest available. This was 2772 kg multiplied 0.021 kg CO_2e per kg.

5.3   Intelligent Network Gateway Lifecycle Parameters

A virtualised Intelligent Network gateway (ING) is envisaged in year 2020. This would be located at a data centre which may be part of the internet point-of-presence (PoP). The power consumption was estimated by taking the power consumption of a server of today and scaling it according to Moore's Law to year 2020. This was estimated at a power of 650 W and a maximum fan-out of 94081 in year 2020. To obtain an estimate of the energy used by the INGs the power per server was multiplied by the maximum number of user terminals (6.4 m) divided by the fan-out.

Upgrade of the terrestrial backhaul network between a satellite gateway and the PoP may be needed to cope with additional traffic. It is not thought that this upgrade would contribute significantly to the GHG emissions, so no allowance for this was included in the model.

5.4   User Modem Lifecycle Parameters

An average satellite modem power of 7.5 W was considered to be representative of hybrid satellite systems over the period 2020–2035 based upon a modem of peak power 22 W and the existence of standby modes which reduce the duty-cycle to around 30 %. The majority of the upstream traffic would be carried by the ADSL2 modem. Whereas downstream traffic will increase over the period, the upstream traffic is not expected to increase significantly. This is because illegal file sharing via peer to peer networking is changing to favour legitimate downloads using paid-for services.

The manufacturing emissions were estimated from the incremental weight 10 kg (including modem cable and antenna) in kg multiplied by a carbon factor of 1.149 kg CO_2e per kg weight of mixed electrical and electronic equipment [9]. The 2014 figure was used as the latest available.

It is assumed that user terminals are not replaced during the life of the BATS system and the waste is disposed of during the final year 2035. The conversion factor used was 0.021 kg CO_2e/kg, from the DEFRA 2014 sheet on 'waste treatment' [9].

5.5   Intelligent User Gateway

The Intelligent User gateway (IUG) is assumed to be functionally similar to a home gateway as described in Section C1 of the EU CoC for broadband equipment [15]. The power consumption target for 2015–16 may be estimated by summing the power consumption of the WAN interfaces (including the central processor and data storage) plus the LAN interface(s) as shown in Table 10. The average power consumption with cache was estimated to be 5.7 W. It is assumed that the manufacturing weight is similar to that of an ADSL modem. This was estimated to be 0.47 kg including power supply. The IUG was assumed installed along with satellite modem as described above. It is assumed that IUGs (and satellite modems) are not replaced during the life of the BATS system and the waste is disposed of during the final year 2035. The conversion factor used was 0.021 kg CO_2e/kg, from the DEFRA 2014 sheet on 'waste treatment' [9].

6   Model Structure and Operation

The model, an Excel spreadsheet which totals the GHG emissions over the life cycle which can be attributed to Europe (28 countries plus Turkey). It includes emissions within GHG protocol Scopes 1, 2 and 3 [16] (Fig. 6).

	A	B	C	D	E	F	G	H	I
1									
2		Year >>			2020				
3	Calculation of market size	value	Unit						
4	Growth rate of households	0.41	%		245500000				
5	Addressable market	14.4	%						
6	Take rate peak	35	%						
7	Addressable market annual profile % relative to peak at 100%	See>>	%		10				
8	Number of households with satellite broadband				1237320				
9	Number of BATS units operating based upon share of available sat	50	%		618660				
10	Number of Units added (or removed) in year				618660				
11	Conversion Factors								
12	Conversion factor-electricity	0.33181	kg CO2e per kWh				0.33181		
13	Conversion Factor - delivery vehicle (light commercial-van)	0.181	kg CO2e per km				0.164		
14	Conversion Factor - truck of capacity 3.5-7 tonnes	0.642	kg CO2e per km				0.6309524		
15									
16	Network Elements and Processes				Number	Number	Conversion	GHGe	
17					added	Operational	factor	kgCO2e	
18	BATS network elements and processes								
19	ING power average	42443	W		1	1	0.33181	1.23E+08	
20	ING manufacture	27	kg		1	1	0.53724	1.45E+01	
21	ING installation	1	km		1	1	0.164	1.64E-01	
22	ING waste treatment	27	kg		1	1			

Fig. 6. Screen shot of the EU GHG emissions model showing key input parameters

The names of key global parameters are entered in cells A4-A14 together with their base value (e.g. for year 2014) in cells B4-B14. Years (of BATS implementation) are shown in Cells E2 and J2 etc. Values for each year are separated by purple columns such as D and I. Key parameters such as number of households and conversion factors for electricity change gradually from year to year as fuel sources change. These have been estimated according to the best available publication or statistics and entered into

	A	B	C	D	E	F	G	H	CG
17					added	Operation factor		kgCO2e	
18	BATS network elements and processes								BATS
19	ING power average	42443	W		1	1	0.33181	1.23E+08	1.75E+09
20	ING manufacture	27	kg		1	1	0.53724	1.45E+01	1.45E+01
21	ING installation	1	km		1	1	0.164	1.64E-01	
22	ING waste treatment	27	kg		1	1			2.15E+01
23	Earth station (gateway) power	4605	W		19	19	0.33181	2.54E+08	6.96E+09
24	Earth station (gateway) manufacture	2722	kg		19		0.53724	2.78E+04	5.56E+04
25	Earth station (gateway) installation	400	km		19		0.164	1.25E+03	2.47E+03
26	Earth station (gateway) waste treatment	2722	kg		19				5.72E+01
27	Satellite weight	6400	kg		1		2.865	1.83E+04	8.55E+04
28	launch vehicle +payload manufacture	21417	kg		1		2.865	6.14E+04	1.23E+05
29	solid fuel booster casing manufacture	39680	kg		1		0.8623	3.42E+04	6.84E+04
30	launch fuel emission butadiene	138880	kg		1		3.26	4.53E+05	9.05E+05
31	launch fuel manufacture aluminium	27392	kg		1		2.865	7.85E+04	1.57E+05
32	launch fuel manufacture hydrogen	22016	kg		1		8.5	1.87E+05	3.74E+05
33	launch fuel manufacture oxygen	26387	kWh		1		0.33181	8.76E+03	1.74E+04
34	launch fuel manufacture chlorine	172848	kWh		1		0.33181	5.74E+04	1.14E+05
35	User satellite modem energy and power amplifier energy	7.5	W		618660	618660	0.33181	1.35E+10	1.45E+12
36	User satellite modem manufacture	10	kg		618660		1.149	7.11E+06	7.37E+07
37	IUG (user hub) power average	5.7	W		618660	618660	0.33181	1.03E+10	1.10E+12
38	IUG (user hub) manufacture	0.466	kg		618660		1.761	5.08E+05	5.26E+06
39	CPE installation 1 visit in light commercial van	34.3	km		618660		0.164	3.48E+06	3.38E+07
40	User equipment waste treatment	10.466	kg						1.40E+06
41	Total BATS							2.41E+10	2.56E+12
								embodied	1.16E+08

Fig. 7. Screen shot of the BATS network element input parameters

the spreadsheet for each of the years 2020–2035. Extensive use was made of the comments field to show the source reference of a parameter value.

The network elements and processes are listed in Column A. Column B shows the parameter in kW, kg or km which is associated with carbon emissions for the network. The number added in year 2020 is shown in column E. The conversion factor to kg CO_2e is in Column H. These are obtained from tables or deduced from them according to the year of emissions.

The total BATS emission for the year 2020 (cell H41) is 2.41 E10 kg CO_2e (24.1 Mtonnes CO_2e). Each year is calculated similarly and is then totalized over the life cycle and plotted.

7   Results

Results appear in numerical form in kg CO_2e in the final column of the spreadsheet. This is column CG in Fig. 7 which shows totals the emissions for each category over the years 2020–2035. The total for the EU-28 plus Turkey is shown as 2.56 Gtonnes in cell CG41 (Fig. 8).

The most significant emissions may be attributed to the use of electricity measured in kWh over the 16 year lifecycle. This includes the earth station energy, user satellite

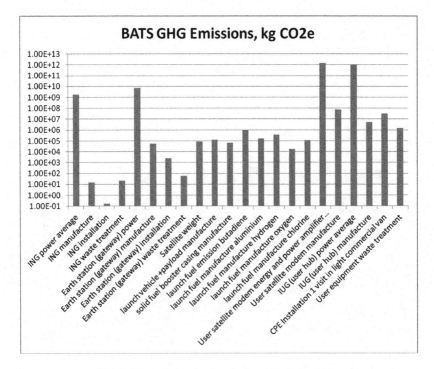

Fig. 8. Emissions of network elements in kg CO_2e

modem (including its power amplifier) and the IUG. The latter two items are the most significant as 6.4 million of these are assumed to be deployed in households in Europe when the service reaches its peak in year 2029. The contribution of all other lifecycle phases including, manufacture, transport, and waste treatment is 116 ktonnes as shown in cell CG42 in Fig. 7. This is only 0.00453 % of the total.

As the most significant contributor, the satellite modem was studied in some detail in an effort to reduce energy usage and emissions. The result presented above is for the case when the satellite modem is operating with a low duty-cycle at a mean power consumption of 7.5 W as compared with an always-on power of 22 W. To make this low duty-cycle possible, traffic is preferentially routed via the ADSL network.

Further modelling work was carried out to compare the hybrid satellite system with terrestrial alternatives: VDSL2, fibre and LTE assuming the same market share number (i.e. number of households) in Europe. A key assumption for the LTE example was the fan-out of the LTE base station which was assumed to be 100 fixed users per base station serving a rural area. This is lower than the average for mobile base stations which is typically 1500. The results of this analysis are shown in Fig. 9.

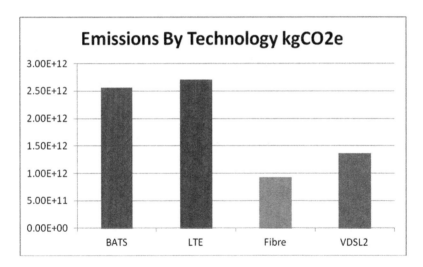

Fig. 9. Comparison of the emissions of four fixed access systems serving rural communities with a downstream rate of approximately 30 Mbit/s

This figure shows that the GHG emissions of the hybrid terrestrial/satellite system (BATS) compares favourably with a terrestrial wireless system based upon LTE technology but the VDSL2 and fibre systems have significantly lower emissions. Note that the hybrid system included operation with a representative low duty-cycle modes which reduced the average power of the modem from 22 W fully-on to 7.5 W average. However, the systems cannot be judged on GHG emissions alone. The hybrid satellite system may turn out to be more economical than the alternatives in rural areas.

8 Conclusion

This paper has shown that, the CO_2e emissions arising from the BATS hybrid terrestrial/satellite system over the lifecycle is 2.56 Gtonnes. This is to a peak market of 6.4 million rural households in year 2029. The lifecycle analysis showed that the electricity used by the customer premises equipment was responsible for the majority of the GHG emissions, assuming that the power plants continue to use fossil fuels. Emissions from manufacture, transport and waste treatment represented only 0.00453 % of the total emissions. Under a 1 % cut-off rule only the in-use emissions from on grid electricity would need to be considered. Manufacture, transport, and waste treatment can be safely ignored. This includes emissions from the manufacture of the satellite launch vehicle and the transport of the satellite into geostationary orbit.

Acknowledgement. The authors wish to acknowledge support and funding of the EU 7th R&D Framework Programme. Without this support the assessments carried out under the BATS Project would not have been possible. The authors acknowledge the work of other partners in the EU FP7 BATS project whose reports have provided input parameters for the inventory. The authors also acknowledge the work of the University of Surrey under Professor Barry Evans who provided guidance on the modelling technique used here.

Acronyms

ADSL	Asymmetric Digital Subscriber Line
BATS	Broadband Access via Integrated Terrestrial and Satellite Systems
CO_2e	Carbon Dioxide equivalent
CPE	Customer Premises Equipment
DEFRA	UK Department for Environment, Food and Rural Affairs
ETSI	European Telecommunications Standards Institute
EU	European Union
FP7	EU 7th R&D Framework Programme
GEO	Geostationary (Satellite)
GHG	Greenhouse Gas
HPA	High Power Amplifier
HVAC	Heating, Ventilation and Air Conditioning
ICT	Information and Communications Technologies
ING	Intelligent Network Gateway
IPCC	Intergovernmental Panel on Climate Change
ISO	International Organization for Standardization
ITU	International Telecommunications Union
IUG	Intelligent User Gateway
LAN	Local Area Network
LCA	Life Cycle Assessment
RES	Renewable Energy Sources
UPS	Uninterruptible Power Supply
VSAT	Very Small Aperture Terminal
WEEE	Waste Electrical and Electronic Equipment

References

1. Dickerson, K., Faulkner, D., Wall, N., Watts, S.: Environmental assessment of hybrid broadband satellite systems. In: Chapter in Book on "Green Services Engineering, Optimization, and Modeling in the Technological Age" to be published by IGI Global (2015)
2. ETSI TS 103 199 (2011). Environmental Engineering (EE); Life Cycle Assessment (LCA) of ICT equipment, networks and services; General methodology and common requirements. www.etsi.org/services/etsi-webstore
3. ITU-T Recommendation L.1410 (2012). Methodology for the assessment of the environmental impact of information and communication technology goods, networks and services. http://www.itu.int/ITU-T/recommendations/index_sg.aspx?sg=5
4. EU Energy Trends to 2030. http://ec.europa.eu/energy/observatory/trends_2030/doc/trends_to_2030_update_2009.pdf
5. Number of private households in Europe. http://www.pordata.pt/en/Europe/Private+households+total+and+by+number+of+children-1615
6. Number of households in Croatia. http://www.dzs.hr/default_e.htm
7. Number of households in Turkey. http://w3.unece.org/pxweb/Dialog/Saveshow.asp?lang=1
8. EU Energy Trends to 2030, Figure 12. http://ec.europa.eu/clima/policies/package/docs/trends_to_2030_update_2009_en.pdf
9. DEFRA (2014). Guidelines to DEFRA/DECC's GHG Conversion Factors for Company Reporting. http://www.ukconversionfactorscarbonsmart.co.uk/
10. Ariane introduction. http://www.arianespace.com/launch-services-ariane5/ariane-5-intro.asp
11. Details of the manufacture of liquid oxygen, IEA "Tracking Industrial Energy Efficiency and CO_2 Emissions". http://www.iea.org/publications/freepublications/publication/tracking_emissions.pdf
12. Liquid hydrogen manufacturing conversion factor, US EPA, "Technical Support Document for Hydrogen Production: Proposed Rule for Mandatory Reporting of Greenhouse Gases' page 2. www.epa.gov/ghgreporting/documents/pdf/archived/tsd/TSD%20HydrogenProduction%20EPA_2-02-09.pdf
13. Chemistry of the solid rocket booster propellant. http://www.americanchemistry.com/s_chlorine/docs/images/ammperch_form1.gif
14. Forecast emission factors for vehicles 'Chapter 5: Reducing emissions from transport'. http://www.theccc.org.uk/wp-content/uploads/2013/12/1785b-CCC_TechRep_Singles_Chap5_1.pdf
15. EU Code of Conduct on Energy Consumption of Broadband Equipment Version 5. http://iet.jrc.ec.europa.eu/energyefficiency/sites/energyefficiency/files/files/documents/ICT_CoC/cocv5-broadband_final.pdf
16. Listing of all tools provided by the GHG Protocol. http://www.ghgprotocol.org/calculation-tools/all-tools

Workshop on Advanced Next Generation Broadband Satellite Systems (BSS) 2

Extending the Usable Ka Band Spectrum for Satellite Communications: The CoRaSat Project

Barry Evans[1], Paul Thompson[1](✉), Eva Lagunas[2], Shree Krishna Sharma[2], Daniele Tarchi[3], and Vincenzo Riccardo Icolari[3]

[1] University of Surrey, Guildford, GU2 7XH, UK
{b.evans,p.thompson}@surrey.ac.uk
[2] University of Luxembourg, Luxembourg, Luxembourg
[3] University of Bologna, Bologna, Italy
{daniele.tarchi,vincenzo.icolari2}@unibo.it

Abstract. Broadband access by satellite in Ka band will become constrained by spectrum availability. In this context, the EU FP7 project CoRaSat is examining the possible spectrum extension opportunities that could be exploited by a database or sensing approach in Ka band via the use of cognitive mechanisms. The database/sensing approach utilises spectrum sharing scenarios between Fixed Satellite Services (FSS), Fixed Services (FS) and Broadcast Satellite Service (BSS) feeder links are considered. Data bases and spectrum sensing have been evaluated to determine white spaces across the shared spectrum for several EU countries. Resource allocation schemes are investigated to place the carriers in the white spaces so as to maximize the throughput of the system. A multi-beam satellite system model has been used to demonstrate the capacity gains that can be achieved by using the cognitive schemes. The overall system is being demonstrated in a laboratory trial.

Keywords: Satellite/terrestrial · Spectrum sharing · Data bases · Spectrum sensing · Resource allocation

1 Introduction

The demand for higher rate and reliable broadband communications is accelerating all over the world. Within Europe the Digital Agenda sets a target for universal broadband coverage of at least 30 Mbps across the whole of Europe by 2020 and 100 Mbps to at least 50 % of the households [1]. Fixed connections and cellular cannot alone meet this target, particularly in the rural and remote areas but also in some black spots across the coverage. In these latter regions satellite broadband delivery is the only practical answer as satellite will cover the whole territory. Some recent studies of the roll out of broadband have shown that up to 50 % of households in some regions will only have satellite available as a means of accessing broadband and thus 5–10 million households are potential

© Institute for Computer Sciences, Social Informatics and Telecommunications Engineering 2015
P. Pillai et al. (Eds.): WiSATS 2015, LNICST 154, pp. 119–132, 2015.
DOI: 10.1007/978-3-319-25479-1_9

satellite customers [2]. Current Ku band satellites do not have the capacity to deliver such services at a cost per bit that makes a business case and thus the satellite community has turned to High Throughput Satellites (HTS) operating at Ka band and above. Examples of early HTS Ka band satellites dedicated to such services are Eutelsats KaSat [3] and VIASAT 1 [4]. These satellites employ multiple (around 100) beams using four fold frequency reuse over the coverage area to achieve capacity of the order of 100 Gbps per satellite. The latter is limited by the exclusive spectrum available to satellite (FSS) of 500 MHz in both the up and downlinks and this limits the feasible user rates to 10–20 Mbps. Thus looking ahead to the increased user demands we have to look to larger satellites (maybe up to a Terabit/s [5,6]) and to more spectrum. Moving up to Q/V bands has already been suggested for feeder links but for user terminals the additional expense is not considered desirable so we return to the problem of getting more usable spectrum at Ka band.

The Ka-Band exclusive bands for satellite are 19.7 to 20.2 GHz in the downlink and 29.5 to 30 GHz on the uplink. In these bands FSS terminals can operate in an uncoordinated manner, which means that they do not have to apply for and be granted a licence by the national regulators, provided they meet set performance characteristics. The issue in other parts of the Ka band is that the spectrum is allocated, not just to FSS but also to fixed links (FS) and to BSS (uplinks for broadcast satellites) as well as mobile services (MS). This spectrum is allocated by the ITU in three regions of the world as shown in Table 1 for Ka band (Europe is Region 1). In these so-called shared bands the different services need to co-exist. Within Europe the CEPT [7] have adopted decisions that expand those of the ITU and produce tighter regulation as follows;

- 17.3–17.7 GHz: the BSS feeder links are determined as the incumbent links but uncoordinated FSS links are also permitted in this band.
- 17.7–19.7 GHz: FS links are considered incumbent but FSS terminals may be deployed anywhere but without right of protection.
- 27.5–29.5 GHz: CEPT provide a segmentation of the band between FSS and FS portions. Within each segment there is a specified incumbent but for instance FSS terminals can operate in FS portions provided they do not interfere with the incumbent FS.

The work reported in this paper has been conducted within the EU FP7 project CoRaSat [8–11] which examines ways in which FSS satellite terminals in the Ka band can co-exist with FS and BSS links given the regulatory regime discussed above. Specifically, a database approach for such coexistence schemes is investigated and demonstrated to exploit the frequency sharing opportunities for uncoordinated FSS terminals and verify its applicability. The aim is to show that future satellite systems can access additional spectrum beyond the exclusive bands that is needed to deliver cost effective broadband services.

2   Scenarios and Database Approach

Within the CoRaSat project, three scenarios have been investigated that reflect the three spectrum components detailed in the previous section. In Fig. 1 we illustrate the interference paths in these scenarios. Two of the scenarios are downlink for the FSS; scenario A, 17.3–17.7 GHz where the potential interference is from BSS uplinks and scenario B, 17.7–19.7 GHz where the potential interference is from incumbent FS transmitters. In both of these cases the FSS is permitted to operate but is not protected by the regulatory regime and thus it is important to ascertain the level of the interference and its affect on the FSS received signal. The third scenario C, is in the transmit band of the FSS from 27.5–29.5 GHz and the interference is from the FSS transmitting earth station into the FS receivers which are protected. The latter is more critical in that we need to demonstrate that the FSS does not contravene interference limits imposed by the regulatory regime. The forward link, e.g. the downlink can be considered more important as it carries more capacity and in addition to this, operation in the downlink bands do not require regulatory changes but merely reassurance to the FSS users that the services need not be impaired. The calculation of interference can be performed if the corresponding accurate database has been obtained, which includes the characteristics and locations of the potential interferers, by using accurate models of the equipment, propagation and the path details. Similar ideas have been employed in TVWS systems [12] to allow UHF frequencies to be used in the gaps between TV transmission regions. For scenario A the number of BSS uplinks in Europe is small and thus a database system is similar in magnitude to that of TVWS. However for scenario B and C the number of FS links runs into the tens of thousands and the database is much more complex. The data on the positions and the characteristics of the BSS and FS are generally held by national regulators and these need to be available for a database system to work.

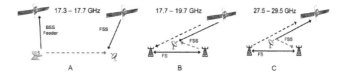

Fig. 1. Scenarios in CoRaSat.

The information of a real interferer database is interfaced to an interference modeling engine which uses ITU- Recommendation P.452-15 [13] procedures plus terrain databases. This is the latest version of this ITU Recommendation that contains a prediction method for the evaluation of path loss between stations. ITU-R P.452-15 includes all the propagation effects on the surface of the Earth at frequencies from 0.1 GHz to 50 GHz. In addition, other factors which affect interference calculation, such as terrain height, bandwidth overlapping are also considered in the proposed database approach, which is illustrated in Fig. 2. The typical interference threshold we determine is based on the long term interference

Table 1. ITU-R table of allocation

Frequency bands	ITU region 1	ITU region 2	ITU region 3
17.3–17.7 GHz (Scenario A)	FSS (space-Earth) BSS (feeder links) Radiolocation	FSS (space-Earth) BSS (feeder links) Radiolocation	FSS (space Earth)BSS (feeder links) Radiolocation
17.7–19.7 GHz (Scenario B)	FSS (space-Earth) BSS (feeder links up 18.1 GHz) FS	FSS (space-Earth) FS	FSS (space-Earth) BSS (feeder links up 18.1 GHz) FS
27.5–29.5 GHz (Scenario C)	FSS (Earth-space) FS MS (Mobile Services)	FSS (Earth-space) FS MS	FSS (Earth-space) FS MS

which can be expected to be present for at least 20 % of the average year and it is set at 10 dB below the noise floor. The interference thresholds for FSS reception and for FS reception are therefore −154 dBW/MHz and −146 dBW/MHz, respectively as given in [14,15].

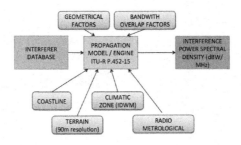

Fig. 2. Interference modelling by ITU-R P.452-15.

Having determined the interference level at the FSS (in scenarios A or B) it can be compared with the regulatory threshold. The action is then taken in the resource allocation at the gateway where a new carrier can be assigned either in another part of the shared band where interference is acceptable or in the exclusive band. For scenario C the situation is different as the interference is caused by the FSS into the FS. Here the database is used to calculate the maximum permissible power that can be transmitted from the FSS in the vicinity in order to retain the threshold condition at the FS receivers

Scenario A

A UK BSS database made available for this study is used for scenario A and contains 442 carriers from a total of 31 BSS uplink earth stations at 8 physical sites, to 12 different satellites. The number of carriers of each BSS earth station ranges from 1 to 42. The carriers span the range 17.3 GHz to 18.35 GHz. The bandwidths of the carriers that belong to the same BSS earth station are the same while those that belong to different earth stations might be different and are typically 26 MHz, 33 MHz, 36 MHz or 66 MHz. The EIRP of these earth station antennas ranges from 69 dBW–84 dBW and all antenna radiation patterns are as defined in ITU-Recommendation S.465 [16] or S.580 [17]. Using the BSS database, area analysis for scenario A in the UK is provided to investigate how much area would be affected by interference from the BSS feeder links. The band of interest is split into 10 × 40 MHz sub-bands (SB1/SB10) and the analysis is then conducted in each sub-band to determine the area of the contours at different cognitive zone thresholds. These mirror the usual 40 MHz channel spacing adopted for BSS satellites. Area analysis is based on BSS database with the full ITU-R P.452-15 model employing the terrain and climatic zones and the FSS terminal evaluated points to a satellite at 53 degrees E longitude. The results are for long term interference (normally 20 %). One example of affected area at difference cognitive zone thresholds is shown in Fig. 3, which represents SB1. It was found that in general across the sub-bands at a −155 dBW/MHz threshold less than for 2 % of the area of the UK is affected by BSS feeder links and thus more than 98 % of the area of the UK can be used by an FSS terminal without the need for any further action We have performed the same analysis for Luxembourg with very similar results and as the UK is the most dense BSS case we would expect the results to be similar or better in other countries.

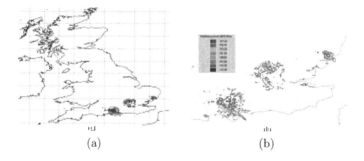

(a)   (b)

Fig. 3. Example of cognitive zones for the sub-band 1 (17.3–17.34 GHz) based on full ITU model.

Scenario B
FS data bases at 18 GHz (17.7 to 19.7 GHz) were used to evaluate the interference. An up to date FS data base was made available to this project by Ofcom UK. This data base for the UK FS in the band 17.7 to 19.7 GHz contains 15,036 carrier records. A French data base has also been examined at 18 GHz and is based on the latest ITU terrestrial services BR IFIC database [18], which contains 16,136 carrier records. Similarly data for Hungary and Slovenia has been obtained from the same source and these contain 2,402 and 1,237 carrier records respectively. Data at 18 GHz for Poland was obtained which is more up to date

than the BR IFIC data and contains 8,323 carrier records. An example of the spectrum occupancy for an FSS terminal placed in the SE of the UK is shown in Fig. 4. Here the calculation based on free space loss only and the full ITU model clearly shows the necessity for inclusion of the terrain effects. It is also clear that white spaces exist for FSS carriers at this location. However the actual position of the white spaces varies with location and thus the data base can be used together with the resource allocation scheme to place the carriers appropriately to the FSS terminals. By analysing the interference results for five data bases for different countries it is possible to get an increased insight into the situation. A CDF is shown for the total occupied bandwidth of the FS interferers at a point over the regions of interest in Fig. 5. This demonstrates that for the majority of locations a large percentage of the 2 GHz spectrum is available for FSS use in all of the countries examined.

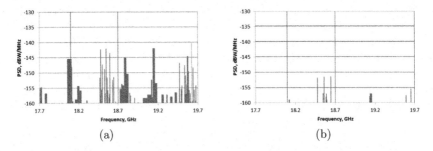

(a)   (b)

Fig. 4. Channel occupancy at an FSS point in SE UK for (a) Free space loss model and (b) Full ITU model.

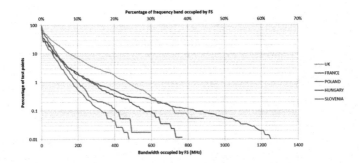

Fig. 5. CDF of FS bandwidth available for threshold of −154 dBW/MHz.

3   Spectrum Sensing

A database approach, although very efficient, requires knowledge on BSS and FS links, which might be confidential for some countries. Moreover, even in countries where such information is available, the database approach does not allow to adapt to short-term variations in spectrum occupancy.

In this scenario, coexistence between FSS cognitive systems and incumbent systems is thus limited by the interference generated from the latter towards the FSS terminal. In particular, a significant amount of aggregate interference may occur at a given FSS terminal due to the side-lobes of the receiving antenna pattern. CR techniques can thus be employed to foster the coexistence between FSS DL and incumbent links, as shown in the following sections. In the following, it is assumed that the receiving chain at the cognitive terminal is used for both sensing and secondary transmissions.

Among several Spectrum Sensing techniques [19], we will firstly focus on energy detection (ED) that is assessed in the considered scenario. Simulation results show that CR-based satellite systems can significantly improve spectrum utilization, which would enable the integration between terrestrial and satellite networks, as well as provide additional spectrum for both systems.

Spectrum sensing (SS) aims at detecting the incumbent user signal by scanning a selected frequency band B [19,20]. It refers to the detection of an unknown or partially known signal, and a trade-off between the probability of false alarm (Pf) and the probability of detection (Pd) is necessary for achieving an accurate degree of certainty in such detection. SS techniques can be modeled as a binary hypothesis test problem, comparing a statistical metric with a given threshold.

An energy detector aims at detecting the presence of incumbent signals based on the energy estimated at the antenna input of the cognitive terminal [21,22]. It is a blind detection technique, as it does not require a-priori knowledge on the incumbent signal, and therefore has a general applicability in CR-based systems. However, it is highly susceptible to Signal-to-Noise Ratio (SNR) wall problem, that prevents from achieving the desired target probabilities Pd or Pf, as the uncertainty in noise power estimation, N, can easily erroneously trigger the detection [23,24].

As an example, in Fig. 6 the comparison between a real interference map, obtained by comparing the real interference values with a certain threshold $(I/N) = -10$ dB, and the interference map obtained through the use of an Energy Detector are shown. In Table 2, the parameters used for obtaining the numerical results of the ED map are listed.

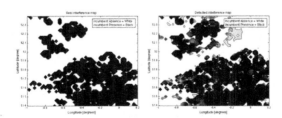

Fig. 6. Comparison between a real interference map and the detected interference map by exploiting the ED.

Table 2. Simulation parameters for the ED.

Frequency [GHz]	17.634
Bandwidth [MHz]	36
FSS terminal latitude	From 51.4 N to 52.4N
FSS terminal longitude	From 1.0 W to 0.2E
FSS satellite longitude	53E
Probability	Pfa = 0.1
Sensing time [us]	20
Noise uncertainty [dB]	0
Number of simulation	10000

Within the proposed scenarios, a Signal-to-Interference plus Noise Ratio (SINR) estimation algorithm has been also proposed. Among different interference estimators available in the literature [25], we rely on the Data Aided SNORE (DA-SNORE) algorithm described in [26]. It is assumed that the cognitive Earth terminal is equipped with a receiving chain able to scan all frequencies of interest with a sensing sub-band equal to 36 MHz, which is the typical bandwidth of DVB-S2 and DVB-S2x standards [27], used by the cognitive satellite system. The algorithm, described in [28], is based on the knowledge of the pilot blocks of the DVB-S2 standard. It is worthwhile highlighting that, as the pilot blocks are the same for both Scenario A and B, the algorithm can be applied with no modification to either of them.

Focusing on Scenario A, the 400 MHz band is split into 11 sub-bands, and on each sub-band the DA-SNORE algorithm is applied to determine the interference level received from BSS feeder links. As the incumbent spectrum utilization is almost constant in time, the sensing operation can be performed with a relatively low duty cycle and when no data transmission is required, so as to lower the computational load. The information gathered during this initial sensing phase can then be reported to the Network Control Center (NCC), which allocates to each user the most reliable sub-band.

The performance of the SINR estimation algorithm has been compared to data extracted from databases. In particular, the potential geographical reuse factor of a specific carrier as a function of the relative location between interferer and interfered terminals has been performed. As an example, Fig. 7 provides the SINR values obtained from the database over a specific geographic area and compares them to the values estimated through the DA-SNORE algorithm. For obtaining the comparison, the same parameters listed in Table 2 have been used, with, additionally, a number of pilots equal to 10 and the SNR value at the DA-SNORE based interference estimator antenna input equal to 4 dB.

The estimated values excellently match the SINR values obtained from the database, and thus the DA-SNORE algorithm provides a valuable solution for spectrum awareness either to complement the information stored in databases or to provide the spectrum occupancy when databases are not available.

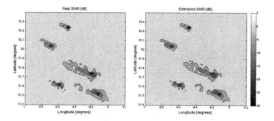

Fig. 7. Comparison between real SINR and estimated SINR by exploiting the proposed DA-SNORE based algorithm.

4   Resource Allocation

After obtaining the spectrum awareness, the available resources need to be allocated among the cognitive terminals. In this section, we provide a numerical evaluation of the resource management techniques presented in the context of [8], which aim at optimizing the allocation of available resources, while employing interference management techniques. Given the similarities between Scenario A and Scenario B, dynamic carrier assignment (CA) techniques based on Signal-to-Interference-plus-Noise Ratio (SINR) are applied in both cases. In Scenario A and B, beamforming (BF) will be considered together with CA in order to mitigate the received interference and enlarge the cognitive zones presented in Sect. 2. Essentially, the level of FS interference at the carrier level is firstly determined based on the available information of FS databases. Having determined the interference level and using the signal level obtained from the FSS system analysis, the SINR is computed for all the FSS terminals considering all the carrier frequencies. Subsequently, we apply BF only in the FSS terminals which suffer excessive interference. Next, the improved SINR is fed to the CA module in order to assign each user to a carrier so as to maximize the total sum-rate of the system.

Here, we present results for Scenario B. Scenario A, as indicated in Sect. 2, is expected to provide higher gains due to a few number of BSS feeder links. For Scenario B evaluation, we consider the country France and the selected beams are depicted in Fig. 8. These beams have been selected based on the potential FS interference receivers and the final results are obtained based on the weighting factor provided in Table 3.

The results shown in this section were obtained after 50 Monte Carlo runs, in which the locations of the FSS terminals were selected uniformly at random for each realization within the considered beam coverage according to the population density database produced by NASA Socioeconomic Data and Applications Center (SEDAC) [29]. The considered system parameters are summarized in Table 4.

The evaluation is made in the following cases:

– Case 1: Exclusive band only: In this case, the SINRs and user rates are calculated using only exclusive carriers.

Table 3. Selected beams for
Scenario B.

	FS links	Weighting factor
1	1522	0.038462
2	1681	0.076923
3	635	0.5
4	906	0.26923
5	1220	0.11538

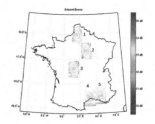

Fig. 8. Selected beams according
to FS antenna density.

Table 4. Simulation parameters for Scenario B.

Simulation parameter	Value
Carrier bandwidth	63.4 MHz
Shared band	17.7–19.7 GHz (32 carriers)
Exclusive band	19.7–20.2 GHz (8 carriers)
Satellite	13°E
EIRP satellite	65 dBW
Re-use pattern	4 color
FSS antenna gain (max)	42.1 dB
FSS Rx noise temperature	262 K
FSS terminal height	15 m
LNBs at FSS terminal	3

- Case 2: Shared plus exclusive band w/o FS presence: In this case, the SNRs
  and user rates are calculated considering both shared and exclusive carriers,
  but without considering the FS system.
- Case 3: Shared plus exclusive bands w/ FS presence: In this case, the SINRs
  and user rates are calculated considering both shared and exclusive carriers,
  and considering the FS system.

The results of the five evaluated beams are shown in Table 5 in terms of per
beam throughput (Mbps). The methodology followed for throughput evaluation
is based on [30] and the employed carrier allocation and beamforming methods
have been discussed in [30]. From Table 5, it can be noted that the throughput
values significantly differ across the considered beams even for the case of exclusive only case, which is due to different beam gains and Carrier-to-Interference
(C/I) values over these beams. The main conclusion that can be extracted from
Table 5 is that the throughput per beam improvement obtained with the proposed CA and BF techniques is 405.92 %. What is most important is that using
the proposed CA and BF we can achieve similar average throughput as if there
were no FS system.

Table 5. Per beam throughput (Mbps) for Scenario B.

Beam no.	Case 1		Case 2		Case 3	
	w/ o CA	w/ CA	w/o CA	w/CA	w/o CA	w/ CA+BF
1	675.10	675.42	3414.17	3419.78	3413.73	3468.05
2	679.13	679.49	3404.98	3410.56	3404.25	3457.66
3	660.42	660.72	3304.87	3309.07	3304.11	3331.52
4	725.76	725.95	3641.67	3646.28	3640.18	3661.03
5	718.47	718.85	3626.94	3646.07	3623.62	3659.71
Average	686.56	686.84	3444.97	3451.16	3443.74	3473.46

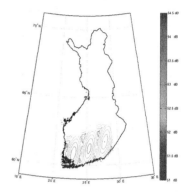

Table 6. Per beam throughput (Mbps) for scenario B.

	Number of Rx FS	Weight
1	32	0.222
2	902	0.111
3	6	0.667

Fig. 9. Selected beams according to FS antenna density.

The applicability of resource allocation techniques in Scenario C was discussed in [31]. Here, we summarize the results obtained for Scenario C in Finland, from which a reliable FS Database obtained from the national regulator is used. Again, we select the most representative beams in terms of FS density. The beams are shown in Fig. 9 and Table 6 summarizes the details.

It can be noted that in Scenario C, the cognitive transmitters create interference to the FS system and, thus, not only the carriers have to be optimally assigned but also the transmit power of the cognitive FSS terminal devices has to be adjusted so that the interference at each of the FS stations is kept below the given threshold. As we did for Scenario B, 50 Monte Carlo runs were averaged in which the FSS terminal locations were determined based on population data. A summary of the most relevant parameters and the FSS link budget details are presented in Table 7.

The results of the three evaluated beams are shown in Table 8. It is important to keep in mind that with the proposed RA we ensure that we never violate the FS interference threshold. On average, the proposed RA provides 400 % gain over the exclusive only band, which coincides with the best results that can be achieved in this particular scenario.

Table 7. Simulation parameters for Scenario C.

Simulation parameter	Value
Carrier bandwidth	7 MHz
Shared band	27.5–29.5 GHz (285 carriers)
Exclusive band	29.5–30 GHz (71 carriers)
Satellite	13E
EIRP	50 dBW
Re-use pattern	4 color
FSS antenna gain (max)	42.1 dB
[G/T] (max)	29.3 dB/k
FSS terminal height	15 m

Table 8. Per beam throughput (Mbps) for Scenario C.

Beam no.	Case 1		Case 2		Case 3	
	w/ o RA	w/ RA	w/o RA	w/RA	w/o RA	w/ RA
1	651.96	651.96	3258.19	3258.33	3253.60	3258.33
2	683.53	683.53	3421.73	3422.31	3252.67	3422.31
3	691.87	691.87	3449.44	3449.94	3449.30	3449.94
Average	682.08	682.08	3403.91	3404.34	3384.03	3404.34

5   Conclusions

It has been demonstrated that a data base system will allow frequency sharing in the 17.3 to 19.7 GHz down link bands and in the 27.5 to 29.5 GHz uplink band between satellite FSS and fixed FS links. Significant additional spectrum in the shared bands is available to satellite FSS and only small areas across Europe would need to adopt additional interference mitigation. Using a carrier resource allocation scheme at the satellite gateway it has been demonstrated that up to four times capacity gains can be achieved over the use of the exclusive band only. It has also been shown that spectrum sensing at the satellite terminals is feasible and can be used where data base information is not available or to augment and hence improve the data base.

Acknowledgement. The authors would like to acknowledge the EU FP7 project CoRaSat which has supported the work herein and in particular the inputs from industrial partners, SES,TAS and Newtec.

References

1. A Digital Agenda for Europe, FCC 02–155, European Commission COM 245,Technical report, Brussels.(2010)

2. EU FP7 Project BATS. http://www.batsproject.eu/
3. Fenech, H., Lance, E., Kalama, M.: KA-SAT and the way forward, Ka-Band Conference. Italy, Technical report, Palermo (2011)
4. Highest-capacity communications satellite. http://www.guinnessworldrecords. com/records-1/highest-capacity-communications-satellite/
5. Thompson, P., Evans, B., Castenet, L., Bousquet, M., Mathiopoulos, T.: Concepts and technologies for a terabit/s satellite. In: Proceedings of SPACOMM-2011 (best paper award in 2011), Budapest, Hungary, April 2011
6. Kyrgiazos, A., Evans, B., Thompson, P., Mathiopoulos, P.T., Pa-paharalabos, S.: A terabit/second satellite system for european broadband access: a feasibility study. Int. J. Satell. Commun. Netw. **32**(2), 63–92 (2014)
7. The European conference of postal and telecommunications adminis- trations. http://www.cept.org/cept
8. EU FP7 Project CoRaSat. http://www.ict-corasat.eu
9. Liolis, K., Schlueter, G., Krause, J., Zimmer, F., Combelles, L., Grotz, J., Chatzinotas, S., Evans, B., Guidotti, A., Tarchi, D., Vanelli-Coralli, A.: Cognitive radio scenarios for satellite communications: The corasat approach. In: 2013 Future Network and Mobile Summit (FutureNetworkSum- mit), pp. 1–10, July 2013
10. Maleki, S., Chatzinotas, S., Evans, B., Liolis, K., Grotz, J., Vanelli-Coralli, A., Chuberre, N.: Cognitive spectrum utilization in ka band multibeam satellite communications. IEEE Commun. Mag. **53**(3), 24–29 (2015)
11. Cognitive radio techniques for satellite communications operating in Ka band, Technical report, ETSI System Reference document. http://webapp.etsi.org
12. Standardization of TV white space systems. http://www.ict-crsi.eu/index.php/ standardization-streams/tv-white-spaces
13. Recommendation P.452-15: Prediction procedure for the evaluation of interference between stations on the surface of the earth at frequencies above about 0.1 GHz, International Telecommunication Union, Technical report (2013)
14. Methods for the determination of the coordination area around an earth station in frequency bands between 100 MHz and 105 GHz, ITU Radio Regulation Appendix 7, International Telecommunication Union, Technical report (2012)
15. Recommendation F.758-5: System parameters and considerations in the development of criteria for sharing or compatibility between digital fixed wireless systems in the fixed service and systems in other services and other sources of interference, International Telecommunication Union, Technical report (2012)
16. Recommendation ITU-R S.465: Reference radiation pattern for earth station antennas in the fixed- satellite service for use in coordination and interference assessment in the frequency range from 2 to 31 GHz, International Telecommunication Union, Technical report (2010)
17. Recommendation ITU-R S.580: Radiation diagrams for use as design objectives for antennas of earth stations operating with geostationary satellites, International Telecommunication Union, Technical report (2004)
18. ITU-R Terrestrial BRIFIC. http://www.itu.int/ITU-R/index.asp?category=terrest rial&rlink=terrestrial-brific&lang=en
19. Haykin, S.: Cognitive radio: brain-empowered wireless communications. IEEE JSAC **23**(2), 201–220 (2005)
20. Hossain, E., Niyato, D., Han, Z.: Dynamic Spectrum Access and Management in Cognitive Radio Networks. Cambridge University Press, Cambridge (2009)
21. Urkowitz, H.: Energy detection of unknown deterministic signals. Proc. IEEE **55**(4), 523–531 (1967)

22. Axell, E., Leus, G., Larsson, E.G., Poor, H.V.: Spectrum sensing for cognitive radio: state-of-the-art and recent advances. IEEE Sig. Proc. Mag. **29**(3), 101–116 (2012)
23. Cabric, D., Mishra, S.M., Brodersen, R.W.: Implementation issues in spectrum sensing for cognitive radios. In: Proceedings of the 38th Asilomar Conference on Signals, Systems and Computers, pp. 772–776, November 2004
24. Kim, H., Shin, K.G.: In-band spectrum sensing in IEEE 802.22 WRANs for incumbent protection. IEEE Trans. Mob. Comput. **9**(12), 766–1779 (2012)
25. Pauluzzi, D., Beaulieu, N.: A comparison of SNR estimation techniques for the AWGN channel. IEEE Trans. Commun. **48**(10), 16811691 (2000)
26. Cioni, S., Corazza, G., Bousquet, M.: An analytical characterization of maximum likelihood signal-to-noise ratio estimation. In: 2nd International Symposium on Wireless Communication Systems 2005, pp. 827–830, September 2005
27. ETSI EN 302 307 v1.3.1, Digital Video Broadcasting (DVB): Second Generation Framing Structure, Channel Coding and Modulation Systems for Broadcasting, Interactive Services, News Gathering and Other Broadband Satellite Applications (DVB-S2), March 2013
28. Icolari, V., Guidotti, A., Tarchi, D., Vanelli-Coralli, A.: An Interference Estimation Technique for Satellite Cognitive Radio Systems, to appear in ICC 2015, June 2015
29. NASA, Socioeconomic Data and Applications Center (SEDAC). http://sedac.ciesin.columbia.edug. Accessed 27 October 2014
30. Sharma, S.K., Lagunas, E., Maleki, S., Chatzinotas, S., Grotz, J., Krause, J., Ottersten, B.: Resource allocation for cognitive satellite communications in Ka-band (17.7-19.7 GHz). In: Workshop on Cognitive Radios and Networks for Spectrum Coexistence of Satellite and Terrestrial Systems, IEEE International Conference On Communications (ICC), London, June 2015
31. Lagunas, E., Sharma, S.K., Maleki, S., Chatzinotas, S., Grotz, J., Krause, J., Ottersten, B.: Resource allocation for cognitive satellite uplink and fixed-service terrestrial coexistence in Ka-band. In: International Conference on Cognitive Radio Oriented Wireless Networks (CROWNCOM), Doha, April 2015

On the Feasibility of Interference Estimation Techniques in Cognitive Satellite Environments with Impairments

Daniele Tarchi[1][✉], Vincenzo Riccardo Icolari[1], Joel Grotz[2],
Alessandro Vanelli-Coralli[1], and Alessandro Guidotti[1]

[1] Department of Electrical, Electronic and Information Engineering,
University of Bologna, Bologna, Italy
{daniele.tarchi,vincenzo.icolari2,alessandro.vanelli,
a.guidotti}@unibo.it
[2] Newtec Cy, 9100 Sint-Niklaas, Belgium
joel.grotz@newtec.eu

Abstract. The increasing demand of wider frequency bands in wireless communications has lead in the last years to the introduction of novel techniques allowing to exploit more efficiently the radio spectrum. In particular, spectrum sharing is considered one of the key technologies for future wireless systems. Cognitive radio is considered the most important technique for efficiently allowing the spectrum sharing among heterogeneous systems. If on one hand the cognitive radio techniques have been extensively assessed in terrestrial communications, they remain quite an unexplored area in Satellite Communications (SatComs). In this paper an Interference Estimation technique for SatComs aiming to reuse the spectrum resources primarily allocated to terrestrial communications is discussed. In particular, its assessment in a realistic scenario, where multiple impairments are considered, is discussed.

Keywords: Cognitive Radio · Satellite communications · Interference estimation · Spectrum sensing · Impairments

1 Introduction

Future satellite systems (2020–2025) are expected to exploit GEO satellites, with capacities ranging from hundreds of Gbps up to Tbps. This will be achieved by means of hundreds of spotbeams, via higher order frequency reuse. In fact, the limited amount of exclusive spectrum that can be accessed by the Fixed Satellite Service (FSS) limits the actual system capacity. Current High Throughput Satellites (HTS) in Ka-band and above have gained momentum to reduce the large cost per bit and allow Ka-band satellites to provide the required capacity.

In this context, access to additional frequency bands through frequency sharing for the user terminal frequency bands would provide additional capacity to

© Institute for Computer Sciences, Social Informatics and Telecommunications Engineering 2015
P. Pillai et al. (Eds.): WiSATS 2015, LNICST 154, pp. 133–146, 2015.
DOI: 10.1007/978-3-319-25479-1_10

further increase the capacity of the satellite system. Cognitive Radio (CR) techniques are seen as the most promising mean to tackle the spectrum scarcity problem [1]. They allow to efficiently share some portions of the spectrum while limiting harmful interference among different communication systems. CRs potential has already been demonstrated in wireless terrestrial services [2], while in SatCom their implementation and study is still in its infancy. SatComs represent a challenging application scenario for CRs, due to, *e.g.*, the geographically wide coverage of the spectrum allocation and the power imbalance among ground and user terminals [3].

In order to enable coexistence between primary and secondary systems, CR techniques can be employed in such scenarios [4]. In particular, spectrum awareness techniques allow the cognitive system to be aware of the primary system presence. The most common techniques that provide this kind of awareness, and that are often proposed also in terrestrial scenarios, are spectrum sensing [5] and databases [6].

These frequencies are defined as *non exclusive* since they are assigned to both services, with a primary use for feeder uplinks (BSS) or fixed services (FS). Aiming at reusing a wider spectrum portion by exploiting those bands in which weak incumbent signal levels may be present, FSS terminals that usually operate in the *exclusive* frequency bands can additionally use the *non exclusive* bands if room is found in order to get additional capacity. It is worthwhile noting that even bands in which incumbent signals are present might be exploited by the FSS, if a proper set of transmission parameters can be found to satisfy the target Quality of Service (QoS).

Rather than typical spectrum sensing techniques, which focus on discovering spectrum holes [5], in this paper the estimation of interfering levels is considered. In fact, while spectrum sensing is usually considered for its detection capabilities, our aim is to have a joint detection and estimation approach that allows to exploit also the underused spectrum intervals where the interference is not harmful.

To this aim, a proper interference estimation technique along the *non exclusive* spectrum should be employed allowing to identify which bands are available for transmission also allowing geographical reuse due to propagation terrain effects, geographical isolation and different direction of the links.

Among other interference estimators proposed in the literature [7], we rely on the Data Aided SNORE (DA-SNORE) algorithm [8] for estimating the signal to interference plus noise ratio (SINR) in each band. The effectiveness of the proposed technique has been described in [9] by assuming a satellite cognitive system based on the DVB-S2 [10] or DVB-S2x standards [11] for downlink transmission and the cognitive earth terminal, equipped with a receiving chain able to scan all the frequencies of interest with a sensing sub-bandwidth, able to perform the data aided estimation algorithm by means of pilot blocks.

In this paper, after recalling the most important characteristics of the SNORE based Interference Estimation algorithm [9], we will assess its performance in a realistic scenario. Indeed, a real cognitive radio SatCom scenario is characterized by presence of several impairments. To this aim, the robustness of the proposed technique in presence of the most typical impairments is evaluated.

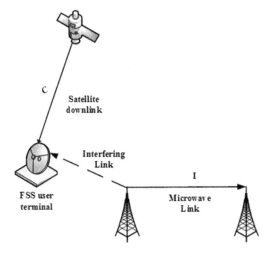

Fig. 1. Scenario representation

2   The Reference Scenario

A Ka-band FSS cognitive system, reusing for downlink transmission the frequency bands allocated to FS links (incumbent systems) as depicted in Fig. 1, is considered for performance assessments of the proposed technique. Since Radio Regulations define limits for emissions from FSS systems [12], it can be assumed that cognitive transmitters do not cause harmful interference towards FS links. Thus, the main focus is on allowing the cognitive system to operate in presence of interfering incumbent systems, as highlighted in Fig. 1. To this aim, the cognitive system shall be able to scan the non exclusive spectrum, estimate the interference levels, and identify the frequencies suitable for the transmission. Therefore, instead of typical cognitive sensing techniques, the system will rely on an estimation technique able to give an increased knowledge of the incumbent presence.

The SNORE based interference estimation algorithm has been proposed in [9]. For the sake of clarity, we summarize here the most important characteristics of the proposed algorithm by focusing more on the impairments in Sect. 3.

We assume that the receiver earth terminal of the cognitive system is equipped with a receiving chain able to scan all the frequencies of interest with a sensing sub-band, B_W, equal to 36 MHz, which is the typical bandwidth of DVB-S2 [10] and DVB-S2x [11] communications. Thus, the 400 MHz *non exclusive* spectrum portion is split into 11 sub-bands, and for each of these bands the DA-SNORE algorithm [8] is used in order to assess whether incumbents activities are present or not.

In the depicted scenario, we can represent the lowpass equivalent expression of the signal at the cognitive receiver denoted as $r(t)$ by

$$r(t) = \begin{cases} s(t) + \sum_k^{N_i} i_k(t) + n(t) & \text{with interferers} \\ s(t) + n(t) & \text{without interferers} \end{cases} \qquad (1)$$

where: (i) $s(t) = \sqrt{P_0}v(t)e^{j\phi_0}$ is the useful signal from the satellite; (ii) $i_k(t) = \sqrt{P_k}\iota_k(t)e^{j(2\pi f_k t + \phi_k)}$ the k-th interferer, with $k = 1, \dots, N_i$ and N_i denoting the number of the interferers; (iii) $P_0, P_k, k = 1, \dots, N_i$, denote the received power on the useful and the k-th interfering links, respectively; (iv) $\phi_0, \phi_k, k = 1, \dots, N_i$, denote the phase of the useful and k-th interfering signal, respectively. We can further assume $\phi_0 = 0$ without loss of generality; (v) $f_k, k = 1, \dots, N_i$, is the frequency shift between the useful signal carrier ($f_0 = 0$ since we have lowpass equivalents) and the k-th interfering signal carrier; (vi) $v(t), \iota_k(t)$, with $k = 1, \dots, N_i$, denote the lowpass complex signals of the cognitive and k-th interferering signals, respectively. We also assume that the interfering signals have white spectral density in their band, and can thus be modeled as a Gaussian random variable with variance $P_k, k = 1, \dots, N_i$; and (vii) $n(t)$ the additive white Gaussian noise (AWGN) with zero mean and power spectral density N_0.

By exploiting the methodology introduced in [13], the explicit expressions highlighting the relationships between W, SINR, the estimation error variance σ_ϵ^2, and the Cramer-Rao bound (CRB) for the design of the proposed cognitive technique has been derived in [9].

It is worth to be noticed that in the SNORE estimator, the CRB normalized with respect to SINR^2 is defined as [13, Eq. (21)]:

$$\frac{CRB}{SINR^2} = \frac{4WN_s}{(2WN_s - 3)^2}\left(\frac{2}{SINR} + 1\right) \qquad (2)$$

and the relationship between the error variance, σ_ϵ^2, and the SINR can be derived by taking into account the number of pilot blocks W that allow to reach the CRB (3) [13, Eq. (20)]:

$$\frac{\sigma_\epsilon^2}{SINR^2} = \frac{4\left(2W^2 N_s^2\left(SINR + 2\right)\ SINR + WN_s\left(1 - 4SINR\right) - 1\right)}{\left(2WN_s - 3\right)^2\left(2WN_s - 5\right)} \qquad (3)$$

2.1 The SNORE Based Interference Estimation

In order to cope with the problem of identifying which band gives best performance among those sensed in the *non-exclusive* spectrum, we propose to perform the scan operation periodically. In the depicted scenario, it is possible to sense with a very low duty cycle, in order to guarantee the desired capacity and satisfy QoS requirements without increasing the computational load on the cognitive system.

In [9] the authors highlighted that the estimation performance strongly depends on W, as the more pilots blocks are accumulated the lower the estimation error will be. On the other hand, increasing the number of pilot blocks to be accumulated also requires longer sensing periods.

A proper value of W for the design of the algorithm can be obtained according to the scenario we have to cope with and by exploiting (2) and (3). From (2), it is possible to derive W, as:

$$W = \frac{1}{N_s} \left(\frac{3}{2} + \frac{SINR^2}{CRB} \left(\frac{2}{SINR} + 1 \right) \left(\frac{1}{2} + \sqrt{1 + \frac{6CRB}{SINR\,(2 + SINR)}} \right) \right) \tag{4}$$

This closed-form expression provides the number of pilot blocks W as a function of the Cramer-Rao bound, the SINR, and the number of pilots per block N_s. The number of pilot blocks required to achieve a target error variance $\sigma_\epsilon^2|_t$ as a function of the SINR can be instead evaluated by solving the third degree equation:

$$W^3 N_s^3 - \left(\frac{11}{2} + \frac{SINR}{\sigma_\epsilon^2}(SINR + 2) \right) W^2 N_s^2$$
$$+ \left(\frac{39}{4} - \frac{1}{2\sigma_\epsilon^2} + 2\frac{SINR}{\sigma_\epsilon^2} \right) W N_s - \frac{45\sigma_\epsilon^2 - 4}{8\sigma_\epsilon^2} = 0 \tag{5}$$

that has been obtained from (3) after a few mathematical steps. The values of W obtained from (4) and (5) represent respectively the lower bound and the minimum number of W that achieve the target error variance. Hence, (4) and (5) give the description of how many pilots are needed to achieve a target estimation error as a function of SINR and can be used for the design of the proposed technique.

3 Interference Estimation Impairments

Even if the SNORE based interference estimation algorithm have promising performance [9] a proper calibration and performance assessment under realistic impairments should be performed. Thus, the proposed spectrum sensing technique based on measuring the SNIR requires a baseline calibration.

In the following we do not refer to any specific knowledge of the incumbent link for this sensing task. The spectrum sensing has to be performed during the first carrier lineup procedure of the terminal. The overall expected link performance is assumed known from planning and previous link budget exercises. This results in an expected signal to noise ratio that has to be met at the installation of the terminal.

During the terminal installation and after the antenna pointing task of the terminal installation, we have a basis of the two values for the expected and measured signal to noise ratio. A residual difference has to be correctly interpreted by the NCC. This difference is measured and expected SNIR may result from

Table 1. Potential perturbations contributing to an incorrect SNIR estimation

Potential contributing factor to measured SNIR delta perturbation	Estimated order of inaccuracy	Potential mitigation measure
Rainfade and atmospheric attenuation during terminal installation	Several dB in Ka-band	Use long term averaging and additional learning procedure in NCC
Inaccurate antenna pointing of the terminal	1 dB p-p	
Cross polarization interference	0.5 dB p-p	May be taken into account and neglected if under control
Bias in expected SNIR value resulting from margins at different levels	1 dB max peak	Reference terminals and an overall system learning of the expected SNIR is required
Interference from other satellite downlinks or adjacent beams of the same system	2 dB max peak	Requires a revised planning tool with potential reference terminals and measurements of expected levels
Interference from terrestrial contributions (scenario A / B) such as BSS feeder links or FS transmitters	Value to be estimated	N.A
Receiver gain variation	Long term variations (seasonal) 1–2 dB p-p	
LNB gain variation over temperature	1–2 dB typical	

different perturbations or inaccuracies in the overall system that needs to be addressed with the NCC integration of the spectrum sensing techniques. These include the aspects listed in Table 1.

The system spectrum sensing requires as a result a combination of planning tools, reference terminals and a system learning mechanism with a feedback to the terminal installation procedure. Additional mechanisms have to be defined in this context to address all possible sources of practical errors to devise a reliable detector for the terrestrial interference.

In this paper, we focus on the effect of perturbations and impairments that may occur in case the spectrum sensing technique proposed in [9], and previously described, is used. Earth terminals can be affected by typical issues that may arise also during set up procedure as:

– Pointing errors
– Fading uncertainties
– G/T and sat. gain uncertainties over coverage area

Impairments that affect the SINR estimation process during terminals installation and the sensing phase, are here introduced and summarized in Table 1. Further details can be found in [14,15]. Imperfect alignment of the transmitting and receiving antennas could cause pointing errors that are sources of additional losses. These losses are due to a reduction of the antenna gain with respect to its maximum and are function of the misalignment of the angle of reception θ_R, and can be evaluated as:

$$L_R = 12 \left(\frac{\theta_R}{\theta_{3dB}} \right)^2 \quad [dB] \tag{6}$$

where θ_{3dB} is the 3 dB beamwidth angle between the direction in which the gain is maximum and that in which it is half of this latter value. Other losses of earth terminals, which are due to non idealities, are feeder losses L_{FRX} between the antenna and the receiver, and polarization mismatch losses L_{POL}.

Atmospheric events cause additional attenuation and variation with respect to the common free space loss propagation. Several effects are present but an overall contribution affecting the received power can be taken into account by adding to the free space loss attenuation A_{FS} the contribution A_P that includes all the atmospheric attenuation:

$$A_{TOT}[dB] = A_{FS}[dB] + A_P[dB]$$

These losses are significant above 10 GHz as in case of the Ka bands, which are used in the considered scenario. In such bands, tropospheric phenomena are the main contributions of the link availability and service quality degradation. These phenomena are (i) attenuation, (ii) scintillation, (iii) depolarization and (iv) increase of the antenna temperature in the receiving earth terminal. A more detailed description of these phenomena is included in [15, Chapter 3]. Link budget is affected by these contribution in many ways. In the downlink case, the carrier to noise ratio can be expressed as:

$$\left(\frac{C}{N_0} \right)_{DOWNLINK} [dB] = (1 - \Delta_1)EIRP_{SAT} - A_{TOT} + (1 - \Delta_2) \left(\frac{G}{T} \right)_{ES} - k_B \tag{7}$$

and can be rearranged in order to separate the ideal, or expected value $\left(\frac{C}{N_0} \right)_{FS}$ calculated in free space loss conditions, from contributions that cause its variation in the following way

$$\left(\frac{C}{N_0} \right)_{DOWNLINK} [dB] = \left(\frac{C}{N_0} \right)_{FS} - \Delta_1 EIRP_{SAT} - A_p - \Delta_2 \left(\frac{G}{T} \right)_{ES} \tag{8}$$

where $EIRP_{SAT}$ is the satellite EIRP, $\left(\frac{G}{T} \right)_{ES}$ the figure of merit of the earth terminal receiver and Δ_1 and Δ_2 a possible decrease of the satellite $EIRP_{SAT}$ and the figure of merit $\left(\frac{G}{T} \right)_{ES}$ respectively. In particular, the figure of merit G/T of the earth terminal at the receiver input, i.e., including also losses of the receiving chain, can be expressed as

$$\left(\frac{G}{T} \right)_{ES} = \frac{G_{R_{max}}/L_R L_{FRX} L_{POL}}{T_{TOT}} [K^{-1}] \tag{9}$$

where T_{TOT} is the total downlink system noise temperature at the receiver input and it is function of the antenna temperature T_A, the feeder temperature T_F, and the effective input noise temperature of the receiver T_{eRX}

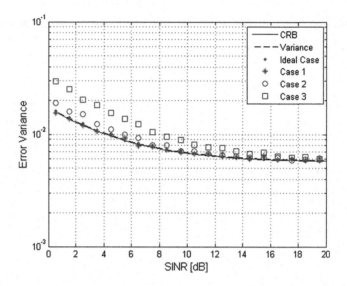

Fig. 2. Error variance as a function of the SNIR in presence of impairments

$$T_{TOT} = \frac{T_A}{L_{FRX}} + T_F \left(1 - \frac{1}{L_{FRX}}\right) + T_{eRX} \tag{10}$$

Temperature variations of the environment cause variation from the nominal value of $\left(\frac{G}{T}\right)_{ES}$ besides other impairments already been addressed as pointing errors. In particular, T_A and T_{eRX} are defined as

$$T_A = \frac{T_{SKY}}{A_p} + T_M \left(1 - \frac{1}{A_p}\right) + T_{GROUND}$$

$$T_{eRX} = (NF - 1)T_0$$

where in the former equation T_{SKY}, T_M, and T_{GROUND} are respectively the sky, the medium and the ground temperatures whereas in the latter equation the NF is the noise figure and T_0 the default noise temperature fixed at 290 K. Variations of these effects can be included in Δ_2.

4 Numerical Results

After having reviewed the impairments that may occur at the receiver during the estimation process, we performed computer simulations in order to verify the robustness of the proposed technique. System reference values taken into account for simulations are reported in Table 2.

As a first result, in Figs. 2 and 3 performance of the SNORE algorithm in presence of three specific impairments cases (Table 3), which have been selected among all the possible for their significance, are depicted.

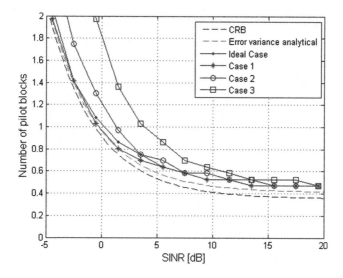

Fig. 3. Minimum number of pilot in presence of impairments

In particular, Fig. 2 shows the normalized error variance $\frac{\sigma^2}{SINR^2}$ as a function of the SINR when 5 pilot blocks of 36 symbols each are used for the estimation, whereas Fig. 3 describes the minimum required number of pilot blocks in order to achieve a target normalized error variance, i.e., $\frac{\sigma^2}{SINR^2}$, of 0.1 as a function of the SINR. In both figures, the solid line represents the Cramer Rao Bound, the dashed one the analytic value derived in [9], and the dots the simulated values under different impairments conditions. Under the ideal case the link budget is calculated without impairments and is considered as reference value, whereas in cases 1, 2, and 3 are introduced respectively the uncertainties shown in Table 3.

The SNORE based technique is also assessed in presence of impairments with respect to its applicability in a specific scenario. Both evaluations in frequency domain for a specific earth terminal and in geographic domain considering a wide region covered by the satellite beam pattern are performed by starting from reference results obtained on databases analysis [16].

The estimation capabilities of the terminal antenna under different impairments as listed in Table 3, are assessed. These simulations aimed at describing how accurate would be the estimation process for a fixed terminal antenna.

We consider a portion of spectrum wide 400 MHz from 18.4 GHz to 18.8 GHz, along which the estimation process is performed in carriers equal to 36 MHz. The FSS terminal is positioned in 47.5 N latitude and 19E longitude. The estimation process is performed under the different impairments listed in Table 3, accumulating 1 and 10 pilot blocks of 36 symbols, in Figs. 4 and 5.

As expected, in each band the SINR value estimated is lower than the real value due to impairments that cause additional losses. However, in case of high SINR values the estimated value even in presence of impairments can be considered reliable while, otherwise, to obtain the desired uncertainty target for lower values of the SINR, more pilot blocks have to be accumulated. In fact,

Table 2. System reference parameters for SNIR based sensing simulations

Parameter name	Value
Sky temperature (T_{SKY})	15 K
Ground temperature (T_{GROUND})	45 K
Temperature of the medium (T_M)	275 K
Downlink frequency	18.4 – 18.8 GHz
Satellite EIRP $(EIRP_{SAT})$	50 – 70 dBW
Carrier bandwidth	36 MHz
Terminal efficiency	0.65
Terminal antenna diameter	0.6 meters
Figure of merit $(G/T)_{ES}$	34.9 dB/K
Additional $(G/T)_{ES}$ variation (Δ_2)	0 – 2 dB
Antenna gain (G_R)	50 62 dB
Polarization losses (L_{POL})	0 – 0.5 dB
Pointing losses (L_R)	0 – 1 dB
Feeder losses (L_{FRX})	0 dB
Terminal antenna temperature (T_A)	78 K
Effective noise temperature (T_{eRX})	262 K
Terminal component temperature (T_F)	290 K
Default temperature (T_0)	290 K
LNA Noise Factor (NF)	1.4 dB
QPSK symbols per pilot	36

considering bands 5 and 6, it can be noticed that in Fig. 4, i.e., in case of 1 pilot block is accumulated, the estimated SINR in presence of impairments is similar to the real value, whereas in Fig. 5, i.e., where 10 pilot blocks are considered, the estimation is more accurate. Thus, an inappropriate estimation of the SINR in presence of impairments causes a misunderstanding in detecting presence of impairments or interference due to incumbent users.

In addition to frequency analysis, geographic assessments are performed to evaluate the SINR estimation process within the beam coverage. Thus, the area from 47 N to 48N of latitude and from 19E to 20E of longitude is considered. Real SINR values that earth terminals experience within the coverage, are pre-

Table 3. The selected impairments cases

	Ideal case	Case 1	Case 2	Case 3
Polarization losses (L_{POL}) [dB]	0	0.5	0.5	0.5
Pointing losses (L_R) [dB]	0	1	1	1
Additional $(G/T)_{ES}$ variation (Δ_2) [dB]	0	2	2	2
Additional atmospheric attenuation (A_p) [dB]	0	0	5	10

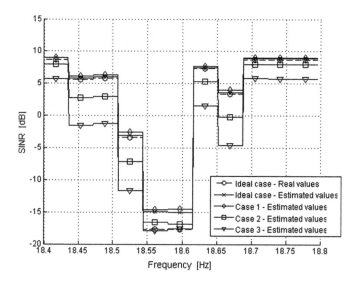

Fig. 4. Assessments on the frequency spectrum in presence of different impairments 1 pilot block

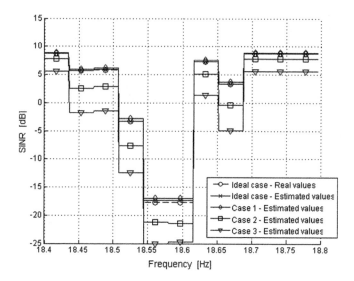

Fig. 5. Assessments on the frequency spectrum in presence of different impairments 10 pilot blocks

sented in Fig. 6. It can be noted the presence of a directive incumbent link and of some incumbent-free regions. Figure 7 shows the results for the cases listed in Table 3, when performing the estimation algorithm with 10 pilot blocks.

Results obtained from geographic simulations also confirm link budget losses due to presence of impairments and a more reliable estimation due to longer

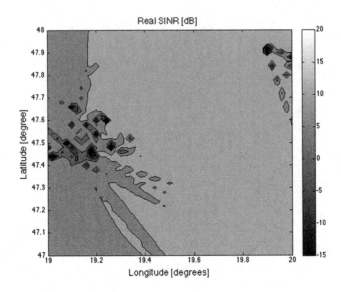

Fig. 6. Real SINR values along the selected geographic region

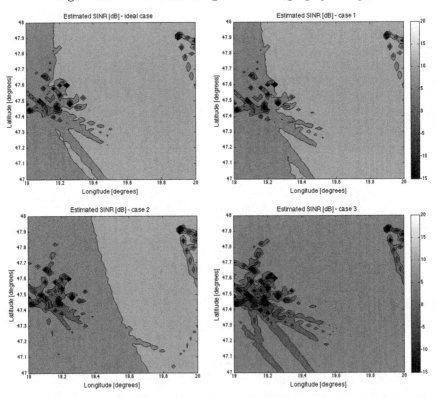

Fig. 7. Comparison between estimated values under different impairments conditions 10 pilot block

Table 4. Geographic assessments results

Case	1 pilot block	10 pilot blocks
Ideal case	0.48 %	0.19 %
Case 1	0.71 %	0.56 %
Case 2	2.42 %	2.46 %
Case 3	3.67 %	3.71 %

observation periods. Percentages of the SINR values estimated that differ from the real value more than $\frac{\sigma^2_{\epsilon|des}}{SINR^2} = 0.1$, where $\sigma^2_{\epsilon|des}$ is the difference between the real value and the estimated value under uncertainties conditions, are shown in Table 4, where the results with 1 pilot block are also reported. Results show that in both cases impairments lead to a degradation of the percentages of points correctly estimated. However, in presence of low impairments losses as for the Ideal case and Case 1, longer estimations provide more reliable estimations whereas in case of higher impairments losses the estimated values do not satisfy the target reliability even in case of longer sensing periods.

5 Conclusions

The success of spectrum sharing is mostly based on the effectiveness of cognitive radio techniques, and, among other components, on the spectrum sensing operations. In this paper, after introducing the SNORE based interference estimation technique, that results particularly effective for the selected heterogeneous terrestrial-satellite scenario, the presence if impairments is introduced. The numerical results shows that the proposed techniques is robust even in the case of communication impairments, resulting in a good candidate for the considered scenario.

Acknowledgment. This work was partially supported supported by the EU FP7 project CoRaSat (FP7 ICT STREP Grant Agreement n. 316779).

References

1. Haykin, S.: Cognitive radio: brain-empowered wireless communications. IEEE J. Sel. Areas Commun. **23**(2), 201–220 (2005)
2. Hossain, E., Niyato, D., Han, Z.: Dynamic Spectrum Access and Management in Cognitive Radio Networks. Cambridge University Press, Cambridge (2009)
3. Liolis, K., Schlueter, G., Krause, J., Zimmer, F., Combelles, L., Grotz, J., Chatzinotas, S., Evans, B., Guidotti, A., Tarchi, D., Vanelli-Coralli, A.: Cognitive radio scenarios for satellite communications: The CoRaSat approach. In: Proceedings of 2013 Future Network and Mobile Summit, Lisbon, Portugal, July 2013

4. Icolari, V., Tarchi, D., Vanelli-Coralli, A., Vincenzi, M.: An energy detector based radio environment mapping technique for cognitive satellite systems. In: Proceedings of IEEE Globecom 2014, Austin, TX, USA, December 2014

5. Yucek, T., Arslan, H.: A survey of spectrum sensing algorithms for cognitive radio applications. IEEE Commun. Surv. Tutorials **11**(1), 116–130 (2009)

6. Nekovee, M., Irnich, T., Karlsson, J.: Worldwide trends in regulation of secondary access to white spaces using cognitive radio. IEEE Trans. Wirel. Commun. **19**(4), 32–40 (2012)

7. Pauluzzi, D., Beaulieu, N.: A comparison of SNR estimation techniques for the AWGN channel. IEEE Trans. Commun. **48**(10), 1681–1691 (2000)

8. Cioni, S., De Gaudenzi, R., Rinaldo, R.: Channel estimation and physical layer adaptation techniques for satellite networks exploiting adaptive coding and modulation. Int. J. Satell. Commun. Netw. **26**, 157–188 (2008)

9. Icolari, V., Guidotti, A., Tarchi, D., Vanelli-Coralli, A.: An interference estimation technique for satellite cognitive radio systems. In: Proceedings of IEEE ICC 2015, London, UK, June 2015

10. ETSI: Digital Video Broadcasting (DVB); Second generation framing structure, channel coding and modulation systems for Broadcasting, Interactive Services, News Gathering and other broadband satellite applications (DVB-S2), March 2013

11. ETSI: Digital Video Broadcasting (DVB); Second generation framing structure, channel coding and modulation systems for Broadcasting, Interactive Services, News Gathering and other broadband satellite applications Part II: S2-Extensions (DVB-S2X), March 2014

12. International Telecommunications Union - Radiocommunication Sector: Radio Regulations (2012)

13. Cioni, S., Corazza, G., Bousquet, M.: An analytical characterization of maximum likelihood signal-to-noise ratio estimation. In: Proceedings of 2nd International Symposium on Wireless Communication Systems, Siena, Italy, September 2005

14. Corazza, G. (ed.): Digital Satellite Communications. Springer, Heidelberg (2007)

15. Maral, G., Bousquet, M., Sun, Z. (eds.): Satellite Communications Systems: Systems, Techniques and Technology, 5th edn. Wiley, New York (2009)

16. Thompson, P., Evans, B.: Analysis of interference between terrestrial and satellite systems in the band 17.7 to 19.7 GHz. In: Proceedings of IEEE ICC 2015 Workshops, London, UK, June 2015

Technology Trends for Ka-Band Broadcasting Satellite Systems

Nader S. Alagha$^{(\boxtimes)}$ and Pantelis-Daniel Arapoglou

European Space Agency, Research and Technology Centre (ESTEC),
Noordwijk, The Netherlands
nader.alagha@esa.int

Abstract. This paper provides an overview of the technology trends pertinent to Ka-band broadcasting satellite systems. Starting from the state-of-the-art digital broadcasting systems, we present technology trends that can further expand the use of Ka-band satellite broadcasting and improve the performance and efficiency. In particular, it is shown that the combination of DVB-S2X features offer significant advances and opportunities to service providers. This is true particularly in geographical areas that are subject to severe atmospheric attenuation.

Keywords: Ka-band broadcasting · Multi-beam satellites · Channel bonding · Variable coding and modulation · Simulcasting

1 Introduction

The Direct to Home (DTH) broadcasting today is by far the largest market in the satellite communications sector. The capability of broadcasting satellites to rapidly setup services and provide multitude of TV channels at relatively low cost and to a large number of subscribers have resulted in a significant growth of the DTH market in the past two decades.

Despite the past success, the opportunity for steady growth of the DTH market in the future is under threat. In many parts of the world consumers are rapidly changing their habits when viewing television, going from passive consumption of delivered content to more on-demand interactive experience. The use of a second screen such as a tablet allows interactions with peers or searching for relevant information while viewing the video content. There is a growing demand for multiple video streaming into a household. These trends are expected to continue, creating new businesses opportunities for highly customized unicast video content distribution. Such services have been creating fierce competition against conventional direct satellite broadcasting services.

On the other hand, the quality of video-on-demand is moving from standard definition quality of high-definition. More bandwidth demanding video contents such as Ultra- High Definition (UHD) and 3D TV create new opportunities for direct broadcasting satellite, particularly targeting home viewing screens.

The cost competitiveness of satellite broadcasting can be maintained or improved through different technical advances. New video compression techniques such as HEVC [1] along with more sophisticated transmission schemes such as DVB-S2X [2]

© Institute for Computer Sciences, Social Informatics and Telecommunications Engineering 2015
P. Pillai et al. (Eds.): WiSATS 2015, LNICST 154, pp. 147–159, 2015.
DOI: 10.1007/978-3-319-25479-1_11

are promising technologies to reduce the required bandwidth per channel. The use of Ka-band broadcasting can help reducing the cost of the bandwidth and allow higher multiplexing gain especially for UHD content broadcasting.

Technical solutions using variable coding and modulation and scalable video coding can provide different quality of service and graceful degradation of the broadcast content. The use of front-end processing together with multiple LNB's can improve the effective antenna size and help maintaining the alignment. New system-on-chip design approach integrating the RF front-end and the baseband processing functionality will allow simultaneous access to full spectrum while lowering the power consumption and possible integration of functionality at the front-end [4].

The satellite video distribution can also be integrated with micro content delivery networks for multi dwelling houses, apartments or neighborhoods and allow the possibility of providing local content. This will help off-loading from the Internet backbones the delivery of bandwidth demanding contents while maintaining many interactivity features that are demanded by users.

The use of new generation of hybrid broadband/broadcast satellite can allow more efficient use of on-board resources and gradual migration from one type of service to the next. However, given the number of existing satellite fleets and their expected lifetime, it is important to rely on ground/user segment technical advancement allowing more efficient use of conventional broadcasting in the next 5 to 10 years while new generation of satellite being designed and put in service.

In this paper we particularly focus on the opportunities and challenges of Ka-band broadcasting. By defining some study cases, we present technology trends both at the space and ground segments that can enable cost-effective content delivery using Ka-band broadcasting satellites even in geographical areas that are subject to severe atmospheric attenuation.

2   Ka-Band Broadcasting: Opportunities and Challenges

Satellite broadcasting in so called "Reverse BSS" (17/24 GHz) [5] Ka-band frequencies is a well-established technology in North America, contributing significantly to the satellite broadcasting business revenue especially for the HDTV program offerings [6].

The deployment of Ka-band broadcasting satellites is also gaining momentum in other regions of the world. Recent advancements in the regulatory situation in ITU Region 1 and 3, as defined in [7], have contributed to this progress. Complementing the original Ka-band broadcasting frequency allocation of 21.4-22.0 GHz by Word Radiocommunication Conference (WRC) in 2007, during WRC-12 a new feeder link (uplink) frequency band was allocated to Ka-band broadcasting (24.65-25.25 GHz). This further improved the feasibility of the frequency planning for Ka-band broadcasting satellite systems in ITU Region 1 and 3.

Current examples of commercially operated Ka-band broadcasting satellites are EUTELSAT 25B at the 25.5°E orbit providing Ka-band broadcasting system coverage to the Middle East and North Africa, Nilesat 201 Satellite with Ka-band coverage of North Africa [8]. Both satellites provide commercial broadcasting services in Ka-band that complements the Ku-band broadcasting content.

The Ka-band broadcast technology provides the means for enhanced TV integrated services where the Ka-band frequency can complement the existing Ku-band services by providing a higher quality video content in addition to the standard quality content offering of Ku-band services. This could be realized by deploying dual-band receivers at the user premises. Technologies such as multi-tuner set-top boxes and dual Ka/Ku band LNBs are today's reality [8].

In some regions including Europe and Far East Asia, the deployment of Ka-band broadcasting faces challenges caused by severe atmospheric attenuation. This could impose a large link margin that would hamper the efficiency and cost-effectiveness of such systems. This issue is recognized and discussed also in recent work as reported in [9]. In the remaining of this paper, we present system and air interface solutions that can address this issue and facilitate the use of Ka-band broadcasting services even in the presence of atmospheric fading conditions.

3   DVB-S2X for Broadcasting

A technical module of DVB (Digital Video Broadcasting) took the mandate to extend DVB-S2 in order to achieve higher spectral efficiencies without introducing fundamental changes to the complexity and structure of DVB-S2. The outcome of this development was approved by DVB in March 2014 as DVB-S2 Extension (herein referred to as DVB-S2X) and was published as Part II of the ETSI Standard [2].

There are number of new features introduced in DVB-S2X, each targeting a particular improvement in the efficiency, flexibility and applicability of the standard to the core or emerging market segments. This paper is concerned with the direct to home broadcasting. The goal is to examine and quantify at system level the benefits offered by the features of DVB-S2X, individually or collectively, in a set of representative system scenarios relevant to Ka-band broadcasting. As a starting point, an overview of the relevant features of DVB-S2X [2] that are considered in our analyses is provided below.

3.1   Broadcasting with Differentiated Channel Protection

A major difference of DVB-S2X compared to DVB-S2 is that it supports the Variable Coding and Modulation (VCM) technique as a normative feature for the broadcasting profile. The VCM can provide different level of signal protection by time sharing different Modulation and Coding (MODCODs) in each physical layer frame [10]. This allows satellite operators to adjust transmission robustness according to the service availability they wish to guarantee, and to differentiate services according to the quality of service requirements. The use of VCM in conjunction with simulcast can be used to guarantee service continuity in the presence of a heavy atmospheric fading, while at the same time offering a high quality service in the absence of rain attenuation. By allowing a tolerable degradation in the picture quality during a heavy atmospheric fading, it is possible to significantly increase the overall system spectral efficiency.

At the receiver, it is assumed that the two streams corresponding to the same content are video decoded independently. However, the time synchronization between the two streams is maintained such that the output stream for the same content can switch seamlessly between the two.

Compared to a scalable video coding with two or more layers of coded content, the use of simulcasting of low and high quality streams offers a simpler video coding and decoding solution and possibly a lower overhead (if the required bit rate for the lower quality stream is considerably lower than that of higher quality stream). It should be noted that the cost of simulcasting, in terms of overhead depends on the assigned bit rates per two video streams.

3.2   Channel Bonding

The DVB-S2 system was designed to carry a single or multiple MPEG Transport Streams (or generic continuous streams) over a single satellite transponder. In a conventional DTH application, HDTV multi-program and MPEG-4 video encoding allows for multiplexing a sizeable number (5-6) of programs per transponder. This could be increased taking into account Statistical Multiplexing gain as shown in Fig. 1. The DVB-S2X aims to integrate with HEVC video coding and UHDTV (four times the definition of HDTV). Assuming that an UHDTV signal requires four times the transmission capacity of HDTV for the same compression system, and that HEVC doubles the compression efficiency versus MPEG-4, we roughly estimate 20 Mbits/s bit rate requirement per each UHD program. This would reduce the number of supported UHD to 2 or 3 programs per transponder, and the statistical multiplexing gain would be significantly reduced (less than 12 %, assuming Fig. 1 is applicable to UHDTV and HEVC), thus not allowing the transmission of an additional program.

In order to increase the statistical multiplexing gain for UHDTV, S2X implements "channel bonding", that is merging the capacity of two or three transponders to transport a single stream. This functionality is available only for multi-tuner receivers,

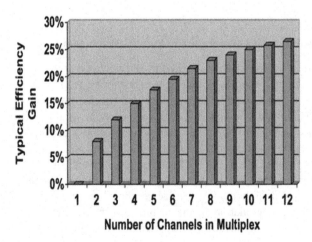

Fig. 1. Statistical multiplexing gain for increasing number of bonded channels (from [11]).

which are anyway finding their way in the market also to implement other functionalities. It should be noted that channel bonding requires simultaneous reception of signal from two or more transponders.

3.3   Higher Order Modulation for Broadcasting Applications

DVB-S2X supports as a normative feature 16APSK (as well as 32APSK) modulation for broadcasting applications. The use of higher order modulations combined with VCM are particularly advantageous for Ka-band broadcasting systems [1], as 16APSK allows a higher spectral efficiency for users in clear sky along with a more protected MODCOD under heavy fading conditions. This physical layer configuration can ensure the target service availability by allowing a graceful degradation of the service quality following the VCM concept.

3.4   New MODCODs and Finer SNR Threshold Granularity

The increasing demand for broadband satellite services has pushed for a more efficient utilization of resource, namely available power and spectrum. Multi-carrier operation is usually envisaged for this configuration, implying that the High Power Amplifiers (HPA) non-linearities can be considered quasi-linear. The new DVB-S2X standard includes additional MODCODs optimally designed to operate on a linear channel and in the presence of phase noise. Furthermore, the additional MODCODs brought as main advantage a finer granularity in the SNR threshold, reducing the gap between thresholds from 1-1.5 dB of the DVB-S2 to 0.5 dB.

3.5   Smaller Pulse Shaping Filter Roll-off Values

The new standard allows for the usage of sharper roll-off values, i.e 15 %, 10 % and 5 %, in addition to the DVB-S2 20 %, 25 % and 35 %. As shown in [3], for DTH applications these narrower roll-offs do not seem to offer significant performance advantage (at least for the channel model that has been considered), while for VSAT networks some performance improvements at system level are expected.

4   Multi-beam Ka-Band Broadcasting

In order to assess the comparative performance of the DVB-S2X with respect to DVB-S2, a broadcasting satellite system deploying the Ka-band broadcasting frequency allocation (21.4-22.0 GHz) is considered.

The satellite is assumed to be located at 10°E of the geostationary orbit providing broadcasting services to eight linguistic beams covering Europe as shown in Fig. 2. The frequency plan associated with 8 beams is defined according to a 4 colour scheme as shown in Fig. 3. The frequency plan allows for a frequency re-use factor of 2 while maintaining reasonable level of carrier to interference ratio.

The Ka-band satellite payload is assumed to have 40 active Ka-band transponders (5 transponders per beam) with single carrier per transponder. The TWTA power amplifier per transponder is assumed to radiate 200 W power at the peak, which agrees with the near term available HPA technology. A nominal bandwidth of 54 MHz per IMUX/OMUX filters and an adjacent frequency spacing of 60 MHz between transponders are assumed. Hence, each beam consists of 5 carriers with a total band-width of 300 MHz.

As a benchmark, a Ku-band broadcasting satellite at the same geostationary orbit with a similar DC power envelope of 15 kW is considered, which typically allows to accommodate up to 64 Ku-band transponders, each with a 120 W TWTA power amplifier.

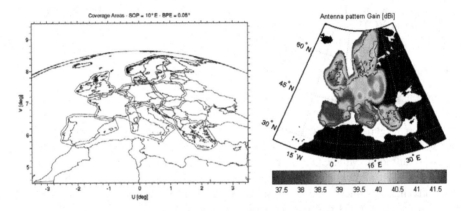

Fig. 2. Broadcasting satellite with 8 linguistic beams in BSS Ka-band

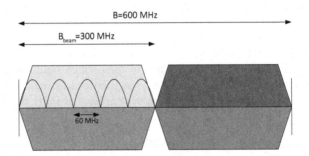

Fig. 3. Frequency Plan – Multi-beam Ka-band

4.1   Receiver Architecture Assumptions

The performance comparison between DVB-S2 and DVB-S2X is based on the following assumptions concerning receiver technologies:

DVB-S2 Legacy Receivers: Implementing DVB-S2 air interface according the broadcasting profile with no added capability to tolerate inter-symbol interference (ISI).

Such receivers are assumed to operate only in CCM mode with QPSK or 8PSK modulation scheme and no capability of digital channel bonding.

DVB-S2 Enhanced Receivers: Implementing DVB-S2 air interface broadcasting profile (similar to the legacy receivers). Additionally, the enhanced receivers are assumed to be capable of mitigating interference caused by band limited channel (e.g. using baseband equalizers) that would allow increasing the symbol rate. It should however be noted that the MODCOD selection is constraint by the broadcasting profile of DVB-S2. The physical layer performance results are shown in [3].

DVB-S2X Enhanced Receivers: Implementing DVB-S2X air interface broadcasting profile and capable of equalizing the linear distortions caused by the band-limited channel. Such receivers are assumed to support digital channel bonding (up to three BBFRAMEs), 16APSK modulation and MODCODSs according to DVB-S2X broadcasting profile. The physical layer performance results are shown in [3].

4.2   Video Service Quality

The target service availability in each scenario is 99.9 % of the time over the entire coverage area. For a DTH service based on DVB-S2 air interface, this service availability is applied directly to each video stream (HD or UHD). For DVB-S2X that is supporting VCM, the service availability of 99.9 % is provided by two video quality levels associated with each video streams as follows:

- The HD service availability of 99.0 %, complemented by a lower quality stream (MPEG-4 coded at 1.0 Mbits/s) of the same program with a service availability of 99.9 %.
- The UHD service availability of 99.0% (or 97.0 %), complemented by a lower quality stream (HEVC coded at 2.0 Mbits/s) of the same video content with a service availability of 99.9 %.

The higher quality (HQ) and lower quality (LQ) streams are decoded independently but the receiver is assumed to be able to maintain the synchronization between the two streams and switch between the two to maintain the quality and service availability.

4.3   Comparative Performance Assessment

Figure 4 shows two locations within the coverage area for which the outage probability as function of atmospheric attenuation for geographical locations is plotted based on ITU-R Rec. P.618 [12]. It is noted that the more severe propagation conditions of the Mediterranean location compared to the North European one are counter balanced by the higher elevation of the former location compared to the latter.

Several study cases are defined to establish a benchmark performance according to the existing DVB-S2 solutions as well as enhanced solutions according to DVB-S2X specifications. The definition and justification of each study case is outlined below. A summary of key parameters associated to the each study case is provided in Table 1.

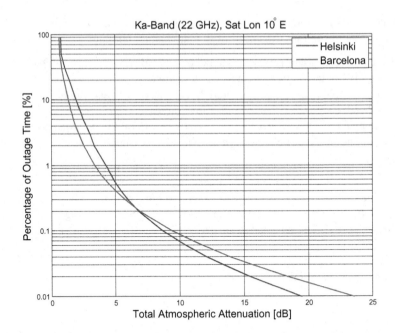

Fig. 4. Examples of atmospheric attenuation statistics in Ka-band

Study Case 1: Ku-band reference system with DVB-S2 and legacy receiver.

This study case serves as a benchmark, outlining the expected performance of legacy DVB-S2 receivers in a conventional Ku-band system with a single beam and wide coverage over Europe using a broadcasting satellite with the same envelope DC power of 15 kW. The payload consists of 64 transponders in Ku-band. Considering the transponder bandwidth and the transponder spacing, a baud rate of 30 MBaud is considered for this study case. The target service case is the broadcasting of UHD quality television channels with an average rate of 20 Mbits/s per channel. The target availability for each channel is 99.9 %.

Study Case 2: Ku-band reference system with DVB-S2 and enhanced receiver.

This study case is similar to Study Case 1 except for the receiver capability of mitigating ISI. In this case, the receiver remains compliant with DVB-S2 protocol according to the broadcasting profile while allowing for the optimization of the symbol rate.

Study Case 3: Ku-band reference system with DVB-S2X with channel bonding.

In this study case, DVB-S2X receivers with channel bonding and CCM transmission mode are considered. Similar to Case 1 and 2, UHD video quality with 99.9 % service availability is assumed.

Study Case 4: Ka-band system with DVB-S2 and enhanced receiver.

This study case examines the use of DVB-S2 together with an enhanced receiver and HEVC video decoder to deliver UHD video quality content to the end users in

Table 1. A summary of parameters for Study cases

Study Case Parameter	Case 1	Case 2	Case 3	Case 4	Case 5	Case 6
Scenario	Ku-band benchmark			Ka-band Multi-beam		
Transponder Frequency Spacing (MHz)	40	40	40	60	60	60
Transponder Bandwidth (MHz)	36	36	36	54	54	54
Total Number of Transponders	64	64	64	40	40	40
Aggregate bandwidth (MHz)	2560	2560	2560	2400	2400	2400
Number of Carriers per transponder	1	1	1	1	1	1
Air Interface (DTH Profile)	DVB-S2	DVB-S2	**DVB-S2X**	DVB-S2	**DVB-S2X**	**DVB-S2X**
Receiver Assumptions	Legacy	Enhanced	Enhanced	Enhanced	Enhanced	Enhanced
Transmission Mode	CCM	CCM	CCM	CCM	**VCM**	**VCM**
Video CODEC	**HEVC**	**HEVC**	**HEVC**	**HEVC**	**HEVC**	**HEVC**
Average bit Rate per stream (Mbits/s)	20	20	20	20	20 (HQ) 2 (LQ)	20 (HQ) 2 (LQ)
Availability	99.9 %	99.9 %	99.9 %	99.9 %	**99.0 %** 99.9 %	**97.0 %** 99.9 %

Ka-band multi-beam broadcasting satellite system. A service availability of 99.9 % is targeted. The DVB-S2 receiver in this case operates in CCM transmission mode.

Study Case 5: Ka-band system with DVB-S2X.

This case examines the use of DVB-S2X air interface and enhanced receiver that supports VCM in broadcasting profile. The receiver is equipped with HEVC decoder to receiver two classes of video qualities; UHD at 99.0 % service availability and a lower quality video HEVC coded at 2 Mbits/s to complement the UHD stream at 99.9 % service availability.

Study Case 6: Ka-band system with DVB-S2X and 97.0 % UHD availability.

This is similar to Study Case 5 except for the service availability of the UHD quality TV program that is relaxed to 97.0 %. It is meant to investigate the sensitivity of the results on the availability required.

Table 2 provides a summary of the key link budget assumptions for each study case.

Table 3 presents a summary of the performance results for cases studied. For each case, the effective bit rate per transponder as well as the total number of video channels delivered by the broadcasting satellite in this example scenario is reported. It can be noted that:

Table 2. A summary of key link budget parameters

Study Case Link budget parameters	Case 1	Case 2	Case 3	Case 4	Case 5	Case 6
Scenario	Ku-band benchmark			Ka-band Multi-beam		
EIRP at Saturation (dBW)	53.5	53.5	53.5	60.7	60.7	60.7
OBO (dB)	0.5	0.3	0.3	0.3	0.5	**1.5**
Fade Margin for 99.9 % availability (dB)	3	3	3	9.4	9.4	9.4
Fade Margin for 99.0 % availability (dB)	N/A	N/A	N/A	N/A	3.4	N/A
Fade Margin for 97.0 % availability (dB)	N/A	N/A	N/A	N/A	N/A	2.2
User Terminal minimum antenna size (cm)	65	65	65	65	65	65
Terminal G/T (dB/K) for 99.9 % of the time	12.7	12.7	12.7	15.5	15.5	15.5
Terminal G/T (dB/K) for 99.0 % of the time	N/A	N/A	N/A	N/A	15.7	N/A
Terminal G/T (dB/K) for 97.0 % of the time	N/A	N/A	N/A	N/A	N/A	16.2
C/N0 (dBHz) for 99.9 % of time	83.7	83.7	83.7	82.5	82.5	82.5
C/N0 (dBHz) for 99.0 % of time	N/A	N/A	N/A	N/A	87.5	N/A
C/N0 (dBHz) for 97.0 % of time	N/A	N/A	N/A	N/A	N/A	89.3
Adjacent Satellite C/I (dB) (Note 3)	13	13	13	22	22	22
Co-Channel Interference (dB) (Note 4)	N/A	N/A	N/A	20	20	20
C/(N+I) (dB) for 99.9 % of the time	7.7	7.1	7.1	5.2	5.1	4.1
C/(N+I) (dB) for 99.0 % of the time	N/A	N/A	N/A	N/A	10.5	N/A
C/(N+I) (dB) for 97.0 % of the time	N/A	N/A	N/A	N/A	N/A	11.1
Baud Rate (MBaud)	30	**34**	**34**	51	51	51
MODCODs	8PSK3/5	**QPSK 5/6**	**QPSK 5/6**	QPSK 2/3	8PSK 5/6 (HQ) QPSK 2/3 (LQ)	16APSK 2/3 (HQ) QPSK 3/5 (LQ)
Required Threshold (dB)	6,6	**5,7**	**5,7**	3,6	10,4 (HQ) 3,6 (LQ)	10,8 (HQ) 2,8 (LQ)
Link Margin (dB)	1.1	**1.4**	**1.4**	1.6	1.5	1.3

Table 3. A summary of performance results

Study Case Parameter	Case 1	Case 2	Case 3	Case 4	Case 5	Case 6
Scenario	Ku-band benchmark			Ka-band Multi-beam		
Transponder IMUX/OMUX Bandwidth (MHz)	36	36	36	60	60	60
Total Number of Transponders	64	64	64	40	40	40
Symbol Rate (MBaud)	**30**	34	34	54	54	54
Air Interface (DTH Profile)	DVB-S2	DVB-S2	**DVB-S2X**	DVB-S2	**DVB-S2X**	**DVB-S2X**
Receiver Assumption	Legacy	Enhanced	Enhanced	Enhanced	Enhanced	Enhanced
Transmission Mode	CCM	CCM	CCM	CCM	**VCM**	**VCM**
Average bit rate per video stream (Mbits/s)	20	20	20	20	20 (HQ) 2 (LQ)	20 (HQ) 2 (LQ)
Assigned MODCODs	8PSK 3/5	QPSK 5/6	QPSK 5/6	QPSK 2/3	8PSK 5/6 (HQ) QPSK 2/3 (LQ)	16APSK 2/3 (HQ) QPSK 3/5 (LQ)
Effective bit rate per transponder (Mbits/s)	54.3	56.5	56.5	67.9	118.1	122.4
Digital Bonding (Note 1)	N/A	N/A	**Yes**	N/A	**NO**	**NO**
Fractional number of video streams per transponder (Note 2)	2.7	2.8	2.8	3.4	5.4	5.6
Statistical multiplexing gain	8 %	8 %	23 %	12 %	16 %	16 %
Effective number of video streams per transponder	3	3	10/3 (Note 3)	3	**6**	**6**
Total number of video streams (TV channels) delivered by the satellite	64x3=192	64x3=192	64 × 10/3 ~ 213	40 × 3=120	40 × 6=240	40 × 6=240

Note 1: For DVB-S2X (CCM), the channel bonding can be applied to up to 3 transponders.
Note 2: this ratio does not include the statistical multiplexing gain.
Note 3: For DVB-S2X CCM, the effective number of video streams is computed for every 3 transponders to take into account the digital channel bonding gain.

- The use of DVB-S2 enhanced receivers (Study Case 2) compared to the legacy receivers (Study Case 1) does not offer any improvement in terms of the number UHDTV channels for this particular system and only provides a marginal improvement (4 %) in terms of the throughput per transponder.
- DVB-S2X channel bonding used in the Ku-band (Study Case 3) offers around 11 % increase in the number of UHDTV channels compared to DVB-S2 (Study Case 2).
- The use of DVB-S2 enhanced receiver with UHD video quality in Ka-band (Study Case 4) shows a significant reduction (37 %) in the number of video streams compared to a Ku-band broadcasting satellite with the same 15 kW DC power envelope (Study Case 2.2). This confirms that Ka-band broadcasting should be realized with DVB-S2X.
- The use of DVB-S2X enhanced receiver with UHD video quality in Ka-band can offer a significant increase in the number of channels compared to DVB-S2 in

Ka-band. The increase in the number of UHD channels is around 100 % (Study Cases 5 or 6) compared to DVB-S2 (Study Case 4). It is worth highlighting that this result is achieved with the same overall power consumption and with fewer HPAs on board.

5   Concluding Remarks

Satellite broadcasting services in Ka-band are readily available in many regions of the world. In this paper, we presented solutions based on multi-beam satellites and new features of DVB-S2X to offer Ka-band broadcasting services even for geographical areas with high atmospheric fading.

For the broadcasting services, the use of the channel bonding feature of DVB-S2X allows for a reasonable gain in the number of offered TV channels on conventional transponders. The performance improvement of the channel bonding is attributed to the statistical multiplexing gain due to the aggregation of two or three multiplexing streams (as opposed to one that is supported by DVB-S2).

The use of VCM together with simulcasting allows for significant improvement in number of channels while maintaining the service availability. In particular, in such scenarios, the use of VCM together with simulcasting, would allow up to 100 % increase in the number of offered TV channels, compared to similar services without VCM (conventional DVB-S2).

Acknowledgments.   The performance results were obtained internally at the European Space Agency. The performance results were submitted to DVB – TM-S working group in support of DVB-S2X development. Opinions, interpretations, recommendations and conclusions presented in this paper are those of the authors and are not necessarily endorsed by the European Space Agency.

References

1. Draft Recommendation ITU-T, H.265: High efficiency video coding (V2). Approved in October 2014. http://www.itu.int/rec/T-REC-H.265-201410-P
2. ETSI EN 302 307-2: Digital Video Broadcasting (DVB); Second generation framing structure, channel coding and modulation systems for Broadcasting, Interactive Services, News Gathering and other broadband satellite applications; Part II (DVB-S2X)
3. ETSI TR 102 376-2 V1.1.1, Digital Video Broadcasting (DVB), User guidelines for the second generation system for Broadcasting, Interactive Services, News Gathering and other broadband satellite applications- Part 2: S2 eXtensions (DVB-S2X)
4. Girault , N., Alagha, N.: Next generation consumer satellite terminal architectures. In: Proceedings of 29th AIAA International Communications Satellite Systems Conference (ICSSC-2011), November 2011
5. Björkman, M.R.: The 17/24-GHz broadcast satellite service. In: 27th AIAA International Communications Satellite Systems Conference, (ICSSC 2009), 1–4 June 2009, Edinburgh

6. DirecTV Annual report 2012. http://investor.directv.com/files/doc_financials/annual/DirecTV_2012_AR.PDF
7. International Telecommunication Union, "Radio Regulations Articles", Edition of 2012. http://www.itu.int/dms_pub/itus/oth/02/02/S02020000244501PDFE.PDF
8. Fenech, H., Tomatis, A., Amos, S.: Broadcasting in Ka-band. In: 30th AIAA International Communications Satellite Systems Conference, (ICSSC 2012), Ottawa, Canada, 24–27 September 2012
9. Shin, M.-S., et al.: A study on the Ka-band satellite 4 K-UHD broadcastingservice provisioning in korea. In: Proceedings of the Eighth International Conference on Mobile Ubiquitous Computing, Systems, Services and Technologies UBICOMM 2014
10. Mirta, S., et al.: HD Video Broadcasting using Scalable Video Coding combined with DVB-S2 variable coding and modulation. In: Advanced Satellite Multimedia Systems Conference (ASMS) and the 11th Signal Processing for Space Communications Workshop (SPSC), pp. 114–121 (2010)
11. McCann, K., Gledhill, J., Mattei, A., Savage, S.: Beyond HDTV: Implications for Digital Delivery: An Independent Report by ZetaCast Ltd Commissioned by Ofcom, July 2009
12. ITU-R Recommendation P. 618–11. Propagation data and prediction methods required for the design of Earth-space telecommunication systems. Geneva, Switzerland 2011

WiSATS Session 1

Evaluating the Performance of Next Generation Web Access via Satellite

Raffaello Secchi$^{(\boxtimes)}$, Althaff Mohideen, and Gorry Fairhurst

School of Engineering, University of Aberdeen,
Fraser Noble Building, Aberdeen AB24 3UE, UK
r.secchi@abdn.ac.uk

Abstract. Responsiveness is a critical metric for web performance. Update to the web protocols to reduce web page latency have been introduced by joint work between the Internet Engineering Task Force (IETF) and the World Wide Web Consortium (W3C). This has resulted in new protocols, including HTTP/2 and TCP modifications, offering an alternative to the hypertext transfer protocol (HTTP/1.1). This paper evaluates the performance of the new web architecture over an operational satellite network. It presents the main features of the new protocols and discusses their impact when using a satellite network. Our tests comparing the performance of web-based applications over the satellite network with HTTP/2 confirm important reductions of page load times with respect to HTTP/1.1. However, it was also shown that performance could be significantly improved by changing the default server/client HTTP/2 configuration to best suit the satellite network.

Keywords: SPDY · HTTP/2 · PEP

1 Introduction

In the last ten years, the types of data exchanged using HTTP has changed radically. Early web pages typically consisted of a few tens of kilobytes of data and were normally not updated frequently (static web). Today, typical web pages are more complex, consisting of (many) tens of elements, including images, style sheets, programming scripts, audio/video clips, HTML frames, etc.[1,2]. Moreover, many web pages are updated in real time (e.g., when linked to live events or dynamic databases). Web interfaces have found use in a wide range of non-browsing applications, e.g., to interact with distributed web applications, where both data and application reside at a remote side.

Raffaello Secchi was funded by the European Community under its 7th Framework Programme through the Reducing Internet Transport Latency (RITE) project (ICT-317700).

Althaff Mohideen was by the RCUK Digital Economy programme of the dot.rural Digital Economy Hub; award reference: EP/G066051/1.

© Institute for Computer Sciences, Social Informatics and Telecommunications Engineering 2015
P. Pillai et al. (Eds.): WiSATS 2015, LNICST 154, pp. 163–176, 2015.
DOI: 10.1007/978-3-319-25479-1_12

The HTTP/1.x suite of protocols [3,4] has dominated web usage for over twenty years. It has become widely used in a variety of contexts beyond its original goals, limitations of the HTTP protocols have become apparent. The original HTTP request/response model, where each web object was requested separately from the others, has been shown to not scale to modern complex web pages and can introduces a significant amount of overhead at start-up. HTTP interacts poorly with the transport layer (Transmission Control Protocol [5]) resulting in poor responsiveness for the web. These performance issues have been accompanied by changing patterns of web.

Changes in the size and structure of web pages now mean that HTTP/1.x can become a significant bottleneck in web performance [6]. HTTP/1.0 [3] allowed only one request at a time on a given TCP connection. HTTP/1.1 [4] added HTTP persistence, i.e., the ability to reuse a TCP connection for subsequent HTTP object requests. HTTP persistence feature avoided the overhead of the TCP connection establishment, but did not resolve the performance gap. The growing number of objects per page, caused web designers to increase the number concurrent TCP connections to decrease down-load time, sometimes scattering the components of a web page over many servers, known as "sharding". While opening many TCP connections can reduce page load times, it is not a network-friendly solution, because it results in web sessions competing aggressively with other network flows. It also does not resolve the head-of-line blocking (HoLB) that occurs when an important object is scheduled over a slow TCP connection and received after a less important one.

Initiatives from Google^{TM} proposed new protocols as an alternative to HTTP 1.1, better suited to new web content, and designed to make HTTP more responsive and more network-friendly. SPDY [7] and later QUIC [8] are prototype protocol implementations of this new approach. In 2009, work on SPDY triggered formation of the HTTPbis working group in the Internet Engineering Task Force (IETF) and related work in Worldwide Web Consortium (W3C). This has since resulted in the definition of a new standard, HTTP/2 [9] in 2015.

This paper seeks to understand the implications for the satellite community of the introduction of HTTP/2 and other TCP/IP stack modifications. We show that HTTP/2 offers significant benefits in terms of reduction of the page load time (PLT) in different scenarios, including cases with performance enhancement proxies (PEPs) and encrypted tunnels. However, we also observed that the default HTTP/2 configuration is optimised for a terrestrial networking and may not be suitable for the satellite scenario, thus negatively impacting HTTP/2 performance. This suggests that HTTP/2 may not be yet mature to support high latency links.

This conclusion is also backed by previous studies that considered the performance of HTTP/2 over terrestrial and satellite networks. An extensive comparison of HTTP(S) and HTTP/2 in terrestrial networks was presented in [6]. This study confirms that HTTP/2 effectively address the problems of HTTP but also highlights some of its potential pitfalls. In particular, the paper observes that a single connection is more exposed to transient loss episodes than multiple HTTP/1.1 connections, and that the HTTP/2 could under-perform due to the current highly *sharded* structure of the web. Performance limitations of HTTP/2 were also pointed out in [10] where PLT was related to the structure of the page. Tests with HTTP/2 over satellite were performed in [11–13]. In [11] it was observed that a proper scheduling discipline and countermeasure to compensate for wireless errors may be needed to achieve good link utilisation with HTTP/2. In [13] HTTP/2 performance were compared to HTTP/1.1 with various demand/assignment multiple access (DAMA) schemes concluding that HTTP/2 can achieve in some scenarios close-to-terrestrial performance.

The remainder of this paper is organised in a series of sections. Section 2 introduces the novelties of HTTP/2 and discusses their usage over satellite. Section 3 surveys the new TCP modification proposals to make it more responsive for the new type of web applications. Section 4 reports the results of the experiments over the satellite platform. A conclusion on the results and the new web architecture is given in Sect. 5.

2   A New Architecture for Web Technology

HTTP/2 maps the entire bundle of HTTP request/response transactions required to retrieve a web page onto a single TCP connection. It introduces the concept of *stream* as an independent data flow within the TCP connection. This allows interleaving of *frames* (stream protocol data units) onto the connection and the prioritisation of streams, letting more important requests complete faster. In addition, HTTP/2 defines a Server Push mechanism, an experimental communication mode where the server can send data to a client without an explicit request. Server Push enables a fully bidirectional client-server channel and may be used by a server to anticipate client requests predicting what data the client is going to request.

2.1   Connection Setup

Although HTTP/2 is designed to replace HTTP/1.x, an initial deployment phase is envisaged where HTTP/2 will co-exist with older HTTP/1.x systems (The httpbis working group therefore also introduced a number of innovations into HTTP/1.1.) Not all servers will immediately change to use HTTP/2, and coexistence with HTTP/1.1 is expected for the foreseeable future.

The design of HTTP/2 chose to reuse the same URI scheme of HTTP and HTTPS to indicate HTTP/2 web resources. The new protocols also use the same port numbers (TCP port 80 and 443). This required a method to signal use of HTTP/2. HTTP/1.1 defined the *Switching Protocol* header, that, when inserted in a request, indicates the intention of the client to change the session protocol. This mechanism, however, costs an extra RTT after connection establishment. If HTTP uses SSL, however, the application-layer protocol negotiation (ALPN) transport-layer security (TLS) extension header could be used to signal the presence of HTTP/2 during the TLS handshake.

ALPN was the preferred method when the HTTP/2 connection is made over SSL, expected to be used in the majority of cases with HTTP/2. Use of SSL also avoid middle-boxes hindrance as the new protocol is deployed. Using SSL requires a SSL handshake before every HTTP/2 session can be established.

2.2 Multiplexing

HTTP/2 allows a session to multiplex transmission of web page resources using a Stream. Streams are independent, bi-directional sequences of frames between the server and the client. Streams can be initiated by both a client and a server. (In HTTP/1.1, the HTTP connection could only be started by the client). A HTTP/2 connection can contain multiple concurrently open streams, with either end-point interleaving frames from multiple streams. This removes the need to open multiple transport connections.

Multiplexing resources can offer important performance benefits for long delay paths, including clients that access the network via satellite. This reduces the amount of traffic over the network removing the overhead of opening several TCP connections and allows the server to send more data per TCP segment. Multiplexing over a single connection could also reduce the need to place HTTP proxies at the edge of a satellite network (i.e., reducing the need for application-layer protocol enhancing proxies - PEPs).

2.3 Flow Control

Flow Control has been introduced in HTTP/2 to regulate contention among the streams that share a TCP connection. This avoids, for instance, a stream from being blocked by another contending stream. Flow Control also seeks to regulate the overall connection rate. This can be used to protect resource-constrained endpoints from being overwhelmed by received data. This addresses cases where the receiver is unable to process data on one stream, yet wants to continue to process other streams within the same connection. In the case of a multi-hop HTTP/2 connection, each hop receiver can control the rate over its hop.

HTTP/2 uses a simple credit-based flow control mechanism. Clients are required to specify an initial size in the SETTINGS frame indicating the amount of data they are prepared to accept for the entire connection and for each stream. During the session, clients use WINDOW_UPDATE frames to increase the size of a stream or the connection. The WINDOW_UPDATE frame specifies the increment in size of the stream.

While Flow Control is an essential feature of HTTP/2, the Flow Control mechanisms are still at an early development stage. The current standard only provides the necessary tools to implement this feature, but does not mandate any particular method.

The choice of the size of the streams can have a direct impact on the performance of the long-delay paths found with satellite networks. Too small values in initial settings may be insufficient to achieve high throughput over a large bandwidth-delay product connection.

2.4   Server Push

HTTP/2 allows a server to *push* responses to a client associated with a previous client request. This could be useful when the server can anticipate which resources the client is going to request based on previous requests. To avoid *pushing* undesired data a client can request that Server Push is disabled when the connection settings are negotiated.

Server Push is implemented using the PUSH_PROMISE frame, which contains the set of requests that would originate a response. The PUSH_PROMISE frame is followed by PUSH_RESPONSE frames that contains the actual pushed data. If the client decides to accept the pushed data, it is required not to send any request for the pushed resources. Otherwise, the client can issue a control frame to stop the transmission of the data.

Server Push could be beneficial for users connected via satellite, because it could potentially reduce the time to load a page by at least one RTT. For example, eliminating the need for a client to send a request for the resources linked to the index page. However, the optimisation may come at a cost, since the client is less able to decide what content uses the satellite capacity. A similar concern has also been expressed by mobile cellular operators, where pro-actively downloading non-needed contents can incur into additional cost [10].

3   Transport Layer Modifications

The effectiveness of HTTP modifications would be limited if not complemented by changes at the transport layer. The awareness that the transport layer needs to be harmonised to the web application layer for higher responsiveness and lower overhead has provoked a series of proposals since 2010. The new protocols update the way web protocols use the Internet by using new transport mechanisms (larger initial window, initial data, updates to congestion window validation, support for thin flows, etc.).

Although most TCP modifications were designed to improve performance over terrestrial networks, they can impact the performance for satellite networks. Removing even a few initial RTTs has macroscopic benefits for the satellite networking. This section surveys key recent proposals to update TCP.

3.1   Fast Open

TCP currently only permits transmission of useful data after the two endpoints have established a connection [5]. This initial three-way handshake (3WHS) introduces one RTT of delay for each connection. For short data transfers, the extra RTT incurs a significant portion of the flow duration [14]. According to [15] the cost of the initial handshake is between 10 % and 30 % of the latency for a HTTP transaction. Data on SYN was initially proposed in RFC793, but later discouraged because this would allow data to be delivered to the application before the connection has been established.

TCP Fast Open (TFO) [16] is an IETF experimental specification published in December 2014 that provides new rules that allow data to be carried in an initial SYN segment and to be consumed by the receiving endpoint during the initial connection handshake. This can save one RTT compared to a standard TCP connection.

In TFO, the server side uses a new security mechanism (based on cookies) to authenticate a client initiating a connection thus addressing previous data integrity concerns caused by dubious SYN packets. This avoids the pitfalls of earlier methods, such as T/TCP [17].

TFO is designed to deliver data safely during the 3WHS without requiring the server to store per-client authentication information. A client obtains a TFO cookie from the server when access the server for the first time. This cookie contains encrypted information that authenticates the client (for example, the client's IP address). When the client wants to open a new connection to the same server, it returns the cookie to the server by inserting it into the payload. The server recognizes the IP address of the client deciphering the cookie, and therefore it can accept the TCP payload of the SYN segment.

TFO can reduce the delay of initial connection establishment, which is non negligible in the overall web page download delay budget as shown in the next section.

3.2 Larger Initial Window

RFC 3390 specifies the TCP initial window to be 3 segments [18]. This has been widely deployed. Motivated by the desire to improve Fast Retransmit, Google proposed further increasing the IW to 10 segments (about 15 kB) to quickly complete a considerable number of short TCP transfers [19].

Experiments by Google [20] showed benefits in reducing web object transfer times at moderate cost in terms of increased congestion and associated packet losses when used over their network. This motivated an experimental update to TCP in 2013 [19].

This specification [19] also recommended TCP implementations to refrain from resetting IW to one segment unless multiple SYN or SYN-ACK retransmissions occurred or congestion losses confirmed. The current standard specifies resetting IW to 1 even when a single control packet is missing. However, considering RFC 6298 reduction of initial RTO from 3 s to 1 s, this would excessively penalise connections with high RTTs (e.g. satellite links).

Current analysis of using a large initial window has not explored the potential collateral damage to other flows that share a bottleneck where the large IW is continuously used.

An adaptive approach to set IW was proposed in [21]. Instead of defining a particular value for IW or prescribing a schedule for increase over time. This proposed an automatic mechanism to increase the IW by objectively measuring when an IW was too large, allowing IW to be adjusted automatically. Generic methods to choose a suitable IW remain an item of future research.

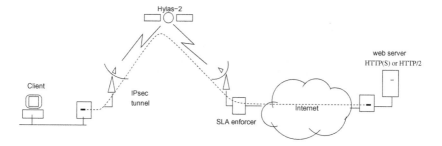

Fig. 1. HTTP/2 experimental testbed topology based on Hylas-II satellite platform

4   Performance Evaluation of HTTP over Satellite

A range of experiments evaluated the page load time (PLT) using both HTTP/1.1 and HTTP/2, and the impact of the different TCP stack modifications. Our tests (Fig. 1) accessed a web server via satellite terminals using the Hylas-II satellite network operated by Avanti^{TM} [22] based on IPoS [23]. Metrics other than the PLT have been used to evaluate web performance. These include the time-to-the-first-paint and the object OnLoad time [10]. However, we preferred to use the PLT which reflects both network performance and user experience.

Avanti^{TM}provided various SLA options for Return Link (RL) and Forward Link (FL) bandwidth provisioning. A traffic shaper co-located with the Satellite Gateway acted as SLA enforcer for the FL. Peak rates in our settings were 512 kb/s for the RL and 2048 kb/s for the FL. The satellite network implemented a range of transport and application layer PEP techniques. Application-layer PEP mechanisms were disabled when using HTTP/2 due to the use of SSL the encryption of the TCP payload.

By default, a transport PEP splits the TCP connection into three parts, one between client and terminal, one between terminal and satellite gateway, and one between gateway and server. To evaluate the impact of Transport PEPs, some tests disabled PEP by using IPsec encryption to hide the TCP header. This allowed us to study a HTTP session with a direct connection between server and client.

We also performed experiments using a network emulator introducing an artificial RTT of 750 ms, which approximated the Internet plus observed satellite round-trip. This was used as a reference case, since it avoided the varying delay from the dynamic bandwidth allocation used by the satellite system.

The web server ran on a Linux-based platform with kernel 3.8. This kernel implemented the recent TCP modifications for fast start-up. In particular, the kernel used RFC6928 by default. All experiments used the Mozilla Firefox (vers. 31) web browser, which was equipped with a protocol analyser to capture the HTTP/2 conversation between the client and server. The Firefox configura-

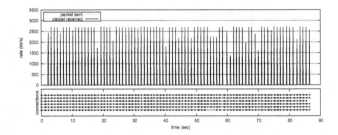

Fig. 2. Dynamics of HTTP traffic rate

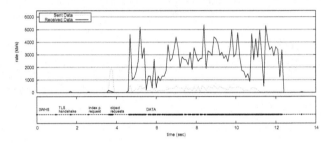

Fig. 3. Dynamics of HTTP/2 traffic rate

tion panel allowed the user to switch between HTTP and HTTP/2. The implementation of HTTP/2 was based on the SPDY/3 module provided by Google[TM].

To explore a range of web page compositions, we considered three cases of web-pages lengths (of 500 kB, 1500 kB, and 2500 kB respectively) and three cases of object sizes (of about 5 kB, 20 kB and 100 kB respectively). This resulted in 9 combinations of pages with homogeneous object size of different length. The number of objects in our tests varied between 5 (when the page size is 500 kB and the object size is 100 kB) and 500 (when the page size is 2500 kB and object size is 5 kB). Although pages with more than 200 objects are unlikely in today's Internet, the number of objects per page has steadily increased in recent years and HTTP/2 should be devised having in mind the extreme cases. This approach was also followed in another HTTP/2 study [10]. Each experiment was repeated 10 times to mitigate the variability of download duration on the average PLT.

4.1 Rate Patterns in HTTP and HTTP/2

Figure 2 illustrates the transport dynamics for HTTP/1 with a web-page of 500 objects of around 5 kB[1]. The figure also plots the activity, i.e., when packets are received/transmitted, of each of the six parallel HTTP/1 client-server connections that Mozilla Firefox opens towards the server. Each connection carries around one-sixth of the total number of objects and each object is downloaded by a separate request/response transaction.

---

[1] Rate samples were taken every 100 ms.

Performance over satellite were influenced by the *connection retry timeout* in the Firefox configuration which is set by default at 250 ms. This parameter sets the time Firefox has to wait after opening a connection before trying to open a new connection. A too small value for the connection retry timeout caused multiple connections being opened and connection errors at start.

Since the objects are small and each connection can carry at most one object per RTT, six connections are far too few to utilise the large bandwidth-delay product path (about 650 kB in this case). This results in a low throughput (230 kb/s) which is much less than the 8 Mb/s available bandwidth. Moreover, HTTP performance cannot be improved by transport PEPs because only a few kilobytes are delivered at each request/response over a TCP connection. The amount of data per transaction is therefore smaller than the TCP IW and could be sent end-to-end by TCP in one RTT.

A very different output is observed instead when HTTP/2 is used (Fig. 3). Indeed, the same 500 HTTP objects can be delivered in about 12 s with an average throughput of 1.7 Mb/s using HTTP/2. The better result is obtained mainly due to object multiplexing. Once that the HTTP/2 client has received the index page, it can generate a sequence of object requests issuing stream synchronisation messages (SYN_STREAM). Each stream is identified by a *stream-ID* which is used by the server to refer to a particular object in the response. The transmission of SYN_STREAMS is shown at around time 4 s when a 2 Mb/s peak rate by the browser. As multiple request messages can be sent simultaneously, the server can receive request messages faster, respond to many more request at the same time and maintain a high throughput. The throughput is however not optimal due to the limited number of concurrent streams allowed by the web server: The default HTTP/2 configuration in Apache allows only up to 100 concurrent streams per session, while the path could sustain up to 150 streams.

A significant amount of time (about 3 s) is spent in connection opening. This includes the TCP three-way handshake (one RTT) and a full TLS handshake (two RTTs). Both TCP and TLS connection establishment require the synchronisation between server and client and hence full satellite RTTs to complete. However, a three RTT connection opening is strictly required only the first time the site is accessed. In a subsequent access the web client could indicate the session-ID of a previous TLS session to resume the session (if session caching, RFC 5246, is supported by the server) or use a session ticket, RFC 5077, which was previously released by the server for a certain TLS session. Both these mechanisms reduce the TLS handshake to one RTT (the *abbreviated* TLS handshake). In addition, the client could use a TCP Fast Open cookie to send data on SYN and eliminate the TCP handshake entirely.

4.2   Impact of Page Composition on Web Performance

Figure 4 illustrates the PLT of test web-pages with respect to the number of objects for HTTP and HTTPS when the pages were accessed through the PEP. The results of our experiments suggest that the number of objects is an important parameter in determining the PLT with HTTP/1 and HTTPS. These results

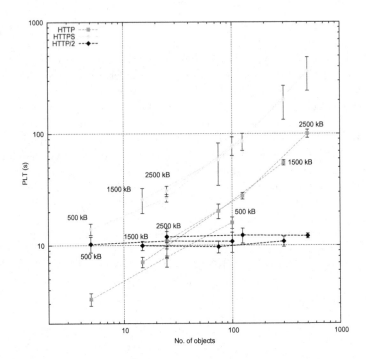

Fig. 4. Page load time (PLT) with respect to the number of objects for HTTP, HTTPS and HTTP/2 with satellite split connection. The object size is the ratio between the page size (label) and the no. of objects.

were obtained with a FL capacity of 8 Mb/s and the default Linux configuration for the TCP/IP stack.

Although the PLT exhibits large variability, an increasing trend of PLT with respect to the number of object is evident in all cases. The number of objects influences the PLT more than the page length. For example, downloading a web page consisting of a hundred objects of 5 kB (the *500 kB* label in the picture) takes around 35 s when using HTTPS with direct access, while a page of 15 100 kB objects (*1500 kB* label) takes only 12 s.

The dependency of PLT on the number of objects with HTTP is not surprising. When a web page is accessed using HTTP/1 or HTTPS, the web browser opens a certain number of connections towards the web server to request and retrieve web objects. Since only one object can be requested on a connection and only after the transmission of a previous object is completed, the throughput of a web session (T_{http}) is limited to transferring one average object size (S_{obj}) per round-trip time (RTT) per connection, i.e., $T_{http} \approx N_c \times S_{obj}/RTT$. This rate limitation imposed by the HTTP/1 request/response model is particularly a limitation for long delay paths, and is one of the main motivations for using application-layer PEP.

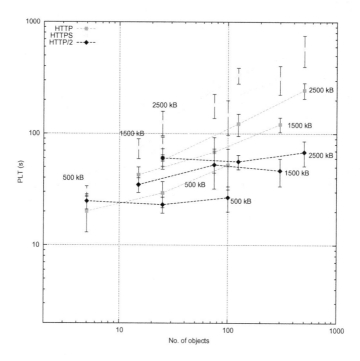

Fig. 5. Page load time (PLT) with respect to the number of objects for HTTP(S) and HTTP/2. The object size is ratio between the page size (label) and the no. of objects.

Figure 4 also shows that the PLT with HTTP/1 is significantly lower than with HTTPS when a page is accessed directly. In reality, this performance is only obtained because the HTTP/1 connection is not persistent and represents the worst case for HTTP/1. If the connection is not persistent, a new TCP and TLS handshake needs to be performed for each object. However, when persistence is enabled, the performance of HTTP/1 and HTTPS are similar.

When HTTP/2 is used (Fig. 4), the PLT is around 10–12 seconds, and is not strongly dependent on the number of objects. This again is not surprising considering that many more objects can be transmitted concurrently. However, we notice that HTTP/2 performance depends weakly also from the page size. This is probably due to the high variability of the rate of the TCP session (as illustrated in Fig. 3). As the throughput can vary substantially from one experiment to the next, the impact of the page size on the PLT is somehow overshadowed by other factors, including the interaction between TCP congestion window (cwnd) and the HTTP/2 request/response mechanism.

4.3   Performance with Large End-to-end Delay (no Transport PEPs)

Figure 5 shows the performance when a client connects to the server without intermediaries. Although the plot exhibits more variability than the one with

PEPs, we can draw similar conclusions. Inspecting the traces, we found that RTT samples are affected by very large variations (on the order of seconds), most likely due to queuing delays at the FL access queue. While these delays are mitigated by the splitting of the connection in the PEP case, using the IPsec tunnel these delays affect the end-to-end RTT. This reflects poorly on the TCP RTO estimation and ultimately on the performance.

The average PLT with HTTP/2 ranges between 11 and 15 s.The PLT with HTTPS spans a much larger interval, taking up to few minutes to download the page when the number of objects is more than one hundred. The performance improvement of HTTP/2 is clearly related to the ability of the protocol to efficiently multiplex many small objects onto the same TCP connection. However, the throughput of HTTP/2 was less than the case with split connection and much less than the available bandwidth (8 Mb/s). This suggests that capacity is not the primary limiting factor on the HTTP/2 performance. Analysing the traces we found that several other factors contributed to reduce the throughput with respect to the split-connection case:

1. The maximum number of concurrent active streams was limited to one hundred (default value in Apache server). This is done as a precautionary measure to constrain the amount of memory used by HTTP/2 streams. While this number is sufficient for a terrestrial wired session, satellite networks could easily host several hundreds of streams if their size is limited to few kilobytes.
2. The memory reserved by the client for the TCP connection receive buffer was about 128 kB. Although TCP Window Scaling Option [24] was used (as in the majority of modern TCP/IP stack) and the application was able to retrieve data quickly from the receive buffer, the TCP receiver was forced to advertise a window that was smaller than the bandwidth-delay product. The small advertised window was due to the limited buffer space reserved by the application.
3. The browser fired a timeout to reset the connection when it was idle for few seconds. In some of our tests, the HTTP session was reset even more than once to finish downloading the objects. At each reset new TCP connections were created.

All these effects are clearly related to the fact that the HTTP session was optimised for a terrestrial Internet path. However, HTTP/2 offers mechanisms (such as the SETTINGS and UPDATE control frame) to configure the HTTP session to the actual client/server requirements. Using these mechanisms it is possible to optimise the session by avoiding parameters statically configured in the browser.

5   Conclusion

This paper reports the results from a series of tests using SDPY/3 (the most recent HTTP/2 implementation from Google^{TM}) over the satellite platform. Our results clearly show that when the number of multiplexed objects is large,

HTTP/2 largely outperforms HTTP(S). In particular, we observed that when the web page is made up of more than hundred objects HTTP/2 completes the web-page transfer in around ten seconds while the HTTP(S) spends several tens of seconds.

On the other hand, with less than ten objects HTTP performs better than HTTP/2. This is because HTTP does not need to wait for the TLS handshake to complete and benefits of the presence of application-layer PEPs. However, HTTP/2 performance could have been improved if protocol parameters had been configured for the satellite delay. For example, the default size of TCP send and receive buffer of the HTTP/2 connection was around 130 kB. While this did not have particular effects when the end-to-end connection was split by the intercepting PEP, it proved insufficient when the client and server were connected directly without the mediation of PEPs. Also, the default initial connection size used in the Flow Control mechanism and the maximum number of permitted parallel streams were inferior to the ones allowed by the bandwidth-delay product. Finally, the HTTP/2 *connection-retry-timeout* was set to 250 ms, which is unrealistic for the satellite case.

PEPs are nowadays essential to achieve satisfactory web performance over satellite. In particular, split connections are required to avoid the long delays created by TCP end-to-end congestion control. However, they have many drawbacks. PEPs break TCP end-to-end semantics leading in some cases to drop TCP options or ignore new options (for instance in our tests split-TCP prevented the use of large window option requested by the client). This somehow limits the ability to upgrade the system when new Internet transport techniques are made available, thus contributing to the *ossification* of the Internet, i.e. the lack of progress due to compatibility issues. Also, PEPs make the system more complex to configure and maintain since transport code need to be run at the terminals and gateway of the satellite network. In other words, PEPs are a *necessary evil*, not loved by network operators.

HTTP/2 may be seen as a way forward to address the performance issue of web traffic over satellite. HTTP/2 offers the ability to send an entire web page over a single TCP connection by introducing methods to multiplex request/response transactions (or more in general web streams) within a persistent connection. HTTP/2 also introduces a mechanism to *push* data to the client without an explicit request, thus saving precious RTTs. The HTTP/2 improvements are complemented at the TCP layer by a series of proposals, such as TCP Fast Open and the larger Initial Window, to improve TCP responsiveness at startup. These recommendations highlight inefficiency of the current network stack and remove inessential RTTs. Combined with the new application-layer protocol, these modifications are expected to speed-up significantly web sessions.

Our tests have shown that HTTP/2 is probably not yet mature technology to definitely replace the traditional satellite web architecture based on PEPs. However, the problem we identified are implementation-related and can probably be overcome by revising HTTP/2 specifications having in mind the issues introduced by the large RTT.

References

1. Domenech, J., Pont, A., Sahuquillo, J., Gil, J.: A user-focussed evaluation of prefetching algorithms. ACM Comp. Comm. **30**, 2213–2224 (2007)
2. Ramachandran, S.: Web Metrics: Size and number of resources (2012)
3. Berners-Lee, T., Fielding, R., Frystyk, H.: Hypertext Transfer Protocol - HTTP/1.0. Internet-Draft draft-ietf-http-v10-spec-05, IETF Secretariat (1996)
4. Fielding, R.T., et al.: Hypertext Transfer Protocol - HTTP/1.1. RFC 2616, RFC Editor (1999)
5. Allman, M., Paxson, V., Blanton, E.: TCP Congestion Control. RFC 5681. RFC Editor (2009)
6. Elkhatib, Y., Tyson, G., Welzl, M.: Can SPDY really make the web faster? In: proceedings of IFIP Networking Conference, Tronhdeim (2014)
7. Peon, R., Belshe, M.: SPDY protocol - Draft 3 (2012)
8. Roskind, J.: QUIC (QUIC UDP Internet Connections), Multiplexed Stream Transport over UDP Google Technical report (2012)
9. Belshe, M., Peon, R., Thomson, M.: Hypertext Transfer Protocol Version 2 (HTTP/2). RFC 7540, RFC Editor (2015)
10. Wang, Z.S., Balasubramanian, A., Krishnamurthy, A., Wetherall, D.: How Speedy is SPDY? In: 11th USENIX NSDI, Seattle, pp. 287–393 (2014)
11. Caviglione, L., Gotta, A.: SPDY over high latency satellite channels. EAI Endorsed Trans. Mobile Comm. Appl. **2**, 1–10 (2014)
12. Luglio, M., Roseti, C., Zampognaro, F.: SPDY multiplexing approach on long-latency links. In: Proceedings of IEEE WCNC, Instanbul, pp. 3450–3455 (2014)
13. Luglio, M., Roseti, C., Zampognaro, F.: Resource optimization over DVB-RCS satellite links through the use of SPDY. In: Proceedingsof WiOpt, Hammamet (2014)
14. Radhakrishnan, S., Cheng, Y., Chu, J., Jain, A., Raghavan, B.: TCP Fast Open. In: Proceedings of ACM Conext 20133, Tokyo (2011)
15. WalkerSand: Quarterly Web Traffic Report. Technical report, WalkerSands Communications, Chicago (2013)
16. Cheng, Y., Chu, J., Radhakrishnan, S., Jain, A.: TCP Fast Open. RFC 7413, RFC Editor (2014)
17. Braden, B.: T/TCP - TCP Extensions for Transactions Functional Specification. RFC 1644, RFC Editor (1994)
18. O'Hara, B., Calhoun, P., Kempf, J.: Configuration and Provisioning for Wireless Access Points (CAPWAP) Problem Statement. RFC 3990, RFC Editor (2005)
19. Chu, J., Dukkipati, N., Cheng, Y., Mathis, M.: Increasing TCP's Initial Window. RFC 6928, RFC Editor (2013)
20. Dukkipati, N., et al.: An argument for increasing TCP's initial congestion window. ACM SIGCOMM Comp. Comm. Rev. **40**, 26–33 (2010)
21. Touch, J.: Automating the Initial Window in TCP. Internet-Draft draft-touch-tcpm-automatic-iw-03.txt, IETF Secretariat (2012)
22. Avanti: Avanti Satellite Communications (2014)
23. Hughes Networks: IP over Satellite (IPoS) - The Standard for Broadband over Satellite (2007)
24. Jacobson, V., Braden, B., Borman, D.: TCP Extensions for High Performance. RFC 1323, RFC Editor (1992)

Multimodality in the Rainfall Drop Size Distribution in Southern England

K'ufre-Mfon E. Ekerete[1(✉)], Francis H. Hunt[1], Ifiok E. Otung[1], and Judith L. Jeffery[2]

[1] Mobile & Satellite Communications Research Group,
University of South Wales, Pontypridd CF37 1DL, UK
{kufre-mfon.ekerete,francis.hunt,
ifiok.otung}@southwales.ac.uk
[2] STFC Rutherford Appleton Laboratory,
Harwell Oxford, Didcot OX11 0QX, UK
judith.jeffery@stfc.ac.uk

Abstract. Mutimodality appears in the modelling of rainfall drop size distributions (DSDs), and the understanding of the distribution in general is important as it helps in the predicting and mitigation of attenuation due to rain of satellite signals in frequencies above 10 GHz. This work looks at the occurrence of multimodality in the rainfall DSDs in southern England, with data captured at the Chilbolton Observatory for a seven year period (2003 to 2009). The investigation looks at the variation in the number of modes against different rain rates and seasons. It shows that multimodality is a relatively common occurrence, and hence there is a need to model this phenomenon when attempting to predict rain attenuation of satellite signals.

Keywords: Probabilistic forecasting · Theoretical modelling · Space and satellite communications · Estimation and forecasting · Probability distributions · Rainfall drop size distribution (DSD) · Multimodality

1 Introduction

Understanding the rainfall drop size distribution (DSD) is important in the understanding of transmissions using frequencies above 10 GHz if optimal use is to be made of bandwidth. There is presently enormous demand for telecommunication services leading to an increasing need to utilize more transmission bandwidth. Lower frequency bands are now congested, and providers presently utilise higher frequencies. The transmission of signals at frequencies above 10 GHz is however much more susceptible to attenuation due to precipitation. The precipitation leads to degradation in the desired quality of service and link availability. Raindrops, in particular, absorb and scatter radio wave energy.

There is the need to properly estimate the attenuation due to rainfall, as over-estimating is wasteful of resources, while under-estimating may lead to system outages. In order for engineers to design dependable systems, there is a need to reliably predict how precipitation, and rain in particular, attenuates transmitted signals.

© Institute for Computer Sciences, Social Informatics and Telecommunications Engineering 2015
P. Pillai et al. (Eds.): WiSATS 2015, LNICST 154, pp. 177–184, 2015.
DOI: 10.1007/978-3-319-25479-1_13

Modelling the rainfall drop size distribution (DSD) is a key ingredient in the prediction of rain attenuation thereby allowing providers to design mitigation techniques to counter attenuation due to these rain events.

This work describes a study of rainfall drop size distributions, particularly the occurrence of multimodal distributions, based on data collected between 2003 and 2009 from a disdrometer located at the Chilbolton Observatory in southern England.

2   DSD Modelling

2.1   Standard Statistical Models

Rainfall drop size distribution, $N(D)$ (in m^{-3} mm^{-1}), is defined as the number of raindrops per unit volume per unit diameter, centred on D (in mm). $N(D)dD$, expressed in m^{-3}, is the number of such drops per unit volume having diameters in the infinitesimal range $(D - dD/2, D + dD/2)$ of size dD centred on D.

Several researchers have proposed various standard classical statistical distributions for the non-negative continuous DSD, $N(D)$. Marshall and Palmer [1] proposed the relationship

$$N(D) = N_0 exp(-\Lambda D),\ 0 < D \leq D_{max} \tag{1}$$

where D_{max} is the maximum drop diameter, where $\Lambda = \alpha R^{\beta}$ (in mm^{-1}) is a function of the rainfall rate R (mm/h). This however fails for small diameters ($D < 1.5$ mm).

Other later researchers also modelled rainfall data using the lognormal distribution [2–5]. The lognormal distribution for the number of drops in a given volume is given in the general form

$$N(D) = \frac{N_T}{\sqrt{2\pi}\sigma_g(D - \theta)} exp\left[-\frac{\left(\ln(D - \theta) - \mu_g\right)^2}{2\sigma_g^2}\right] \tag{2}$$

Ulbrich and Atlas [6] showed a gamma distribution yielded better rainfall rate computations when combined with radar data. They used a gamma distribution expressed in the form

$$N(D) = N_T D^{\mu} \exp(-\Lambda D) \qquad 0 \leq D \leq D_{max} \tag{3}$$

with Λ, μ, and N_T as the slope, shape and scaling parameters respectively, and these allow for the characterization of a wide range of rainfall scenarios, though in the paper they use $\mu = 2$, and the exponential distribution is a special case of the gamma distribution with $\mu = 0$. Ulbrich and Atlas [6] do not actually claim the DSD is a gamma distribution, but simply that a gamma distribution yields more accurate rainfall rate computations. They accept that other distributions might serve equally well. It should be noted that these three standard models for DSDs are all unimodal.

A typical example of rainfall drop size distribution modelled with lognormal, gamma (with the method of moments) and gamma (with the method of likelihood estimates) is shown in Fig. 1a below for 6th January, 2007 at 14:24, with 11.1 mm/h rain rate. A goodness of fit is done with the data and the various statistical distributions using Pearson's chi square, and the data did not reject any of the distributions tested. Figure 1b, however shows a clear case of multimodality, and the chi square statistic show that none of the distributions can be fitted to the data.

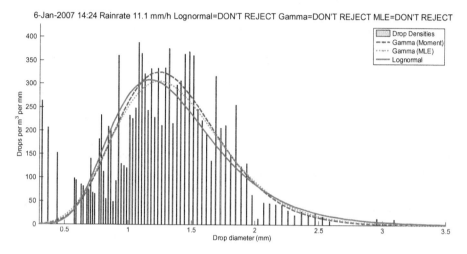

Fig. 1a: DSD for 6th Jan, 2007 at 14:24, modelled with three statistical distributions.

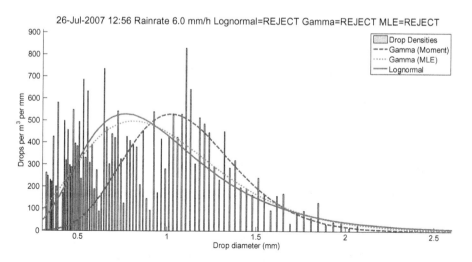

Fig. 1b: DSD for 26th July, 2007 at 12:56, modelled with three statistical distributions.

2.2 Multimodality in the DSD

Despite the standard models for DSDs all being unimodal, raindrop data often strongly suggest underlying multimodal distributions instead [7–10]. Steiner and Waldvogel [8] define an interval of drop sizes to be a mode if "… the concentration of raindrops per unit volume and unit diameter interval of a given interval was significantly larger than the concentrations of the two neighbouring diameter intervals". Sauvageot and Koffi [9] and Radhakrishna and Rao [10] simply implement this as $N(D_{i-1}) < N(D_i) > N(D_{i+1})$.

The challenge in detecting modes in the underlying distribution is that the data collected are necessarily a finite sample – if we have fine enough diameter measurements then every drop will appear in an interval by itself, with no drops in the intervals on either side, so there will be as many modes as there are drops. This is clearly unsatisfactory. In practice drop size data is collected using an instrument such as an impact disdrometer (see Sect. 3.1 for details) that counts the number of drops in a set interval of drop diameters or "bin". The counts in a number of adjoining bins are merged to produce smoother data; however the optimal number to merge is not clear.

Sauvageot and Koffi [9] attribute the presence of multimodality in DSDs to the overlapping of different rain shafts resulting from cloud volumes at different heights, and they also show that the number of peaks, N_m, of a DSD depends on the rain rate variations, and not on the mean rain rate, but this was for rain with $D_i > 2$ mm, where large N_m are inversely related to values of the slope parameter, λ and with large values of the intercept, N_0 of the exponential distribution. Steiner and Waldvogel [8] equally studied multimodal behaviours in DSDs and report that these modes existed for different drop size diameters in convective rain regimes. Radhakrishna and Rao's [10] study indicated that the appearance of multimodal distribution in the DSDs are dependent on the height, and varies with different rain systems, with multimodal distributions frequently encountered in convective rain systems. They classified the rain systems as convection, stratiform, and transition. This work however classes rain regimes as light, moderate, heavy and very heavy, as shown in Sect. 4.

3 Data and Procedure Used in This Study

3.1 Data Collection

This work utilised data captured by the RD-69 Joss-Waldgovel Impact Disdrometer connected to an ADA90 analyser at the Chilbolton Observatory in southern England (51.1° N, 1.4° W) between April 2003 and December 2009. Data was not captured from July 2005 to May 2006 and for 73 other days in this period. The disdrometer works by converting the vertical momentum of an impacting raindrop into an electrical pulse, and estimating the diameter of the raindrop from the amplitude of the pulse. The disdrometer has a surface area of 50 cm^2 and measures raindrop diameters from 0.3 mm to 5.0 mm in 127 gradations, or bins, sampling at 10 seconds intervals. The 127 size classes are distributed more or less exponentially over the range of drop diameters and the accuracy rate of the readings is ± 5 % of measured drop diameter [11]. A picture of the disdrometer at chilbolton (with a co-located rain guage) is shown in Fig. 2. This study aggregated six 10 seconds samples into a one minute samples to achieve a larger sample. This implicitly assumes that the underlying distribution is approximately stationary over a 1 min timescale. This is the approach taken by previous workers [12–14].

Furthermore, five adjoining bins were merged to smooth the data. This number was adopted as the best value after experimenting with different numbers – merging fewer bins results in too noisy a sample, whereas merging more bins results in too great a loss of detail. *Radhakrishna and Rao*, [10] merged over five minute interval, instead of five bins. This however requires stronger assumptions on the stationarity of the distribution in order to be confident that a multimodal distribution is not simply the result of merging different unimodal distributions (Fig. 3).

Fig. 2. The disdrometer and a co-located rain gauge at Chilbolton Observatory.

This work however looks at a different method in the determination of modes in a multimodal distribution by identifying each individual cluster (or distribution) from the troughs separating each cluster, here assuming that each cluster has a peak (mode). A trough (the end of a cluster, with an assumed peak) is thus defined as $N(D_i)$, when $N(D_{i-1}) > N(D_i) < N(D_{i+1}) < N(D_{i+2})$. This ensures a steady rise to determine the beginning of the next cluster. To ascertain the reliability of the method compared to that used by others [8–10], an inspection of the data for 27th July, 2007 showed that from 170 one-minute samples, 40 samples showed the same number of modes for both methods, while a visual inspection seem to agree with 38 samples for the earlier method and 93 for the current method.

4 Results

Using the data and procedure described in Sect. 3, distributions were fitted to a total of 166,065 one-minute samples, with 5 consecutive bins merged. Figure 3 shows a typical one-minute sample of data (26th July, 2007 at 17:45) which clearly suggests a trimodal distribution. The bar chart shows the distribution of the rain drop sizes, the drop densities in mm^{-1} m^{-3} plotted against the log of the drop diameters in mm. A log scale has been used simply to provide equally spaced bars (since as explained in Sect. 3.1, the bin sizes increase exponentially) – the use of a log scale has no effect on the multimodality.

The work investigated the variation of the modes with the rain rate and seasons of the year. Table 1 shows a summary of both unimodal and multimodal distributions for different rain rates. When the rainfall is classified as light (less than 2 mm/h), moderate (2 mm/h to 10 mm/h), heavy (10 mm/h to 50 mm/h), and very heavy (more than 50 mm/h) then it is noticeable that multimodality is more common at higher rain rates.

Table 1. Summary of results for unimode and multimode at different rain rates

Rain type	Rain rate	Unimode	Multimode			Unimode Total	Multimode Total	Grand Total
	(mm/h)	1	2	3	4+			
Light	<2	53 %	38 %	8 %	1 %	72,249	64,520	**136,769**
Moderate	2-10	32 %	48 %	18 %	2 %	8,872	18,446	**27,318**
Heavy	10-50	18 %	48 %	29 %	4 %	349	1,567	**1,916**
Very Heavy	> 50	16 %	47 %	34 %	3 %	10	52	**62**
Total		**81,480**	**66,395**	**16,653**	**1,537**	**81,480**	**84,585**	**166,065**

Results show that of the 166,065 samples, 136,769 (82 %) were classified as light rain (with rain rates less than 2 mm/h) and 72,249 samples in this rain regime were unimodal whilst 64,520 samples (39 %: 38 % bimodal, 8 % trimodal, and 1 % with 4 modes and above) were multimodally distributed as shown in Table 1. Considering all rain rates, 81,480 (49 %) samples were unimodal, whilst 84,585 (51 %) samples were multimodal.

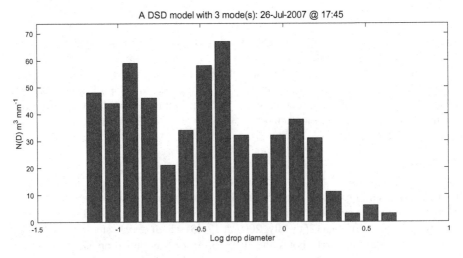

Fig. 3. DSD with three clusters for 26th July, 2007 at 17:45.

Considering the seasons; Spring (March-May), Summer (June-August), Autumn (September-November), and Winter (December to February), there does not seem to be a marked variation between the seasons. Results (Table 2) show that 22,794 (50 %) samples were unimodal in Winter, whilst 22,373 (39 % bimodal, 10 % trimodal, and 1 % with four modes and above) were multimodal in the same season. In all modes, the seasons were distributed as follows: Winter (27 %), Spring (24 %), Summer (24 %), and Autumn (24 %). The detailed distributions are as shown in the table below.

The fundamental result here is that multimodality does occur significantly often, particularly at higher rain rates.

Table 2. Summary of results for unimode and multimodes at different seasons

Season	Unimode	Multimode			Unimode Total	Multimode Total	Grand Total
	1	2	3	4+			
Winter	50 %	39 %	10 %	1 %	22,794	22,373	**45,167**
Spring	46 %	43 %	11 %	1 %	18,320	21,633	**39,953**
Summer	49 %	40 %	10 %	1 %	19,833	20,455	**40,288**
Autumn	51 %	39 %	10 %	1 %	20,533	20,124	**40,657**
Total	**81,480**	**66,395**	**16,653**	**1,537**	**81,480**	**84,585**	**166,065**

5 Interpretations, Further Work and Conclusions

This work has shown that whilst there is no discernible variation between the seasons, the different rain regimes however show a high occurrence of unimodality in light rains and high bimodality in the other rain types (moderate, heavy and very heavy). However, we should note that a large portion of the data was classified as light rain, and very little of the sample was classified as very heavy.

Although moderate, heavy and very heavy rains are less common than light rain, it is precisely these types of rain that cause the most attenuation to signals at frequencies above 10 GHz. Hence it is of concern that multimodality is common here, but none of the standard DSD models are multimodal. Ekerete et al. [7] highlighted the fact that the standard models often do not fit the data well, as assessed using the chi-square goodness of fit test. It would appear that a good explanation for this is that the data is multimodal but the models unimodal. More work is needed to find appropriate multimodal models for these multimodal situations.

References

1. Marshall, J.S., McK, P.W.: The distribution of raindrops with size. J. Meteorol. **5**, 165–166 (1948)
2. Levine, L. M.: The distribution function of cloud and rain drops by sizes, Doklady Akad. Nauk., SSSR 94, No. 6, 1045–1048, 1954. (Translated by Assoc. Tech. Services Inc., East Orange, NJ) cited in Mueller, Eugene Albert (1966), Radar Cross Sections from Drop Size Spectra, Ph.D. thesis in Electrical Engineering, Graduate College of the University of Illinois, Urbana, Illinois, United States

3. Markowitz, A.H.: Raindrop size distribution expressions. J. Appl. Meteorol. **15**(9), 1029–1031 (1976)
4. Feingold, G., Levin, Z.: The lognormal fit to raindrop spectra from frontal convective clouds in Israel. J. Appl. Meteorol. **25**(10), 1346–1363 (1986)
5. Owolawi, P.: Raindrop size distribution model for the prediction of rain attenuation in Durban. Prog. Electromagnet. Res. **7**(6), 516–523 (2011)
6. Ulbrich, C.W., Atlas, D.: Assessment of the contribution of differential polarization to improved rainfall measurements. Radio Sci. J. **19**(1), 49–57 (1984)
7. Ekerete, K'.E., Hunt, F.H., Agnew, J. L., Otung, I. E.: Experimental study and modelling of rain drop size distribution in southern England, IET Colloquium on Antennas, Wireless and Electromagnetics, May 27th 2014. doi:10.1049/ic.2014.0016
8. Steiner, M., Waldvogel, A.: Peaks in raindrop size distributions. J. Atmos. Sci. **44**, 3127–3133 (1987)
9. Sauvageot, H., Koffi, M.: Multimodal Raindrop Size Distribution. J. Atmos. Sci. **57**, 2480–2492 (2000)
10. Radhakrisna, B., Narayana Rao, T.: Multipeak raindrop size distribution observed by UHF/VHF wind profilers during the passage of a mesoscale convective system. Mon. Weather Rev. **137**, 976–990 (2008)
11. Distromet Ltd.: Disdrometer RD-80 Instruction Manual, Switzerland (2002)
12. Montopoli, M., Marzano, F.S., Vulpiani, G.: Analysis and synthesis of raindrop size distribution time series from disdrometer data. IEEE Trans. Geosci. Remote Sens. **46**(2), 466–478 (2008). doi:10.1109/TGRS.2007.909102
13. Townsend, A.J., Watson, R.J.: The linear relationship between attenuation and average rainfall rate for terrestrial links. IEEE Trans. Antennas Propag. **59**(3), 994–1002 (2011)
14. Islam, T., Rico-Ramirez, M.A., Thurai, M., Han, D.: Characteristics of raindrop spectra as normalized gamma distribution from a Joss-Waldgovel disdrometer. Atmos. Res. **108**, 57–73 (2012)

Wireless Sensor Networks and Satellite Simulation

P.-Y. Lucas[1]($\boxtimes$), Nguyen Huu Van Long[2],
Tuyen Phong Truong[2], and B. Pottier[1]

[1] Université de Bretagne Occidentale, Brest, France
pottier@univ-brest.fr
[2] Can Tho University, Can Tho, Vietnam

Abstract. Connecting wireless sensor networks (WSN) by the air becomes attractive due to advances in Satellites and Unmanned aerial vehicle (UAV). This work focus on specification and simulation of situations where several distant WSN have gateways visited periodically by a mobile on a static path. To develop and evaluate collection and control services, it is first needed to run system level simulation. This paper reports on producing automatically representations of complex topology where mobile cooperate with sensor fields with respect to timing constraints from both sides. Simulation programs are produced for graphic accelerators (GPU), and concurrent process architectures.

Keywords: Simulation · Satellite · Sensor network · Distributed algorithm

1 Introduction

Wireless Sensor Networks (WSN) are known to solve practical problems in emergency and environment monitoring. Most of WSN are deployed for managing aspects of *smart cities*: parking allocation, transportation, pollution, home services. They can group thousand of nodes in places where communication systems are abundant. They are of less common use in distant areas, such as shores, deserts, mountains, polar regions, due to the lack of energy and communication supports. In these cases mobile visits, including satellites, enable to collect data periodically, to data center, or to provide remote control on sensor fields.

Beside geostationary satellites, low earth orbit (LEO) satellites systems are operated for purposes such as observation, positioning, rescuing or commercial global communications. Well known LEO systems include manned orbiters and the International Space Station (ISS), geo-positioning system (GPS) variants, Iridium constellation, or Argos services. Very small satellites called *CubeSats* were initiated in USA, in 2003 [3]. They are more and more considered to build valuable experimentations for at least two reasons: energy budget and solution cost. The low altitude, from 160 to 2000 km, means low energy budget for launching and for communications. LEO are also associated to high travel speed, bound

© Institute for Computer Sciences, Social Informatics and Telecommunications Engineering 2015
P. Pillai et al. (Eds.): WiSATS 2015, LNICST 154, pp. 185–198, 2015.
DOI: 10.1007/978-3-319-25479-1_14

to short orbital periods from 1 to 2 h, and short distances suitable for remote sensing. The compactness, low cost, and standardization of devices, have motivated several international projects such as CubeSat, *QB50*, or *Outernet*.

1.1 Simulation Definitions

This paper presents preliminary *system investigations*, with the perspective to simulate access to *large sensor fields* grouping thousands of nodes distributed in several remote area. Unmanned mobiles, such as LEO satellites, are visiting these fields. They control remote data collection, distribute synthetic information to ground stations, and they control sensing operations.

Work Hypotheses. We take the hypothesis of sensor systems permanently sampling physical processes, communicating with neighbours periodically, using a mesh organization. This is common in 802.15 wireless solutions. Sensor fields group any number of sensors, anywhere, with any connectivity, but they are required to share a common abstract clocking system to schedule communications. Mobiles travelling on predictable paths, at predictable speed will visit the sensor fields, establishing communications with one or several gateways.

Objectives. They are to establish specification methods to describe sensor fields from geographic environments. From this, we can infer or tune radio ranges, deduce ground network topology. From mobile path, we can infer meeting schedules with the ground. This enable exploration and design of cooperative algorithms for remote control and data collection. Expected metrics are correctness taking account of ground and air relative speeds, latency, risk of failures, energy budget for communications.

Work Contents. Simulation requires high performance computing and enough flexibility to associate mobiles, sensor field and mobile interaction algorithms. Intermediate level models for these activities are produced with automatic translation of system organization for parallel execution as MIMD and SIMD programs. General Purpose Graphic Processing Units (GPGPU) have been found to be a solution to simulate the sensing activity, the network activities, and interactions with satellites.

According to this context, the paper will firstly describe the WSN-Satellite problem, then will describe the simulation flow applied to 3 collection algorithms.

1.2 Simulation Flow

The flow for simulation has 4 main steps shown Fig. 1.

1. Using geographic tools for specification of satellite path and sensor fields.
2. Building an abstract network and formal representation for the sensor fields. The abstract network hold information such as node names, communication ranges, channels, geographic positions.

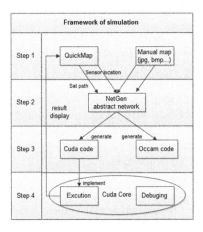

Fig. 1. Framework for simulation

3. Generating concurrent code in CUDA or Occam that represents network topology of sensor fields (no behaviours).
4. Implementing or reusing behavioural distributed algorithms for activities inside WSN, and with satellites. The result is a simulation program that will be executed on powerful platforms enabling step by step analysis.

The objective is to enable ambitious services for a promising domain [1].

2   Geographic Specifications

2.1   Orbiters and Radio Links

Two reference commercial satellite systems are Argos and Iridium. Argos data collection and location system focus on surveillance as well as protection of environment and wildlife, meanwhile, the Iridium network is a global satellite communication service for subscribers from government agencies and public citizen. Each system was established as a global satellite constellation of the low Earth orbiting (LEO) satellites flying at about 700–900 km above the Earth. A typical satellite's footprint is thousands of kilometers in diameter. Besides, both systems have already supported all three links namely up-link, down-link and cross-link to provide the high reliability of the communications network and to remain unaffected by natural disasters such as hurricanes, tsunamis and earthquakes, etc. The systems are often designed with L-band antennas to meet the requirement on high performances, low power supply, compactness, low cost.

Orbits. Due to the Earth's rotation, the swath shifts around the polar axis at each revolution, as shown Fig. 2. As a result, the overlap between successive swaths allows satellites enough time to visit ground stations in footprint vision for sending and receiving data several times per day. As a complement, ground infrastructure ensures 24/7 for controlling and monitoring on all components: transceivers, gateways, interconnections, and terminals. Even with these obvious advantages, the services are still expensive and closed.

Satellite Prediction. Several software packages allow to retrieve and interpret satellite path information from public repositories. As a reference example, GPredict is a public domain software with lot of operational capabilities based on the *prediction* of satellite paths: real time track characteristics, schedule table, footprint, communication establishment, control of antenna. Computation of positions and speed is based on keplerians elements of its orbit, these parameters being provided by public servers.

Miniature Systems. Now, unmanned aerial vehicles (UAVs), airships, balloons and small satellites are experimented for collecting and delivering data. In a common scenario, a CubeSat pico-satellite carries sensors for significant scientific research, and drifts along low attitude polar orbit resulting in easy monitoring the Earth's surface. The velocity of most CubeSat and the corresponding diameter of footprint are about 10 km/s and around 500 Km at attitude 100–200 km respectively. Initially, each CubeSat just performs individually, but then came projects for interconnections and to establish satellite constellations. This also aims to achieve better performance in severe conditions, e.g. magnetic storms, and natural catastrophes [9].

Radio Links. In previous years, the CubeSats have had two radios for different links VHF- 2 m band for up-link and UHF-70 cm band for down-link because of the limit of receiver load, power resource on board. Many digital modulations have been proposed to reduce the noise and to gain higher effective communication. CubeSat communication systems recently tends to perform on S-band with frequencies range from 2–4 GHz using smaller size of antennas in both air and ground segment.

2.2 Sensor Fields

The application deployments cover targets in wide distant area or behind obstacles (mountains, oceans).

A sensor field works like a hierarchical two-stage synchronous network. The first stage is for sampling physical processes, as example for environment monitoring. Then a local distributed decision algorithm take place using radio communications. Practical implementations use radio transceivers for communication standards such as IEEE 802.15, or Zigbee. Several frequencies are available that allow short (100 m) or medium range (10 km) communications between nodes. The first stage produces synthetic information to gateways that participate in the second stage. Several sinks can be embedded in a sensor field to feed visiting mobiles, and their activity can be coordinated. Gateways are expected to have energy support and hardware for receiving and transmitting to the satellite.

Field control implies serious observation of timings since communication can take place only when neighbours have schedule a rendez-vous. The normal status of a sensor is the sleep status awaiting for sensing and communication periods. A good image for this is the cycle for synchronous Time Division Multiple Access super-frames from IEEE 802.15.4.

2.3   Map Browsers and Satellite Tracks

Map browsers are now of common use, and it is an evidence that sensor distributions necessitate such tool to ease sensor field description for preliminary exploration. When interactions with mobile flights are considered, additional supports include mobile path description and radio link connectivities,

Maps are downloaded from servers (OpenStreetMap or GoogleMap) that supply tiles of raster georeferenced images. Map browsers are intuitive: the user moves the map with the mouse, by click-and-drag, and zoom level is changed by the mouse wheel. It is therefore possible to display different level of details. The maximum zoom level depends of the service resolution: OpenStreetMap offers 16 levels, while GoogleMap get down to 22 levels. The number of tiles t is a function of the zoom level z: $t = 2^{2z}$.

Map Projection and Quickmap Tool. The Mercator projection is used with the purpose to convert the round Earth with angular coordinates, longitude λ and latitude ϕ, into a flattened map with metric coordinates (x, y). Formula are given in [6] that allows to compute conversions, with applications to sensors and satellite representation on flat surfaces.

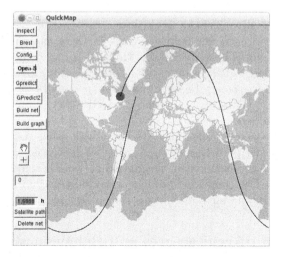

Fig. 2. QuickMap showing an LEO satellite path. This one-period satellite footprint was recorded during approximately 1 h and a half. The shift in horizontal position represents the earth rotation during the period.

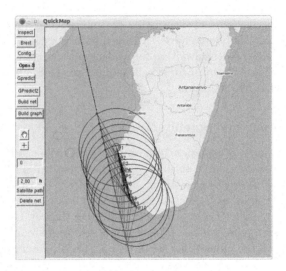

Fig. 3. The satellite path, and a sensor field. The green lines and the red lines represent respectively radio link establishment and deletion (Color figure online).

Quickmap, shown Figs. 2 and 3 was developped to support sensor layout, radio ranges computations, network topology presentation, as well as satellite tracks and further functionalities such as cellular system synthesis. Its main role is to facilitate sensor interactive location specification. Sensors positions are sent to a companion tool NetGen (step 2 from Fig. 1), that computes the network topology and feed back resulting drawings according to radio range.

Sensors and Satellite Tracks Relations. Geographical information, and time sequencement are the key points to consider these relations.

Practically, satellites have predictable track, that can be described by algorithms or by discrete sequences of events ((time, position) ...). To bind mobiles to sensor fields, it is necessary to share time references and geographic positions, For this reason, QuickMap was interfaced to GPredict, grouping mobile and sensor networks models.

Mobiles to fixed communications are computed using the mobile *discrete timed path*. For each point in the path, a set of reachable ground sensors is computed. For each new connexion, the ground network is completed by and air link, while the loss of a connexion produce a link destruction. A graphical view of a ground network and air radio links is shown Fig. 3.

3   Simulation : Architectures and Behaviours

3.1   Model for System Architecture and Distributed Behaviour

The final objective is defining distributed services, performances, costs and risks.

Several entities appeared at different level such as geographic distribution, interaction with physical processes, distributed behaviour, communications, mobility, and local sensing. The situation is similar to basic separation of concerns proposed and design methods for CAD tools, in particular disjoining organization (circuits) from behaviours (algorithms).

According to general principles in the domain of distributed algorithms [5], the design framework should propose supports for two orthogonal abstractions:

– **the system architecture** describe nodes and their communication links,
– **the behaviours** are local programs cooperaing by message exchanges

When applied to simulation, a general idea is to produce concurrent programs that separate almost completely the algorithmic and organization contributions, enabling to use and control huge number of nodes from distributed algorithms, whatever is the architecture topology.

Mobiles are considered as part of the system itself, having their own logic, as it is the case for satellites. They are moved by specific simulation threads, perhaps statically, perhaps dynamically. Clocking systems within WSN and time continuity of satellite moves help considerably to handle synchronization of ground and air activities.

3.2   Architecture Representations and Concurrent Programs

Network topology are expressed internally on a simple network model, as *named nodes* and *communication links*. Few attributes can basically be added, such as the physical location of a node, or the expected communication range.

A Network is therefore a large data structure grouping nodes and their associated links. Several separated graphs generally coexist in this structure, mobiles being nodes that follow static or dynamic trajectory under external control, eventually taken from the real world.

Given this central data structure produced by front end tools, translators allow to produce equivalent representation in terms of communicating processes either in Occam syntax and CUDA syntax. Occam uses micro threads and blocking channels ([8]). It is suitable for mono and multi-processors simulation of WSN (as shown in [4]), for distributed execution, and for execution on sensors, either in the form of virtual machines or native code. CUDA is a programming language associated with GPGPU. GPGPU are using the notion of *Kernels* executing in parallel a sequential procedure. Nodes behaviour can be represented inside these procedures, providing that the networks are acting synchronously [2].

While Occam represent communications by point to point blocking message sending, CUDA is operated by synchronized exchanges in GPGPU memory. These exchanges are executed by automaton based on a description of copy operations to and from neighbour buffers. The list of these operations is produced automatically by NetGen CUDA translator to represent and use the connectivity. The size of the communication program is adapted to the maximum fan-out inside the network (see [7] for technical details).

A nice effect of NetGen software approach is that the *node behaviours* can be specified, validated, and stored in the form of Occam or CUDA libraries. We have been able to simulate very large system concurrently, at the level of 1000 processes and more than 10000 processes respectively (see [2,4]).

Behaviours. Node behaviours group several activities for *sensing*, for making local decisions that will appear as a change in the node state, and for exchanging with neighbours. This also follow the more common way to describe distributed algorithms, and particularly synchronous, self timed behaviours [5]. In this model each node executes cycles for change of state (C_i), message out (M_i), and message in (N_i).

Our simulation program will execute similarly sensing and sleeping, communication phases, and making of decisions. There is only one stage for communications since they are executed by broadcast, each node sending at its turn.

Such behaviours are strongly bound to the synchronous model, the handling of communications on links being executed by procedure calls. The programming pattern is a loop grouping communications, buffer analysis and sense data analysis to obtain local change of state, then starting of sensing/sleeping activities.

Occam simulations do not explicitly make use of time, but are only sequenced by inter-process synchronizations. By removing the delays, this enable respect of causalities inside networks, and asynchronism between separated networks. Small set of nodes are running faster than large ones because the diameters and communication work load are fairly smaller. The simulation progresses are observed by sending node status to a trace process that is embedded automatically in the architecture. Graphic presentation can be observed by filtering the trace flow to produce annotations.

4   Mobile to WSN Collection Algorithms

Three algorithms for satellite and sensor fields cooperation have been experimented.

- **anti**-cipated. In the first case, the proximity of the arrival of a satellite is known. Therefore, nodes anticipate the interaction, preparing data to be sent. The nature of the computation can be static, or passed from the previous satellite visit.

  We will not detail this algorithm which is simply implemented from an downward and upward *breadth first search* algorithm started from the expected visited gateway.
- **trans**-action. In the second case, The satellite send a command at its first contact with the sensor field, which propagate the command and execute distributed computation dynamically. This time, the command must flow inside the sensor field with results sent back to the satellite. This takes at least two diameters period to execute, and the satellite speed becomes a larger constraint.

– **flow**. In the third case, the satellite send a command which is processed on the fly in the sensor network, with the results forwarded to an exit node where the satellite can receive it.

Algorithms have been initially coded into Occam. The mobile is not present, being represented by WSN gateways. Many synchronous loops have been executed for random networks of different size $10, 50, 100, 200, 500$. This number allows to compare the elapsed time in the network with the satellite visit delay. This comparison is critical for verification of system correctness.

As these algorithms are expressed following the synchronous model, the number of loops executed to achieve complete network computation is at least *diameter* steps. The following sections describes algorithms **trans** and **flow** in terms of Occam automaton.

4.1   Transaction

The Problem. Mobile arrives over a sensor field and obtains a connection with one gateway. It sends a command to the gateway asking the execution of a global computation implying a visit of every reachable node. After a moment, it is expected that a result message will be sent back by the gateway to the mobile. Each node contributes to the computation in a commutative way, as examples: computation of the field bounding box, global status, maximum.

Algorithm Informal Description. A breadth first search (BFS) is built dynamically using forward (downward a tree) and then backward (upward a tree) propagation to implement a global computation.

Downward Visit. During this stage, a root node sends a search message carrying the command. Each other node will receive the command at some step of the synchronous loop. At the first time and only at this time, one parent (and only one) is elected and informed. Each other neighbour will receive "search" at the next step. After diameter step, all the node in network have received and propagated successfully the search message. In addition, the relationship between nodes: parent index and list of children is established for next backward algorithm to propagate the result.

Upward Back Propagation. The upward operation starts in nodes after search propagation, with no parent notification from any outgoing link. At this time, a local result is produced and sent to the parent node. As soon as the parent node receive the partial results from children, it computes its local contribution and waits for next result It will send the its final contribution upward as it collected and computed all the results from children. By this way, after diameter step in computing, the Root will receive and update the result to make a global result before sending it to mobile.

Obviously, combining two algorithms, upward and downward, the total time Root node will receive the global computation result is at most $diameter \times 2 + 1$ steps. *As consequence, connection between mobile and gateway must be active over this delay.*

Protocol and Messages. Messages in the abstract algorithm are $null, request,$ $parent, result$. For Occam implementation, a protocol construct associates data to typed messages. With a customization for bounding box computation it comes: *BBoxRequest BBoxParent BBoxResult SendNull.*

State Definition. Nodes retain their local state in a set of variables: index or identities for possible parent and children, Boolean to recall previous visits, etc... The sensing status is figured by the geographic position in an initially unknown bounding box rectangle. In Occam, the state is retained in a set of variables for neighborhood description and automaton stages.

Downward Stage Automaton. The local automaton state is kept in a variable state, with possible values:

StateInit, StateSendRequest, StateReceiveParent, StateEnd, initially $state = StateInit$.

Upward Stage Automaton. the possible values are:

StateReceiveResult, StateSendResult, StateIdle, initially $state = StateReceive$ *Result*. Following the current state, the local automaton fill message buffers to the neighbours according to the computation status. In the initial state, nodes are waiting for *BBRequest*. Upon reception, output buffer are filled with BBParent, or BBRequest messages. To illustrate how the algorithm works in communication rounds, an example network with 50 nodes is presented Fig. 4 The root node R1 of this tree has id = 9. The red link is the downward part and the blue link is the upward part. The satellite sends the command signal at R1 and receive back the global result at R1 also.

Obviously, this algorithm proposed a better solution than the first **anti** algorithm in receiving command from a gateway, executing the command and sending back the result for satellite. However, it has the critical drawback: the race between forward-backward computing and movement of satellite.

4.2   Flow

The Problem. In the previous algorithm, a critical problem is time constraint. A better solution, is to use a traversal propagation in computation and transaction. The strategy is to merge two BFS to receive command and propagate result to a sink. As result, we have more time related to satellite crossing sensor field.

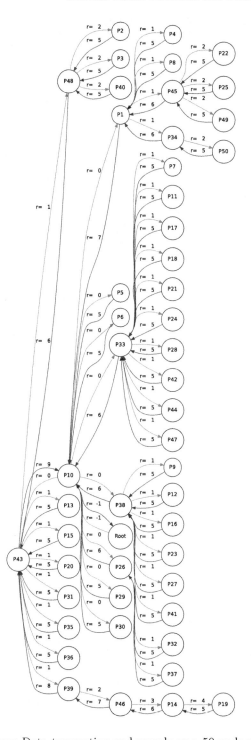

Fig. 4. tran: Data transaction and rounds on a 50 node network

Algorithm Informal Description. Thus two BFS trees are built inside the network. According to the BFS trees, the first Root is elected as the first gateway to receive the command signal of satellite when it passes over the node field. In contrast, the second Root is elected as the second gateway to send back the result of computation just before satellite leaves the node field. We also consider that we have already built both BFS similarly as for second algorithm before satellite enters the field. Therefore, the structures of protocol, messages are similar to **tran** algorithm. The maximum number of steps is *diameter* × 4.

State Definition. The state set is similar to **tran** with some changes. Instead of forwarding result to Root of first tree, we forward the partial result to Root of second tree. Therefore, we need a other collections for list of children and the index for parent node in second tree. In addition, the local automaton state is kept in a variable `state`, with possible values:

$StateWaitForSignal, StateRequestComputing, StateReceiveResult,$
$StateSendResult, StateIdle, initially state = StateInit.$

Message Production and Automaton Management for Downward and Upward Propagation. According to the state, the local automaton will fill message buffers to the neighbours according to the following patterns. In the initial state, nodes are waiting for *BBRequest*. Upon reception, output buffer are filled with BBRequest messages. In special case, the state change can be optimized when a node is a leaf for both tree. The leaf node can send back the result in next round by change from StateWaitForSignal to StateSendResult.

As a result, the third algorithm appears as a best solution for transaction protocol for satellite and sensor network. This algorithm is adaptable and reliable for a real deployment. In next section, we analysis these presented algorithms by testing the performance in execution time based on different scenarios. In addition, we also prove why the third algorithm is adaptable solution to deploy for transaction protocol.

4.3   Simulation Performance

Performance for distributed computations depend on the network topology and maximum number of channel in/out for nodes. These factors can be controlled inside NetGen and QuickMap by checking parameters such as Network zone definition, Node range and Number of node in network. Several scenarios were used in Occam simulation to obtain random distribution according to these parameters. The result were checked to be correct. As an assessment, (Fig. 5) displays performance with a fixed range of node of 800 meters with a varying number of random nodes covering an instrumented region.

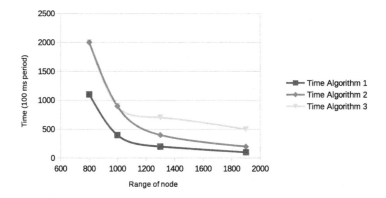

Fig. 5. Test results in time execution

5   Environment, Results and Perspectives

Simulations can be run as Occam or CUDA programs. CUDA code is distributed into few files related to the CPU and GPGPU behaviour description. One file is automatic generation from NetGen and includes the communication description table. The development platform is a variant of Smalltalk supporting dynamic generation and use of shared libraries. It enable to manipulate network structure stored on GPGPU to reflect changes such as mobile to field connectivity.

We have run simulation on Linux machines with NVIDIA graphic cards (GTX480, GFX680) providing respectively, 500 and 1500 processing elements. Both machines are equipped with Intel(R) i7 processors. As for Occam, random distributions over rectangular surface were generated to reflect deployments over wide area such as deserts, polar regions, countries or oceans.

For each set, the distributed simulation begins by producing a number of connected sensors, representing ground collective computations. Experimental behaviour is to elect leaders thus providing the number of isolated sub-networks. By observing the maximum fan-out, the simulation provide an additional information on the credibility of the network in terms of communication load (balance between sensor density and communication range) (Table 1).

Finally the execution time provides an idea on what can happen in long simulations (several orbits over the earth). Here the satellite path is controlled from Gpredict runs with a period of 1 s, during 15 min.

The outline of this simulation framework reveal the interest of two major components. **QuickMap** handles maps and navigation. It also allows specification of sensors or gateways over geographic presentations. **NetGen** handles network models, code generation, and control of simulation including the mobile moves. NetGen also send back annotations to QuickMap, and notably network and mobile communication graphic display.

System simulations are activated by exchange of messages between nodes and mobiles. The paper has provided a simple example with the presentation of a

Table 1. Results on GFX680

Number sensors	100	200	400	600
Real participants	91	186	331	434
Number of networks	17	19	69	123
Max fanout	6	7	6	7
Execution time (s)	52.8619	143.016	467.402	688.439

synchronous model Bounding Box computation. A number of distributed algorithm have been developed and tested over the Occam language giving confidence in the possibility to produce ambitious toolbox for Satellite to WSN problem. An evidence is the need to design protocols for uploading and downloading data appearing in gateways with systems that includes ground stations and final users. These simulation tools help considerably in understanding geographic deployment properties, time constraints, and even communication energy budgets.

The footprint of a LEO satellite is defined by the altitude of its orbit. Its velocity is in inverse proportion of the altitude. As a result, access windows time depends on both satellite's altitude and evaluation angle of ground station. Selecting proper radio frequency and protocols, high gain antenna and mechanic structure for satellite are also key factors in satellite communications. These factors are to be considered to design direct links between sensors fields and small satellites, with the capability of ground networks to elaborate synthetic data collectively. Ground speed is bound to the frequency of sensor networks, and number of hops, if any.

References

1. Celandroni, N., et al.: A survey of architectures and scenarios in satellite-based wireless sensor networks: system design aspects. Int. J. Satell. Commun. Netw. **31**, 1–38 (2013)
2. Dutta, H., Failler, T., Melot, N., Pottier, B., Stinckwich, S.: An execution flow for dynamic concurrent systems: simulation of WSN on a Smalltalk/CUDA environment. In: DYROS/SIMPAR 2010, Darmstadt (2010)
3. Heidt, H., Puig-Suari, J., Moore, A., Nakasuka, S., Twiggs, R.: CubeSat: a new generation of picosatellite for education and industry low-cost space experimentation. In: Proceedings of the AIAA/USU Conference on Small Satellites (2000)
4. Iqbal, A., Pottier, B.: Meta-simulation of large WSN on multi-core computers. In: DEVS 2010, in SpringSim SCS Conference, Orlando, USA (2010)
5. Lynch, N.: Distributed Algorithms. Morgan Kaufmann, San Mateo (1996)
6. Pridal, K.P.: http://www.klokan.cz/projects/gdal2tiles/
7. Thibault, F.: (UBO): NetGen : un générateur de code pour CUDA : principes, implémentation et performances. Technical report (2010)
8. Welch, P.H., Barnes, F.R.M.: Communicating mobile processes: introducing occam-pi. In: Abdallah, A.E., Jones, C.B., Sanders, J.W. (eds.) Communicating Sequential Processes. LNCS, vol. 3525, pp. 175–210. Springer, Heidelberg (2005)
9. Wright, D., Grego, L., Gronlund, L.: The Physics of Space Security. American Academy of Arts & Sciences, Cambridge (2005)

Mission Analysis for the Optimization of the GPS Coverage for an Earth Observation Satellite

Roberta Falone[1]([⊠]), Pietro Ivan Chichiarelli[2],
Ennio Gambi[1], and Susanna Spinsante[1]

[1] Dipartimento di Ingegneria Dell'Informazione,
Universitá Politecnica delle Marche, Ancona, Italy
r.falone@pm.univpm.it
[2] Satellite Operations, Telespazio S.p.A., Fucino Space Centre, Avezzano, Italy

Abstract. The goal of this work is to present an advanced study of the GPS coverage capability of a Low Earth Orbit (LEO) satellite. The motivation for this arises in the satellite control environment, during the operations preparation phase, by the need of predicting the time necessary to the on-board GPS receiver to fix the first orbital position, after the on-orbit injection at LEOP (Launch and Early Operational Phase). Once the first acquisition is performed, the satellite will be ready to enter the normal mode. The availability of a proper GPS coverage at LEO altitudes should be evaluated not only during the LEOP - clearly more critical - but also in the entire routine life of the satellite. Different approaches to optimize the GPS coverage, proposed by manufacturers and satellite operators, are considered, showing key strengths and weak points of the selected strategies. Operational and structural constraints must be taken into account when aiming at the proposal of different approaches to the GPS coverage issue, such as occultation of the on-board star trackers and the GPS antennas.

Keywords: LEO guidance · GPS · Coverage · Optimization · Satellite operations

1 Introduction: The Operational Context

Nowadays, GPS receivers are widely used for LEO spacecraft applications, in particular for precise orbit determination and geo-referencing the acquired images for Earth Observation purposes. Recent satellite LEO platforms rely so deeply on Global Navigation Satellite Systems (GNSS) that the on-board GPS receiver is switched on at launch phase and it is never switched off, apart from some contingency cases. GNSS integration has become such a well-established standard and a recognized support for LEO navigation that unavailability periods of the on-board receiver are always subject to an attempt of minimization, with any kind of proper strategy.

© Institute for Computer Sciences, Social Informatics and Telecommunications Engineering 2015
P. Pillai et al. (Eds.): WiSATS 2015, LNICST 154, pp. 199–211, 2015.
DOI: 10.1007/978-3-319-25479-1_15

Nevertheless, performances of spaceborn GPS receivers are deeply affected by the geometry of the subset of visible satellites in the GNSS constellations and by the variability of the ionospheric layer, in terms of residual altitude (above the LEO satellite) and electronic density. These factors influence the Time To the First Fix (TTFF) too, a critical parameter to be carefully considered after the first switch on of the on-board GPS receiver. TTFF includes a structural time needed to the receiver to perform a *cold start*, rather than a *warm start*. Furthermore, the switch-on operation can be performed only through a direct Telecommand (TC) sent by the Ground Control Centre (GCC) during a visibility opportunity between the satellite and the Main Control Station (MCS). In order to reduce the convergence time needed to the GPS receiver to supply a fixed position, we can operate on the deterministic constraint mentioned above, i.e. the available GPS constellation geometry. The first attempt is to provide the GPS receiver switch-on at the most favorable orbital sections, in terms of GPS coverage and geometric visibility from the MCS.

The proposed analysis and the relevant simulations have been performed by distinguishing between LEOP/contingency phase (two very different phases but with similar constraints from the point of view of the on-board GPS receiver), and the routine phase of the satellite mission life.

Simulations have been implemented and run by means of *FreeFlyer* (a.i. Solutions' software tool) dedicated to the space mission design, analysis and operations for the satellite environment.

The paper is organized as follows: Sect. 2 introduces basic knowledge about the satellite platform and the orbit features. Section 3 presents the simulations performed on GPS coverage, that are discussed, together with possible improvements, in Sect. 4. Finally, Sect. 5 draws the main conclusion of the work.

2 The Satellite Platform: Architecture and Orbit Features

2.1 The Spacecraft Architecture

In this section we will briefly show the general architecture of the considered platform. We will not refer to a particular spacecraft architecture; the proposed one can be considered as a features' "merge" of several platforms. The following features are quite general and common among Earth Observation satellites. This is the reason why we can consider the following platform model without loss of generality.

Let us clarify the reference frame used to set up the satellite axes and the location of the useful sensors, i.e. the GPS antennas and the Star Trackers.

As the Fig. 1 shows, the satellite body axes are arranged in the following manner:

- X axis: the satellite main body axis; the camera for images acquisition is arranged along this direction (shaded red cone);
- Y axis: solar array panels direction;
- Z axis: it completes the right-handed orthogonal reference frame.

Then the GPS antennas are arranged in the following disposition:

– GPS Antenna 1: along $-X$ direction, perpendicularly to solar array panels;
– GPS Antenna 2: along $-Z$ direction.

They feature a Field Of View (FOV) of 40 degrees half cone. See Fig. 1(a).

In the end, the Star Trackers are arranged with the following LOS (Line Of Sight) main directions:

– Optical Head 1: $-X/-Y/+Z$
– Optical Head 2: $-X/+Y/-Z$
– Optical Head 3: $-X/-Y/-Z$

They feature a FOV of 12.5 degrees half cone. See Fig. 1(b).

2.2   The Navigation Guidance Description

The navigation guidance is different for each phase of the satellite's mission life. For the purpose and the type of the satellite analyzed within this work, two main phases have been considered: LEOP and routine. They impose different control laws, attitude and management over the spacecraft. In the following subsections we will describe them in detail.

LEOP Guidance. Once the satellite has been released and injected into the orbit, a "stabilization" phase begins. For safety and charging batteries reasons, the satellite is oriented with its main body axis (X) aligned to the Sun vector, while it is spun around the same axis. Such a configuration is also known as *barbecue*, for clear analogies. In the Fig. 2, the spacecraft axes are shown, with the Sun Vector in yellow, and the useful sensors with their FOV. During the LEOP phase only the GPS Antenna 1 (yellow shaded cone) is active, while the telescope is not used, though displayed here (red shaded cone) for visualization scope of the attitude. Nominally, the satellite control remains in the barbecue phase until the stabilization of the spinning attitude, the first GPS fixing is reached and the star trackers are activated.

Routine Guidance. This phase is entered by the MCS with a proper TC, when the already mentioned conditions on attitude, GPS fixing, and star trackers have been gained. Once the normal-routine mode has been activated on-board, both GPS Antennas and Star Trackers are active and the guidance of the satellite changes from the barbecue to the Autonomous Navigation. This type of guidance is defined autonomous since it is self-managed by the satellite on-board software when some specific orbital events are triggered: the *eclipse entry*, the *eclipse exit*, and the *ascending node*. Clearly, the Autonomous Guidance is active only when no image acquisition activity is scheduled in the sun-lightened part of the orbit. The actual guidance of the spacecraft is featured by:

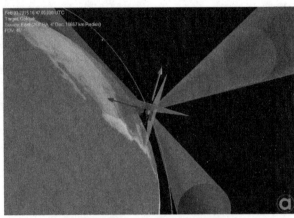

(a) GPS Antennas in shaded yellow

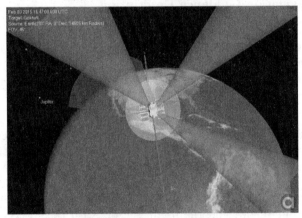

(b) Star Trackers in shaded blue

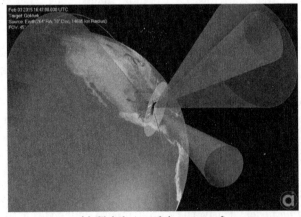

(c) Global view of the spacecraft

Fig. 1. Different perspective of the spacecraft's sensors (Color figure online)

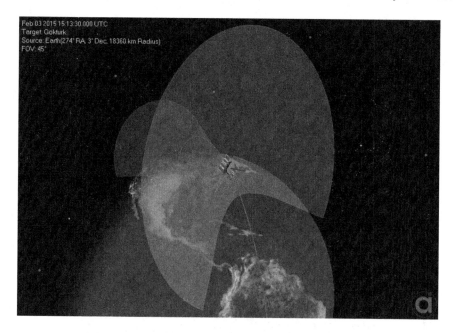

Feb 03 2015 15:13:30.000 UTC
Target: Gokturk
Source: Earth(274° RA, 3° Dec, 18360 km Radius)
FOV: 45°

Fig. 2. Spacecraft attitude during the barbecue phase

- **Earth Pointing:** In the Earth shadow orbit, the spacecraft has its main body axis (X) always oriented toward the center of the Earth, so the satellite constantly rotates along the orbital path. This guidance is triggered by the eclipse entry event (See Fig. 3(a)).
- **Sun Pointing:** In the sun-lightened part of the orbit, the spacecraft autonomously re-orient its main body axis toward the Sun, without spinning around the Sun Vector. Since the solar array panels of the modeled spacecraft are fixed (they are not able to rotate around their own axis - Y), the whole platform is forced to keep its panels perpendicular to the Sun Vector, in order to maximize the power generation by the panels. This guidance is triggered by the eclipse exit event (see Fig. 3(b)). Furthermore, the satellite exits the eclipse phase with the GPS Antenna 2 (on $-Z$) pointed toward the deep space. This orientation features the orbital path until the ascending node. Few minutes before reaching the ascending node, the satellite prepares for a yaw-flip maneuver: it is aimed to rotate the satellite around its Y axis (see Fig. 3(c)). In this new attitude, in fact, the GPS Antenna 2 is now pointed toward the North direction, allowing the platform to benefit from better GPS coverage, without the almost-total Earth occultation. This represents a first possible strategy to optimize the GPS coverage.

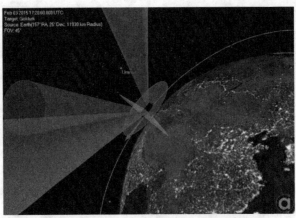

(a) Earth Pointing Attitude during eclipse phase

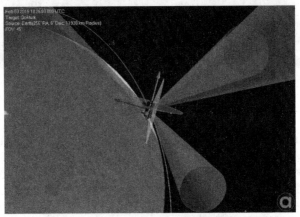

(b) Sun Pointing attitude after the Ascending Node

(c) Sun Pointing attitude before the Ascending Node

Fig. 3. Different attitudes during the normal autonomous guidance

3   GPS Coverage Simulations: Run and Results

In this section we will show the results of the orbit propagation for the LEO satellite and the GPS constellation. Synchronizing their epochs and running a propagation of one week, two different setups have been implemented relevant to the two mission scenarios, LEOP and Routine phase. The model used to propagate the orbit of the LEO satellite is quite standard for any mission at low altitudes: a fixed-step size (30 s) Runge-Kutta 8(9) integrator, adding a classical disturbing force model. We considered the effects of Earth, Moon and Sun, with Zonal and Tesseral Earth potential with solid tides, and the atmospheric drag, with the MSIS-2000 atmospheric density model. For the GPS constellation we used TLE files provided by NORAD.

3.1   LEOP Phase

We already mentioned the features of this mission phase: LEOP is characterized by a Sun Inertial Spinning attitude and the use of just one GPS Antenna to accomplish the first position fixing. A preliminary analysis has been performed on the coverage opportunity between the GPS Antenna 1 and the whole GPS constellation. In Fig. 4 some meaningful results can be gathered. Clearly, the polar diagram shows a statistical description of the number of the visible GPS satellites over 7 days of simulation.

The Fig. 4 is obtained by running the propagation in Sun Inertial Spinning attitude for 7 days and evaluating the number of visible GPS satellites at each propagation step (30 s). No data have been removed from the chart during the propagation, since the main scope of that was to evaluate which "orbital sections" benefit, on average, of the best GPS coverage. Clearly, the GPS geometry changes quite fast over the LEO satellite and it may be very different from one orbit to another.

Nevertheless, some peculiar orbital points feature quite a characteristic behavior of the middle-long term GPS coverage. With regard to this aspect, in order to aid in the readability of the results, in Fig. 4 two orbital sections are shown (eclipse and enlightened sections of the orbit) as function of the common reference theta angle (the argument of latitude, i.e. the angle between the position of the satellite on the orbit and the ascending node). Actually, for the aim of a better presentation of the chart, we have recorded the values of the argument of latitude plus 180 degrees. This was just to report the North Pole in the upper part of the chart. We can infer some important results from the polar diagram.

- The orbital points which statistically experience the best GPS coverage are reasonably comprised between intermediate latitudes over the ascending node, while gradually decreasing (the goodness of the coverage) when approaching to the poles: as is known, the GPS coverage at very high latitudes may experience a slight degradation. In general the entire section within this zone is a good candidate to schedule the switch-on of the on-board GPS receiver, considering that the Fucino Ground Station (involved in our simulation and

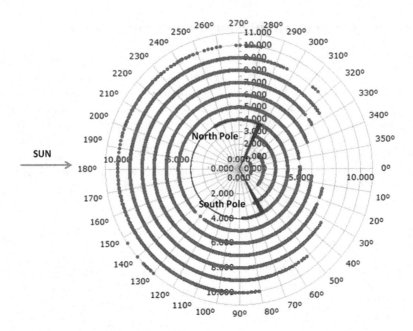

Fig. 4. Number of visible GPS satellites from Antenna 1. On theta angle the Argument of Latitude, on radial axis the number of visible GPS satellites. The solid line is added to show the relation between orbital section (high value in eclipse, zero value in the enlightened part of the orbit) and the statistical coverage.

deputy to the control operations) is also well located in terms of latitudes in order to command the satellite in direct visibility: this is necessary to achieve a real-time control and monitoring during the switching-on of the GPS receiver. Clearly, not every orbit may be as good as expected from this point of view, since the variability of the geometry of the constellation may cause a poor configuration in terms of number and quality of the geometry (DOP values). In Fig. 5 a demonstrative case is shown: this is an extract of the time diagram, where it has been possible to find out that only the shown opportunities give an unavailability to track in the enlightened arc. In this case, it may be necessary to wait for successive orbits in order to match the adequate conditions. Nevertheless, excluding the previous finite circumstance, in the enlightened orbital arc the coverage capability never decreases under the minimum required number of GPS satellites, i.e. four. This observation gives us good expectations about the possibility to even perform a cold start in this orbital arc.

- The polar diagram gives more information and confidence about the orbital arcs which must not be taken into account, regardless the presence of the Ground Station in visibility with the satellite. As expected, the sections to be avoided are those at high latitudes (in absolute value) and those in eclipse: within these sections, the variance of the number of available GPS satellites is significant in a limited timing window, which implies that the on-board GPS receiver may be unable to track the same 4 satellites for a sufficient time to achieve a FF.

– The central eclipse arc is another "hole" for GPS visibility. The diagram clearly shows that the maximum number of satellites in visibility may be 4 (or less): nevertheless, these satellites are visible with very low elevations with respect to the GPS Antenna 1 FOV, and the combination of them changes quite rapidly. This prevent the on-board receiver to have time enough to receive the navigation signal from the same 4 satellites, i.e. enough time to accomplish a trilateration.

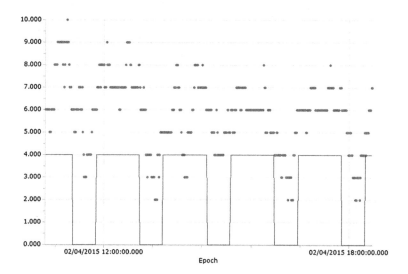

Fig. 5. Time Diagram. The number of visible GPS satellites is shown as a function of the time. This is the only opportunity of "unavailability" of the GPS system encountered during 7 days simulation, limited to the enlightened arc of the orbit.

There are other important observations which cannot be inferred by a diagram like the previous one. Nevertheless, they are equally important in the analysis we are proposing (statistical description parameters).

3.2   Routine Phase

We have already described the routine phase of the satellite, in terms of attitude and navigation. In the diagram of Fig. 6, the results are reported on the helpful polar plot.

Clearly, the switch-on and the actual utilization of the second GPS Antenna, introduces significant improvements. The minimum number of visible GPS satellites never decreases under the value of 8 satellites; the maximum reaches even

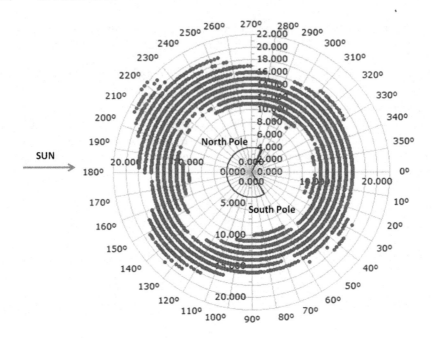

Fig. 6. Polar diagram. Cardinality of the visible GPS satellites observed during the routine phase.

22 satellites in particular favorable conditions of visibility and orbital phase of the LEO satellite. Once the satellite has entered the nominal navigation phase, only severe conditions affecting the on-board GPS receiver (internal failures, platform issues, channel propagation errors or interference) could compromise the tracking and consequently the fixing, since the opportunities to choose the better subset of incoming ranging signals become considerable.

Nevertheless, from the point of view of the Satellite Operations, a yaw-flip maneuver at the ascending node means an increasing complexity to the management of the satellite, especially when acquisition images' activities are scheduled in the arc close to the ascending node itself. In fact when an image is planned to be acquired at latitudes close to zero, the autonomous guidance has to be disabled at least some (about 5) minutes before the ascending node crossing, otherwise the satellite will autonomously start the preparation of the maneuver. Once the preparation has started, no image or other activity's request can be submitted on board. Operationally speaking, the yaw-flip maneuver is very useful to improve the GPS coverage (keeping at the same time at least two star tracker's optical heads not obscured), but it is tedious in some way for the satellite's activities planning.

4   Variations over the Nominal Attitude: Room for Improvement

For the reasons mentioned above, the goal of this work is to justify (or refuse) the actual benefit coming from the yaw-flip maneuver. If another "stationary" attitude could replace the solution shown before, we might encounter a little underperformance of the GPS coverage, with the benefit of completely avoiding the maneuver at the ascending node. The simplification of the operational life of the satellite is the main purpose of the proposed solution.

Nevertheless, when introducing a fixed attitude which keeps the satellite pointed toward the Sun, the constraint requested for the Star Tracker (two optical heads out of three must be not occulted) fails, as the Fig. 7 shows. The solid line with radial value of 5 represents the orbital arcs during which two or even three optical heads are occulted. The only exploitable degree of freedom is the rotation around the X axis: in fact, a complete rotation has been simulated and tested in terms of GPS coverage. No significant coverage variation has been found, only the introduction of issues related to the occultation of the Star Tracker. The Fig. 7 clearly shows the trend of the occultation as a function of the rotation around the X axis: as the satellite attitude varies from the reference attitude, some orbital arcs are featured by the occultation of the Star Tracker. It is interesting to note that the trend is clearly linear as a function of the rotation: once the attitude is established, the orbital arcs featured by occultation remain constant in the total sum, but they change in the "distribution" along the enlightened orbital section. The geometrical arrangement of the optical heads plays a strategic role to these results.

From the operational point of view, the possibility to forgo one occulted optical head's support, in favor of a good GPS coverage together with a simpler autonomous guidance management, seemed to be a good trade-off. This strategy is a potential solution considering that during the autonomous guidance the acquisition of the images is not allowed, hence a highly precise geo-referencing of the images (given by the attitude information in addition to the navigation information) is not strictly needed. Based on these considerations, other simulations have been proposed and performed assuming the new following conditions:

- Research of X axis rotated attitude which minimize the occurrences (duration) of one or two occulted optical heads. No instance of total occultation is allowed;
- Evaluation of GPS coverage is clearly taken into account.

In the same Fig. 7 the results of the simulation are shown. The condition of three optical heads occulted is never verified, whatever the attitude is rotated. This means that at least one optical head is available to provide information about the spacecraft attitude. If we accept this type of availability of the attitude sensor, only the optimization of the GPS coverage remains to be considered in the selection of the best attitude for the autonomous guidance. It is easily remarkable that not every rotation is a good solution, for a simple reason: the

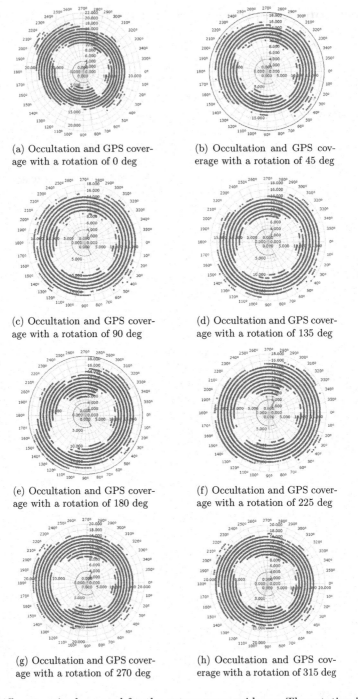

(a) Occultation and GPS coverage with a rotation of 0 deg

(b) Occultation and GPS coverage with a rotation of 45 deg

(c) Occultation and GPS coverage with a rotation of 90 deg

(d) Occultation and GPS coverage with a rotation of 135 deg

(e) Occultation and GPS coverage with a rotation of 180 deg

(f) Occultation and GPS coverage with a rotation of 225 deg

(g) Occultation and GPS coverage with a rotation of 270 deg

(h) Occultation and GPS coverage with a rotation of 315 deg

Fig. 7. Different attitudes tested for the autonomous guidance. The rotation is stated with respect to the reference attitude coincident with the Sun Pointing Reference Frame with the $-Z$ axis toward the North direction.

symmetry that features the arrangement of the optical heads does not affect the GPS antennas, too. We identified in the configuration shown in Fig. 7(c) and (g) or (h) the best trade-off taking into account all the considerations presented along the dissertation.

5   Conclusions and Future Works

This work presents a mission analysis oriented to the study of the GPS coverage by the on-board GPS Receiver on a LEO spacecraft. The main issues, specific for an Earth Observation mission, have been explained and a further approach to the yaw-flip maneuver has been proposed, supported by the simulation results. These ones allow us to accept the validity of an alternative solution to the problem of the GPS coverage, simplifying, at the same time, the operational management of the spacecraft.

The study will be successively deepened by simulating other features of the on-board GPS receiver and trying to build a cost-function shaped in order to optimize the DOP values along the time (rather than the mean value of the GPS visible satellites), considering a model of the upper atmosphere layer affecting the GPS signal, a statistical description of the phenomena and the possibility of a slight de-pointing of the spacecraft with respect to the Sun Vector during the enlightened orbital arc.

References

1. Hauschild, A., Markgraf, M., Montenbruck, O.: GPS receiver performance on board a LEO satellite. Inside GNSS
2. Liu, J., Gu, D., Ju, B., Yao, J., Duan, X., Yi, D.: Basic performance of BeiDou-2 navigation satellite system used in LEO satellites precise orbit determination. Chin. J. Aeronaut. **27**(5), 1251–1258 (2014). ISSN 1000–9361
3. Wermuth, M., Hauschild, A., Montenbruck, O., Kahle, R.: TerraSAR-X precise orbit determination with real-time GPS ephemerides. Adv. Space Res. **50**(5), 549–559 (2012). ISSN 0273–1177
4. Misra, P., Enge, P.: Global Positioning System: Signals, Measurements, and Performance

WiSATS Session 2

Proposal of Time Domain Channel Estimation Method for MIMO-OFDM Systems

Tanairat Mata[1(✉)], Pisit Boonsrimuang[2], Kazuo Mori[1], and Hideo Kobayashi[1]

[1] Graduate School of Engineering, Mie University, Tsu, 514-8507, Japan
`tanairat@com.elec.mie-u.ac.jp`,
`{kmori,koba}@elec.mie-u.ac.jp`
[2] Faculty of Engineering, King Mongkut's Institute of Technology Ladkrabang,
Bangkok 10520, Thailand
`kbpisit@kmitl.ac.th`

Abstract. This paper proposes a time domain channel estimation (TD-CE) method for Multi Input Multi Output-Orthogonal Frequency Division Multiplexing (MIMO-OFDM) systems. The feature of proposed TD-CE method is to estimate channel frequency responses for all links between transmit and receive antennas in MIMO-OFDM systems by using one scattered pilot preamble symbol in the time domain. The proposed method can achieve higher channel estimation accuracy even when the number of transmission antennas is larger and the transmission signal is sampled by a non-Nyquist rate in which the number of IFFT points is different from the number of data subcarriers due to the insertion of null subcarriers (zero padding) at the both ends of data subcarriers. This paper shows various computer simulation results in the time-varying fading channels to demonstrate the effectiveness of the proposed channel estimation method as comparing with the conventional channel estimation methods.

Keywords: Time domain channel estimation (TD-CE) · MIMO · Channel impulse response · Channel frequency response · Non-Nyquist rate

1 Introduction

Orthogonal Frequency Division Multiplexing (OFDM) has been widely adopted in the current wireless communications systems as the standard transmission technique such as the Digital Audio and Video Broadcasting (DAB [1] and DVB [2]), Broadband Wireless Access (IEEE 802.16) [3] and Wireless Local Area Network (WLAN) [4] because of its efficient usage of frequency bandwidth and robustness to the multipath fading. Furthermore, the Multi Input Multi Output (MIMO) technique is employed with OFDM technique (MIMO-OFDM) [5] to achieve higher data transmission rate and higher signal quality in various wireless communication systems.

In MIMO-OFDM systems, the receiver needs to estimate the channel frequency responses (CFR) precisely for all links between transmit and receive antennas which are

© Institute for Computer Sciences, Social Informatics and Telecommunications Engineering 2015
P. Pillai et al. (Eds.): WiSATS 2015, LNICST 154, pp. 215–228, 2015.
DOI: 10.1007/978-3-319-25479-1_16

used in the demodulation of information data with the MIMO detection. From this fact, the channel estimation (CE) method which can achieve higher estimation accuracy is essential in MIMO-OFDM systems to achieve higher data transmission rate with keeping higher signal quality. Up to today, many CE methods have been proposed for MIMO-OFDM systems including the Discrete Fourier Transform interpolation-channel estimation (DFTI-CE) method [6] and the Maximum Likelihood-channel estimation (ML-CE) method [7]. The DFTI-CE method can achieve higher estimation accuracy only when the sampling rate of transmission signal is the Nyquist rate *i.e.* the number of IFFT points (N) is equal to the number of data subcarriers (M). However the estimation accuracy of DFTI-CE method would be degraded relatively in the practical MIMO-OFDM systems in which the sampling rate is taken by the non-Nyquist rate. In the practical OFDM system, the null subcarriers (zero padding) are usually inserted at the both ends of M data subcarriers in every transmission OFDM symbol to reject the aliasing occurring at the output of digital to analogue (D/A) converter. From this fact, the channel estimation accuracy of using this method would be degraded especially at around the both ends of data subcarriers which are the borders between the data and null subcarriers due to the mismatching of sampling rate between the transmission signal with N samples and received signal with M samples. This phenomenon is called the border effect [8]. To solve the above problem, the ML-CE method was proposed for the MIMO-OFDM systems which can achieve higher estimation accuracy than the conventional DFTI-CE method [7]. However its estimation accuracy at the both ends of data subcarriers would be degraded when increasing the number of transmit antennas.

To solve the above problems on the conventional CE methods, this paper proposes a novel channel estimation method which can achieve higher estimation accuracy even when the non-Nyquist rate and increasing the number of transmit antennas in the MIMO-OFDM systems. The salient feature of proposed TD-CE method is to employ the superimposed time domain received scattered pilot preamble (SPP) symbol sent from all the transmit antennas in which pilot subcarriers sent from each transmit antenna are assigned cyclically including the both ends of transmission frequency band for data subcarriers. In the proposed method, the channel frequency responses for all links between transmit and receive antennas are estimated over the frequency band corresponding from the first to the end pilot subcarriers of which frequency band is also used in the transmission of data information. This paper is organized as follows. Section 2 presents the conventional CE methods for MIMO-OFDM systems and their problems on the estimation accuracy at the non-Nyquist rate and when increasing the number of transmit antennas. Section 3 proposes the TD-CE method for MIMO-OFDM systems which can solve the problems on the conventional methods. Section 4 presents various computer simulation results to verify the effectiveness of proposed method and Sect. 5 draws some conclusions.

2   Conventional Channel Estimation Method

2.1   System Model for MIMO-OFDM Systems

For simplicity, we consider the MIMO-OFDM systems employing the Space Division Multiple Access (SDMA) technique with N_T transmit and N_R receive antennas which enables the transmission of separate information data from each transmit antenna. Figure 1 shows a block diagram of MIMO-OFDM system with the SDMA. At the transmitter, the information data encoded by FEC is modulated and separated by M subcarriers into each transmit antenna. The zero padding are added at the both ends of M data subcarriers then converted into the time domain signal by N points IFFT at each transmit antenna. The time domain signal b_k^i at the k-th time sample ($0 \leq k \leq N\text{-}1$) transmitted from the T_i transmit antenna ($1 \leq i \leq N_T$) is sent to the receive antenna R_j ($1 \leq j \leq N_R$) after adding the guard interval (GI) to avoid the inter symbol interference (ISI). At the receiver, the data information transmitted from N_T transmit antennas are demodulated by using the MIMO detection with the estimated channel frequency response matrix consisting of all links between transmit and receive antennas.

In the MIMO-OFDM systems, ($N_T \times N_R$) channels between transmit and receive antennas are required to estimate by using one common SPP symbol. Figure 2 shows an example of pilot subcarriers assignment in the SPP symbol sent from the i-th transmit antenna. In the Fig. 2, N is the number of IFFT points, M is the number of data subcarriers, ($N\text{-}M$) represents the total number of null subcarriers (zero padding) and the half of the null subcarriers ($N\text{-}M$)/2 are inserted at both ends of M data subcarrier. Each transmit antenna can use P ($=M/K_f$) pilot subcarriers within M subcarrier which are inserted cyclically with an interval of K_f subcarrier. J represents the first data subcarrier number ($J=(N\text{-}M)/2$) within N subcarriers including the zero padding. The pilot subcarriers for the i-th transmit antenna are assigned from $J+(i\text{-}1)$ to $J+(i\text{-}1)+sK_f$ ($0 \leq s \leq P\text{-}1$) with the interval of K_f subcarriers. Here it should be noted that the pilot subcarriers sent from each transmit antenna are assigned cyclically so as to avoid a collision among the pilot subcarriers sent from all transmit antennas. The time domain SPP symbol b_k^i transmitted from the i-th transmit antenna is given by,

$$b_k^i = \frac{1}{N} \cdot \sum_{s=0}^{P-1} B_s^i \cdot e^{j\frac{2\pi k}{N}(J+(i-1)+sK_f)}, \quad (0 \leq k \leq N-1) \tag{1}$$

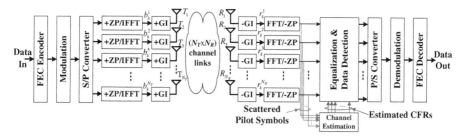

Fig. 1. Overview of MIMO-OFDM system.

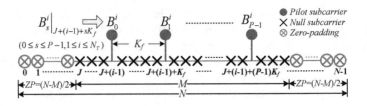

Fig. 2. Assignment of pilot subcarriers in SPP symbol.

Where B_s^i is the s-th pilot data for the i-th transmit antenna of which subcarrier number is $J+(i-1)+sK_f$ within N subcarriers. The channel impulse response (CIR) of multipath fading between the i-th transmit and the j-th receive antennas can be expressed by,

$$h_k^{i,j} = \sum_{q(i,j)=0}^{N_p-1} \rho_{q(i,j)}^{i,j} \cdot \delta(k - q(i,j)) \tag{2}$$

where N_P is the number of delay paths and $\rho_{q(i,j)}^{i,j}$ represents the complex amplitude of CIR for the $q(i,j)$-th delay path occurred in the channel between the i-th transmit and j-th receive antennas. At the j-th receive antenna, the superimposed received SPP signal r_k^j sent from all transmit antennas after removing the GI can be given by,

$$r_k^j = \sum_{i=1}^{N_T} \left\{ b_k^i \otimes h_k^{i,j} \right\} + z_k^j = \sum_{i=1}^{N_T} \sum_{q(i,j)=0}^{N_p-1} \left\{ \rho_{q(i,j)}^{i,j} \cdot b_{k-q(i,j)}^i \right\} + z_k^j, \quad (1 \le j \le N_R) \tag{3}$$

where z_k^j is the additive white Gaussian noise (AWGN) at the k-th time sample of j-th receive antenna and $\otimes$ represents the convolution. The CFR at the pilot subcarriers between the i-th transmit and j-th receive antennas can be estimated independently by using the superimposed frequency domain received SPP symbol which is converted from the time domain signal in (3), because the pilot subcarriers sent from all transmit antennas are assigned without a collision as described above. By performing FFT to (3), the received frequency domain SPP symbol at the $J+(i-1)+sK_f$-th pilot subcarrier is given by,

$$R_{J+(i-1)+sK_f}^j = B_s^i \sum_{q(i,j)=0}^{N_p-1} \rho_{q(i,j)}^{i,j} e^{-j\frac{2\pi q(i,j)}{N}(J+(i-1)+sK_f)} + Z_{J+(i-1)+sK_f}^j = B_s^i H_{J+(i-1)+sK_f}^{i,j} + Z_{J+(i-1)+sK_f}^j \tag{4}$$

where $H_{J+(i-1)+sK_f}^{i,j}$ and $Z_{J+(i-1)+sK_f}^j$ are the CFR at the $J+(i-1)+sK_f$-th pilot subcarrier between the i-th transmit and j-th receive antennas and AWGN at the j-th receive antenna both in the frequency domain. From (4), the CFR $H_{J+(i-1)+sK_f}^{i,j}$ can be estimated by using the pilot data B_s^i known at the receiver which is given by,

$$\hat{H}^{i,j}_{J+(i-1)+sK_f} = \frac{R^j_{J+(i-1)+sK_f}}{B^i_s} = H^{i,j}_{J+(i-1)+sK_f} + \frac{Z^j_{J+(i-1)+sK_f}}{B^i_s} \tag{5}$$

The next section presents the conventional DFTI-CE and ML-CE methods by using the estimated CFR at the pilot subcarriers given in (5).

2.2 Conventional DFTI-CE Method

The DFT-CE method has been proposed for MIMO-OFDM systems [6] which can achieve higher channel estimation accuracy by reducing the noise component in the time domain CIR. By performing P points IDFT to (5), the CIR $\hat{g}^{i,j}_u$ is given by,

$$\hat{g}^{i,j}_u = \frac{1}{P} \cdot \sum_{s=0}^{P-1} \hat{H}^{i,j}_{J+(i-1)+sK_f} \cdot e^{j\frac{2\pi su}{P}} = g^{i,j}_u + w^{i,j}_u, \quad (0 \le u \le P-1) \tag{6}$$

where $g^{i,j}_u$ in (6) is the ideal CIR for $H^{i,j}_{J+(i-1)+sK_f}$ in (5) and $w^{i,j}_u$ is the noise component both at the u-th time sampling. By using (4), the ideal CIR $g^{i,j}_u$ is given by,

$$g^{i,j}_u = \frac{1}{P} \cdot \sum_{q(i,j)=0}^{N_P-1} \rho^{i,j}_{q(i,j)} \cdot e^{-j\frac{2\pi q(i,j)}{N}(J+i-1)} \cdot \sum_{s=0}^{P-1} e^{-j\frac{2\pi s}{P}(\frac{PK_f}{N}q(i,j)-u)} \tag{7}$$

When N is equal to M $(=PK_f)$ at the Nyquist rate, the last term in (7) is given by [9],

$$\sum_{s=0}^{P-1} e^{-j\frac{2\pi s}{P}(\frac{PK_f}{N}q(i,j)-u)} = \sum_{s=0}^{P-1} e^{-j\frac{2\pi s}{P}(q(i,j)-u)} = \begin{cases} P, & q(i,j) = u \\ 0, & q(i,j) \ne u \end{cases} \tag{8}$$

By inserting (8) into (7), the CIR $g^{i,j}_u$ exists only from $u=0$ to N_P-1 which is the same as (2). From this fact, the noise components from $u=N_P$ to $P-1$ can be removed by adding zeros and then the CFR over M data subcarriers can be estimated precisely by performing M points DFT to (6). However when N is larger than M $(N>M)$ due to the insertion of zero padding at the both ends of M data subcarriers which corresponds to the non-Nyquist rate, the last term of (7) can be given by,

$$\sum_{s=0}^{P-1} e^{-j\frac{2\pi s}{P}(\frac{PK_f}{N}q(i,j)-u)} = \frac{1 - e^{-j2\pi(\frac{PK_f}{N}q(i,j)-u)}}{1 - e^{-j\frac{2\pi}{P}(\frac{PK_f}{N}q(i,j)-u)}} \tag{9}$$

By inserting (9) into (7), the CIR $g^{i,j}_u$ exists over P time sampling points from $u=0$ to $P-1$ which is completely different from the ideal CIR given in (2). From this fact, the channel estimation accuracy of using the DFTI-CE method at the non-Nyquist rate would be degraded relatively especially at the borders between the zero padding and data subcarriers.

2.3 Conventional ML-CE Method

To improve the border effect in the DFTI-CE method at the non-Nyquist rate, the ML-CE method was proposed [7]. In the ML-CE method, the CIR $\hat{\rho}_{q(i,j)}^{i,j}$ can be estimated by the following Maximum Likelihood (ML) equation [10].

$$L_{ML}\left\langle \hat{\rho}_{q(i,j)}^{i,j} \right\rangle = \underset{\hat{\rho}_{q(i,j)}^{i,j}}{\arg\min}\left[\left| \sum_{s=0}^{P-1} \left| \sum_{q(i,j)=0}^{Ng-1} \rho_{q(i,j)}^{i,j} \cdot e^{-j\frac{2\pi q(i,j)}{N}(J+(i-1)+sK_f)} - \hat{H}_{J+(i-1)+sK_f}^{i,j} \right| \right|^2 \right] \quad (10)$$

The ML equation given in (10) can be expressed by the following simultaneous equations.

$$\left\| \hat{\rho}_{q(i,j)}^{i,j} \right\|_{Ng\times1} = \dagger \left\| D_{q(i,j),J+(i-1)+sK_f}^{i,j} \right\|_{p\times Ng} \cdot \left\| \hat{H}_{J+(i-1)+sK_f}^{i,j} \right\|_{p\times1} \quad (11)$$

where $\dagger$ denotes the Moore-Penrose pseudo inverse matrix, $\left\| \hat{\rho}_{q(i,j)}^{i,j} \right\|$ is the matrix of the complex amplitude of CIR with the matrix size $[Ng\times1]$, and $\left\| D_{q(i,j),J+(i-1)+sK_f}^{i,j} \right\|$ with the matrix size $[P\times Ng]$ can be given by,

$$D_{q(i,j),J+(i-1)+sK_f}^{i,j} = e^{-j\frac{2\pi q(i,j)}{N}(J+(i-1)+sK_f)}, \quad (0 \le q(i,j) \le Ng-1) \quad (12)$$

Since the number of actual delay paths N_P is unknown at the receiver, the length of GI (Ng) is used in the estimation of CIR in (10). After performing DFT to (11), the CFR can be estimated over M data subcarriers. Although the ML-CE method can improve the border effect and achieve higher channel estimation accuracy, its estimation accuracy at the both ends of data subcarriers would be degraded when increasing the number of transmit antennas [7]. When increasing the number of transmit antennas, the interval of pilot subcarriers K_f becomes larger and the number of subcarriers which are not covered by pilot subcarriers at the both ends of M data subcarriers are increased as shown in Fig. 2. This is the reason that the estimation accuracy of ML-CE method obtained by (10) would be degraded when increasing the number of antennas.

3 Proposal of Time Domain Channel Estimation Method

To solve the above problems having the conventional DFTI-CE and ML-CE methods, this section proposes a time domain-channel estimation (TD-CE) method by using one SPP symbol which can improve the border effect even when the non-Nyquist rate and increasing the number of transmit antennas.

3.1   Proposed Pilot and Data Subcarriers Assignment Method

Figure 3(a) shows the proposed pilot subcarriers assignment for the SSP symbol sent from the i-th transmit antenna. In the Fig. 3, the pilot subcarriers are inserted with the interval of K_f subcarriers starting from the $J+(i-1)$-th to the $J+(i-1)+(P-1)K_f$-th subcarriers which covers the bandwidth of $M_p=M-N_T+1$. Here it should be noted that the frequency band covered by both ends of pilot subcarriers for the i-th transmit antenna is different from other transmit antennas and the data subcarriers are assigned within the same bandwidth of M_p subcarriers between the first to the end pilot subcarriers as shown in Fig. 3(b). In the proposed pilot assignment method, the active number of data subcarriers becomes $M_p=M-N_T+1$ which is smaller number than M for the conventional method as shown in Fig. 2. However, it can be expected that channel estimation accuracy at around the borders between the null subcarriers and data subcarriers could be better than the conventional method because the pilot subcarriers are inserted at the both ends of data subcarriers.

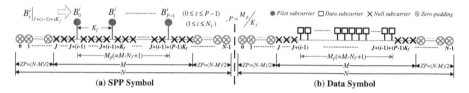

(a) SPP Symbol   (b) Data Symbol

Fig. 3.   Proposed pilot and data subcarriers assignment for i-th transmit antenna.

3.2   Proposed Time Domain Channel Estimation (TD-CE) Method

This section proposes the TD-CE method by using the proposed pilot and data assignment method as shown in Fig. 3 which can achieve higher estimation accuracy at around the borders between the data and zero padding subcarriers even when the Non-Nyquist rate and when increasing the number of transmit antennas. Accordingly, it can be expected that the proposed TD-CE method can achieve better BER performance than those for the conventional DFTI-CE and ML-CE methods.

The superimposed received SPP symbol r_k^j at the j-th receive antenna can be given by the following equation in which the received SPP symbol sent from each transmit antenna to the j-th receive antenna is expressed separately.

$$r_k^j = \underbrace{\sum_{q(1,j)=0}^{N_g-1} \rho_{q(1,j)}^{1,j} \cdot b_{k-q(1,j)}^i + \cdots}_{\text{from } T_1 \text{ to } R_j} + \underbrace{\sum_{q(N_T,j)=0}^{N_g-1} \rho_{q(N_T,j)}^{N_T,j} \cdot b_{k-q(N_T,j)}^{N_T}}_{\text{from } T_{N_T} \text{ to } R_j} + z_k^j \tag{13}$$

The length of GI N_g in (13) is usually taken by longer than the N_P so as to avoid the Inter-symbol interference (ISI). By using the matrix multiplication, (13) can be rewritten by,

$$r = \underbrace{\begin{bmatrix} \ddot{?}_0^1 & \ddot{?}_{-1}^1 & \cdots & \ddot{?}_{1-N\geq}^1 \\ \ddot{?}_1^1 & \ddot{?}_0^1 & \cdots & \ddot{?}_{2-N\geq}^1 \\ \vdots & \vdots & \ddots & \vdots \\ \ddot{?}_N^1 & \ddot{?}_{N-?}^1 & \cdots & \ddot{?}_{N-N\geq}^1 \end{bmatrix}}_{b_1} \cdot \underbrace{\begin{bmatrix} \ell_0^{?,} \\ \ell_1^{?,} \\ \vdots \\ \ell_{N\geq-1}^{?,} \end{bmatrix}}_{c_{1,j}} + \cdots + \underbrace{\begin{bmatrix} \ddot{?}_0^{N_T} & \ddot{?}_{-1}^{N_T} & \cdots & \ddot{?}_{1-Ng}^{N_T} \\ \ddot{?}_1^{N_T} & \ddot{?}_0^{N_T} & \cdots & \ddot{?}_{2-Ng}^{N_T} \\ \vdots & \vdots & \ddots & \vdots \\ \ddot{?}_N^{N_T} & \ddot{?}_{N-1}^{N_T} & \cdots & \ddot{?}_{N-Ng}^{N_T} \end{bmatrix}}_{b_N} \cdot \underbrace{\begin{bmatrix} \rho_0^{N_{T'}} \\ \rho_1^{N_{T'}} \\ \vdots \\ \rho_{N\geq-1}^{N_{T'}} \end{bmatrix}}_{c_{N'}} + z \tag{14}$$

where $\mathbf{r}_j$ is the matrix of received signal r_k^j with the matrix size $[N\times1]$, $\mathbf{b}_i$ and $\mathbf{c}_{i,j}$ are the matrix of time domain SPP symbol b_k^i with the matrix size $[N\times Ng]$ and the matrix of complex amplitude of CIR $h_k^{i,j}$ with the matrix size $[Ng\times1]$ and $\mathbf{z}_j$ is the matrix of AWGN z_k^j with the matrix size $[N\times1]$. The matrix operation of (14) at the j-th receive antenna is rewritten by,

$$\mathbf{r}_j = \underbrace{\mathbf{b}_1\cdot\mathbf{c}_{1,j}}_{from\,T_1\,to\,R_j} + \cdots \underbrace{\mathbf{b}_{N_T}\cdot\mathbf{c}_{N_Tj}}_{from\,T_{N_T}\,to\,R_j} + \mathbf{z}_j = \underbrace{[\mathbf{b}_1 \cdots \mathbf{b}_{N_T}]}_{\mathbf{B}} \cdot \underbrace{[\mathbf{c}_{1,j} \cdots \mathbf{c}_{N_Tj}]^T}_{\mathbf{G}_j} + \mathbf{z}_j = \mathbf{B}\cdot\mathbf{G}_j + \mathbf{z}_j \tag{15}$$

where $[\ ]^T$ is the transpose matrix operation. $\mathbf{B}$ is the matrix of time domain SPP symbol $\mathbf{b}_i$ sent from all transmit antennas with the matrix size $[N_R\times N_T Ng]$ and $\mathbf{G}_j$ is the matrix of CIR $\mathbf{c}_{i,j}$ for all links from all transmit antennas to the j-th receive antenna with the matrix size $[N_T Ng\times1]$. From (15), the CIRs in matrix $\mathbf{G}_j$ can be estimated by using the Moore-Penrose pseudo inverse $\mathbf{B}^\dagger$ which can be given by,

$$\hat{\mathbf{G}}_j = \mathbf{B}^\dagger\cdot\mathbf{r}_j = \mathbf{G}_j + \mathbf{B}^\dagger\cdot\mathbf{z}_j \tag{16}$$

All CIRs can be estimated together by using (16) in the time domain at the j-th receive antenna. Since all the time domain SPP symbols sent from all transmit antennas are known at the receiver, $\mathbf{B}^\dagger$ can be calculated in advance at the receiver. From this fact, the estimation of CIRs for all links can be estimated by simple matrix multiplication as given in (16) which leads the considerable reduction of computational complexity in the proposed TD-CE method. In the time varying fading channels, the SPP symbols are inserted with the interval of K_t symbols in one frame and the CIRs over one frame can be estimated by applying the cubic interpolation method to the CIRs estimated at the SPP symbols with the interval of K_t symbols by (16). The CFRs for all links over one frame can be obtained by performing DFT to the estimated CIRs. Finally, the information data in one OFDM frame can be demodulated precisely by the MIMO detection of using the estimated CFR matrix even in higher time-varying fading channel.

4 Performance Evaluations

This section presents various computer simulation results to verify the effectiveness of proposed TD-CE method in the time-varying fading channel. Table 1 shows the simulation parameters used in the following evaluations.

Table 1. Simulation parameters.

Parameters	Values
Modulation method for data subcarriers	16QAM
Modulation method for pilot subcarriers	QPSK
No. of FFT points (N)	128
No. of data subcarriers in conv. methods (M)	96
No. of data subcarriers in proposed methods (M_p)	$M-N_T+1$
Length of GI (Ng)	11
OFDM frame length (L-Symbols)	17
Interval of pilot in freq. (K_f) and in time (K_t) axes	$K_f = N_T$ and $K_t=4$
No. of transmit and receive antennas ($N_T \times N_R$)	4×4 and 8×8
OFDM occupied bandwidth (BW)	5 MHz
Radio frequency (f_c)	2 GHz
Forward Error Correction (FEC) Code	
-Encoding	Convolution ($R=1/2$, $K=7$)
-Decoding	Hard-decision withViterbi
-Interleaver	Matrix with one frame (L)
-Packet length (information bits/packet)	512 bits
Multipath Rayleigh fading channel model	
-Delay profile	Exponential
-Decay constant	-1 dB
-No. of delay paths (N_P) in all links	10
-No. of scattered rays	20

Figure 4 shows the normalized mean square error (NMSE) performances of estimated CFR evaluated at each subcarrier for the proposed TD-CE, Conv.DFTI-CE and Conv.ML-CE methods when the sampling rate are taken by the Nyquist rate ($N=M$) and non-Nyquist ($N \neq M$) rate. The number of transmit (N_T) and receive (N_R) antennas are 8×8. From the figure, it can be seen that there is no border effect in the NMSE performances for all CE methods when the sampling rate is the Nyquist rate. When the sampling rate is the non-Nyquist rate, the proposed TD-CE methods has no border effects while the Conv.DFTI-CE and Conv. ML-CE methods have the border effects at around the both ends of data subcarriers. From these results, it can be concluded that the proposed TD-CE method can achieve higher estimation accuracy even when the non-Nyquist rate and the larger number of transmit antennas.

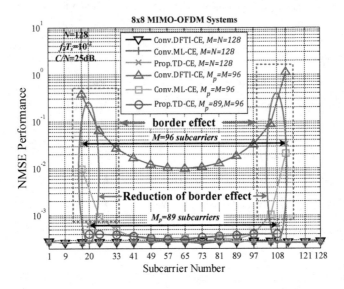

Fig. 4. CFR estimation accuracy for proposed method at each subcarrier.

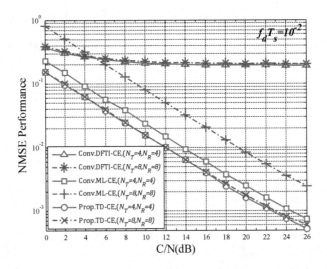

Fig. 5. CFR estimation accuracy for proposed method versus C/N(dB) at non-Nyquist rate.

Figure 5 shows the NMSE performance of estimated CFR when changing the carrier to noise power ratio (C/N) for the proposed TD-CE, Conv.DFTI-CE and Conv.ML-CE methods at the non-Nyquist rate. In the evaluation, the number of transmit and receive antennas $(N_T \times N_R)$=4×4 and 8×8, and the normalized Doppler frequency $(f_d T_s)$ is 10^{-2} where f_d is the maximum Doppler frequency and T_s is the OFDM symbol duration including the guard interval (GI). From the figure, it can be seen that the NMSE performance for the Conv.DFTI-CE method is degraded relatively due to the border effect at the non-Nyquist rate. The performance for the Conv.ML-CE method is degraded when increasing the number of transmit antennas. The proposed TD-CE method shows higher estimation accuracy regardless of the number of transmit antennas than that for the Conv.ML-CE method.

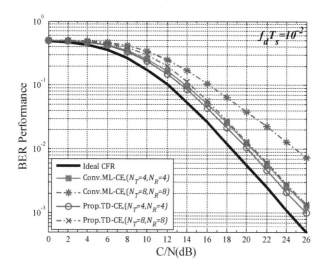

Fig. 6. BER performance for proposed method versus C/N(dB) at $f_d T_s$=10^{-2}.

Figure 6 shows the BER performances when changing C/N at $f_d T_s$=10^{-2} for the Conv.ML-CE and proposed TD-CE methods at the non-Nyquist rate. From the figure, it can be seen that both 4×4 and 8×8 MIMO-OFDM systems with the proposed TD-CE method shows better BER performance than those for the Conv.ML-CE method especially when the number of transmit antennas is 8. The degradation of BER performance for the Conv.ML-CE method as compared with the proposed TD-CE method is come from the degradation of channel estimation accuracy as shown in Fig. 5.

Table 2 shows the comparison of computation complexity required in one CFR estimation for the proposed and conventional methods when $(N_T \times N_R) = 8 \times 8$. From Table 2, it can be observed that the computation complexity for the proposed TD-CE method is slightly larger than the Conv.DFTI-CE and Conv.ML-CE methods. From Table 2 and Figs. 5 and 6, it can be concluded that the proposed TD-CE method can achieve higher estimation accuracy and accordingly better BER performance than the conventional DFTI-CE and ML-CE methods with keeping almost the same computational complexity.

Table 2. Comparison of computational complexity for channel estimation methods.

Channel Estimation Methods	Computational complexity required in one channel estimation (N_T=8)	
Conv.DFTI-CE	$N \bullet log_2 N + P + P^2 + M \bullet P$	2,204
Conv.ML-CE	$N \bullet log_2 N + P + P^2 + M \bullet Ng$	2,108
Prop.TD-CE ($M_p = M - N_T + 1$)	$N \bullet Ng + M_p \bullet Ng$	2,387

The throughput (*Thp*) performances defined by bps/Hz for the proposed and conventional methods are evaluated by using the following equation.

$$Thp = n_{bit} \cdot R \cdot N_T \cdot \frac{N}{N + Ng} \cdot \frac{M_p}{M} \cdot \frac{(L-1)(K_t - 1)}{L \cdot K_t} \cdot (1 - PER), \quad (bps/Hz) \qquad (17)$$

where n_{bit} is the modulation level (n_{bit}=4 for 16QAM), M_p ($=M-N_T+1$) is the number of active data subcarrier for the proposed method ($M_p=M$ for conventional methods), R is the FEC rate, K_t is the interval of SPP symbol in the time axis, L is the frame length and *PER* is the packet error rate when the packet length (information bits/packet) is 512 bits. From (17) it can be seen that the throughput performance for the proposed method would be decreased by (M_p/M) due to the smaller number of active data subcarriers M_p ($=M-N_T+1$) which depends on the number of transmit antennas (N_T). However the NMSE estimation accuracy of proposed method is higher than the conventional ML-CE method as shown in Fig. 5 and accordingly the proposed method can achieve better BER performances as shown in Fig. 6. Consequently, the PER performance in (17) for the proposed TD-CE method becomes better than the conventional ML-CE method. From these reasons, it can be expected that the proposed TD-CE method can achieve higher throughput efficiency even when the number of active data subcarriers M_p for the proposed method is smaller than M that for the conventional ML-CE method.

Figure 7 shows the throughput performances for the Conventional ML-CE and proposed TD-CE methods evaluated by using (17). In the figure, the Ideal throughput performance which is given by assuming the ideal CFR for the Conv.ML-CE method is also shown as for the purpose of comparison with other methods. From the figure, it can be seen that the throughput performance for the proposed TD-CE method is almost the same as the Conv.ML-CE method when ($N_T \times N_R$) = 4×4. This is the reason that the improvement of PER performance by using the proposed TD-CE method is not enough to compensate the reduction of ratio (M_p/M =93/96) for active data subcarriers in (17). However the throughput performance for the proposed method is much higher than the Conv.ML-CE method when ($N_T \times N_R$) = 8×8. The throughput performances for the Conv.ML-CE and proposed TD-CE methods at C/N=30 dB are 2.6bps/Hz and 7.6bps/Hz, respectively which corresponds to the improvement of 5bps/Hz by the proposed TD-CE method.

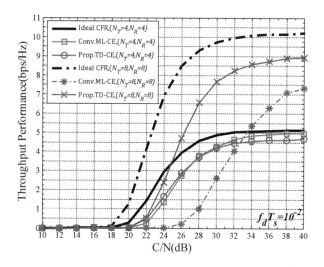

Fig. 7. Throughput performance for proposed method versus C/N(dB) at $f_dT_s=10^{-2}$.

This means that the proposed TD-CE method can achieve the transmission data rate of 38Mbps which is higher transmission data rate by 25Mbps than the Conv.ML-CE method of 13Mbps when the allocated frequency bandwidth is 5 MHz.

5   Conclusions

This paper proposed the time domain channel estimation (TD-CE) method for the MIMO-OFDM systems. The salient feature of proposed TD-CE method is to employ the scattered pilot preamble symbol in which pilot subcarriers sent from each transmit antenna are assigned cyclically including the both ends of transmission frequency band for data subcarriers. In the proposed method, the channel frequency responses for all links between transmit and receive antennas are estimated over the frequency band between the first to the end pilot subcarriers assigned for each transmit antenna. The same frequency band assigned for pilot subcarriers in each transmit antenna is also used in the transmission of data information. From the computer simulation results, it was confirmed that the MIMO-OFDM systems with the proposed TD-CE method can achieve higher throughput performance in higher time-varying fading channel with keeping almost the same computational complexity as comparing with the conventional methods even when the non-Nyquist rate and the larger number of transmit antennas.

Acknowledgments.   The authors would like to thank to the Japanese Government (Monbukagakusho:MEXT) Scholarships who supported this research.

References

1. Radio Broadcasting Systems; Digital Audio Broadcasting (DAB) to mobile, portable and fixed receivers, ETS 300 401, v1.4.1 (2006)
2. Digital Video Broadcasting (DVB); Implementation guidelines for the second generation digital terrestrial television broadcasting system (DVB-T2), ETSI TS 102 831, v1.2.1 (2012)
3. IEEE Standard for Local and Metropolitan Area Networks Part 16: Air Interface for Fixed Broadband Wireless Access Systems. IEEE Std. 802.16 (2004)
4. Kim, J., Lee, I.: 802.11 WLAN history and new enabling MIMO techniques for next generation standards. IEEE Commun. Mag. **53**(3), 134–140 (2015)
5. Jiang, M., Hanzo, L.: Multi-user MIMO-OFDM for next generation wireless systems. Proc. IEEE **95**(7), 1430–1469 (2007)
6. Sure, P., Bhuma, C.M.: A pilot aided channel estimator using DFT based time interpolation for massive MIMO-OFDM systems. Int. J. Electron. Commun. AEU **69**(1), 321–327 (2015)
7. Mata, T., Boonsrimuang, P., Mori, K., Kobayashi, H.: Time-Domain channel estimation method for MIMO-OFDM systems. In: IEICE General Conference 2015, B-1-155 (2015)
8. Diallo, D., Helard, M., Cariou, L., Rabineau, R.: DFT based channel estimation methods for MIMO-OFDM systems. In: Vehicular Technologies: Increasing Connectivity, pp. 97–114 (2011)
9. Mata, T., Hourai, M., Boonsrimuang, P., Mori, K., Kobayashi, H.: Time domain channel estimation method for uplink OFDMA system. In: the 22nd International Conference on Telecommunications, ICT2015 (2015)
10. Kobayashi, H., Mori, K.: Proposal of channel estimation method for OFDM systems under time-varying fading environments. IEICE Trans. Comm. (Japanese edition) **J90-B**(12), 1249–1262 (2007)

An OFDM Timing Synchronization Method Based on Averaging the Correlations of Preamble Symbol

Yunsi Ma, Chaoxing Yan[✉], Sanwen Zhou, Tongling Liu, and Lingang Fu

Beijing Research Institute of Telemetry, Beijing 100076, China
chaoxingyan@foxmail.com

Abstract. In this paper, we propose a novel timing synchronization method for orthogonal frequency division multiplexing (OFDM) systems with a single-symbol preamble. The proposed method has an impulsive timing metric and outperforms the conventional methods in multipath fading channels by averaging M_0 differential correlations of the preamble. Using this method, the system will achieve accurate timing synchronization and keep low out-of-band radiation. Performances of the proposed estimator with different M_0 values are evaluated in terms of bias and mean square error (MSE). Simulation results validate the effectiveness of the proposed timing synchronization method.

Keywords: OFDM · Timing synchronization · Differential correlation

1 Introduction

Orthogonal frequency division multiplexing (OFDM) technique has received much attention in the last decade owing to its effective transmission capability and robustness over multipath fading channels. Currently, OFDM is employed intensively in various satellite and mobile communication systems, such as digital video broadcasting-Satellite to Handheld systems (DVB-SH) [1, 2], and wireless local area network (WLAN) systems [3].

The principal weakness of OFDM is its sensitivity to carrier and frequency offsets. OFDM exhibits some tolerance to timing offsets when the guard interval length is longer than the maximum channel delay spread. However, when the timing error positions the fast Fourier transform (FFT) window to include samples of either preceding or succeeding symbols, it will result in inter-symbol interference (ISI). Several timing offset estimation schemes for OFDM systems have been investigated in [4–8]. The conventional OFDM timing synchronization methods can be divided into four categories [4]: those that use the periodic structure of cyclic prefix (CP), those that utilize null subcarriers, those that take the advantage of the special structure of preamble symbols, and those that use a preamble symbol but work independent of its structure.

© Institute for Computer Sciences, Social Informatics and Telecommunications Engineering 2015
P. Pillai et al. (Eds.): WiSATS 2015, LNICST 154, pp. 229–238, 2015.
DOI: 10.1007/978-3-319-25479-1_17

Schmidl proposed using the autocorrelation of a preamble containing two repetitive patterns to estimate timing offset in the time-domain [5]. However, the timing metric plateau inherent in this method leads to a large variance of the timing estimate. Minn presented two methods as modifications to Schmidl's method to avoid the timing metric plateau [6]. Minn's methods achieve a smaller estimator variance than that of Schmidl's method. In [7], for having an impulse-like timing metric, Ren multiplied the constant envelop preamble containing two identical parts by a pseudo-random noise (PN) sequence. Exploiting the differential cross-correlation of a randomized sequence [8], Ren's method improves the accuracy of the timing offset estimator and achieves a smaller MSE than that of Minn's method. It is noted that the weighted preamble will increase the out-of-band radiation of OFDM signals and produce interference to the adjacent band users.

Considering those above problems in the literature, we propose a novel timing synchronization method in the time-domain for the OFDM system with a single-symbol preamble. Performance of the proposed estimator is evaluated in terms of bias and MSE. Computer simulations are employed to validate the effectiveness of the proposed timing method.

2 System Model and Timing Synchronization

In this section, we describe briefly an OFDM system and then analyze the advantages and weaknesses of conventional timing synchronization methods. Thereby, the motivation of the novel timing synchronization is drawn.

2.1 OFDM System Model

In the OFDM system, the samples of complex-valued baseband OFDM symbol in the time-domain can be expressed by

$$x_i(n) = \frac{1}{\sqrt{N}} \sum_{k=0}^{N_u-1} X_i(k) \cdot e^{j2\pi nk/N} \, , \, n = -N_g, \, \cdots, \, N-1, \tag{1}$$

where $x_i(n)$ denotes the ith OFDM symbol, $X_i(k)$ represents the data sequence modulated on the kth subcarrier, which may assume any modulation format such as quadrature amplitude modulation (QAM) or quadratic phase-shift keying (QPSK), N is the size of inverse fast Fourier transform (IFFT) with N_u active subcarriers, and N_g is the CP length.

In the OFDM digital transmitter, firstly, the preamble sequence is inserted into the channel-coded data according to the frame structure. Then, an OFDM signal is generated by taking the IFFT of QAM or PSK symbols and it is preceded by a cyclic prefix that is longer than the channel impulse response. Due to the equivalent rectangular pulse shaping in modulation, OFDM has high levels of out-of-band (OOB) radiation [9]. In order to make the amplitude go smoothly to zero at the symbol boundaries, windowing is often applied to individual OFDM symbols [10],

$$\bar{x}_i(n) = \begin{cases} c_1(n + N_g) \cdot x_i(n) + x_{i-1}(n + N_g) \cdot c_2\,(n + N_g)\,, & -N_g \leq n < -N_g + \alpha N \\ x_i(n)\,, & -N_g + \alpha N \leq n \leq N - 1 \end{cases},$$

$$\tag{2}$$

where $c_1(n)$, $c_2(n)$ represents respectively the prefix and postfix coefficients of window function. The raised-cosine window is usually employed and expressed by

$$c_1(n) = 0.5 + 0.5\cos\left(\pi + \frac{n\pi}{\alpha N}\right)\,, c_2(n) = 0.5 + 0.5\cos\left(\frac{n\pi}{\alpha N}\right), \tag{3}$$

where α is the roll-off factor of the raised-cosine window.

In the OFDM digital receiver, after removing the cyclic prefix, the nth sample of the received OFDM baseband signal $r(n)$ [11] is given by

$$r(n) = y(n - \varepsilon)e^{j(2\pi v n / N)} + w(n) = \sum_{m=0}^{L-1} h(m)x(n - \varepsilon - m)e^{j(2\pi v n / N)} + w(n), \tag{4}$$

where ε represents the integer-valued unknown arrival time of a symbol, v represents the frequency offset normalized by the subcarrier spacing, $h(m)$ is the channel impulse response whose memory order is L-1, and $w(n)$ is the zero-mean complex additive white Gaussian noise (AWGN).

2.2   Timing Synchronization

Schmidl proposed a preamble-aided timing synchronization method [5]. Ren presented a synchronization method based on a constant envelope preamble [7]. We briefly describe these two methods as follows.

A. Schmidl's Method. Schmidl exploits a preamble containing two identical halves in the time-domain, and the form of preamble is described as

$$a_1 = [A_{N/2}\,A_{N/2}], \tag{5}$$

where $A_{N/2}$ represents a random sequence consisting of $N/2$ samples in the time-domain. Schmidl's preamble can be generated via direct repetition of a suitable pseudo-noise (PN) sequence in the time-domain or an IFFT of the odd-numbered subcarriers in the frequency-domain. This method estimates the starting point of the received signal at the maximum point of the timing metric given by

$$M_1(d) = \frac{|P_1(d)|^2}{\left(R_1(d)\right)^2}, \tag{6}$$

where

$$P_1(d) = \sum_{n=0}^{N/2-1} r^*(d+n) \cdot r(d+n+N/2), \tag{7}$$

$$R_1(d) = \sum_{n=0}^{N/2-1} |r(d+n+N/2)|^2, \tag{8}$$

where $r^*(n)$ represents the complex conjugate of the received OFDM signal $r(n)$. The timing offset of Schmidl's method can be estimated from

$$\hat{\varepsilon}_1 = \arg\max_d \left(M_1(d) \right). \tag{9}$$

B. Ren's Method. The timing metric of Schmidl's method suffers from a plateau which has a length equal to the length of the guard interval minus the length of the channel impulse response in frequency selective channel. In order to reduce some uncertainty of the timing estimate, Ren proposed a timing offset estimation method with a scrambled preamble weighted by PN sequence, which can be defined as

$$x'(n) = s(n) \cdot x(n), \ n = 0, \ \cdots, \ N-1, \tag{10}$$

where $s(n)$ represents the PN sequence weighted factor of the nth sample of the original preamble and the value is $+1$ or -1. The timing metric in [7] is defined as

$$M_2(d) = \frac{|P_2(d)|^2}{\left(R_2(d)\right)^2}, \tag{11}$$

where

$$P_2(d) = \sum_{n=0}^{N/2-1} s(n) \cdot s(n+N/2) \cdot r^*(d+n) \cdot r(d+n+N/2), \tag{12}$$

$$R_2(d) = \frac{1}{2} \sum_{n=0}^{N-1} |r(d+n)|^2. \tag{13}$$

The timing offset of Ren's method can be estimated from

$$\hat{\varepsilon}_2 = \arg\max_d \left(M_2(d) \right). \tag{14}$$

The timing metric of Ren's method has an impulsive shape only at the exact timing point [7]. Therefore, Ren's method eliminates the plateau of Schmidl's method under the same channel condition, and achieves a smaller MSE than Schmidl's estimator (9). However, the preamble weighted by PN sequence will lead to high levels of out-of-band radiation, as illustrated by the simulation results in Fig. 1. In the following section, we will address this problem and develop a novel timing synchronization method.

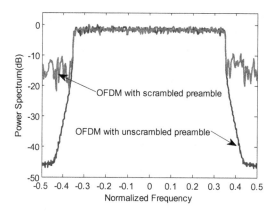

Fig. 1. Power spectrum of OFDM signals with the scrambled/unscrambled preamble

3 Correlation-Based Timing Synchronization

In this section, we will propose a timing synchronization method which works independent of the preamble structure in the time-domain and discuss a frequency offset estimator when the proposed timing method is used.

3.1 Proposed Timing Synchronization Method

Considering that the preamble $x_i(n) = a(n)$ is known at the OFDM receiver, we can make full use of the characteristics of the correlation to eliminate the modulation information of the preamble. We multiply the samples of the received signal $r(n)$ by the samples of the known preamble $a(n)$, which can be described as

$$r_0(n, d) = r(n + d)a^*(n) \ , n = 0, 1, \cdots, N - 1, \tag{15}$$

where d is the timing index corresponding to the first sample in a window of N samples. Then, define the differential correlation function of $r_0(n, d)$ as

$$p(m, d) = \sum_{k=m}^{N-1} r_0(k, d) \cdot r_0^*(k - m, d), \ m = 1, \cdots, M_0, \tag{16}$$

where N-m is the number of summation items in the differential correlation function, and M_0 is an adjustable parameter, which should be carefully evaluated. For $M_0 = 1$, a single-lag differential correlation function can be directly used to produce the timing metric. For $M_0 > 1$, M_0 differential correlation values can be summed together with different weighted factors. Therefore, we derive a reasonable solution with their weights as

$$P(d) = \sum_{m=1}^{M_0} \frac{(N-m) \cdot |p(m,d)|}{M_0 N - M_0(1+M_0)/2}. \tag{17}$$

For simplicity, when M_0 is small, we suggest averaging M_0 differential correlation values, that is, each of weighted factors equal to $1/M_0$. The corresponding timing metric of the proposed method is derived as

$$M(d) = \frac{P(d)^2}{(R(d))^2}, \tag{18}$$

where

$$P(d) = \frac{1}{M_0} \sum_{m=1}^{M_0} |p(m,d)| = \frac{1}{M_0} \sum_{m=1}^{M_0} \left| \sum_{k=m}^{N-1} r_0(k,d) \cdot r_0^*(k-m,d) \right|, \tag{19}$$

$$R(d) = \sum_{n=0}^{N-1} |r(n+d)|^2. \tag{20}$$

The timing offset of proposed method can be estimated from

$$\hat{\varepsilon} = \arg\max_d (M(d)). \tag{21}$$

3.2 Timing Metric with Different Preambles

The timing metric proposed in (19) for OFDM systems will be examined with two different preamble structures: (1) preamble-1 is a single symbol preamble with no repetitive patterns; (2) preamble-2 is a particular preamble containing two identical halves, which can be generated in the frequency-domain by mapping a PSK PN sequence to the odd numbered subcarriers. For the OFDM system, we assume using $N = 256$ subcarriers and the signal-to-noise ratio being SNR = 10 dB in AWGN channel.

In Fig. 2, we can find that, for the larger value of M_0 (e.g. $M_0 = 5$), the noise of the timing metric will get smaller for both kinds of preambles. There is a major peak in the middle at the starting point of the preamble corresponding to a full-symbol pattern match. However, for the preamble-2 with repetitive structure, besides this major peak, there are two minor peaks at the $N/2$ samples left and right corresponding to a half-symbol pattern match. In complicated wireless transmission channel, the values of the minor peaks may exceed the major peak, which will result in a large timing error.

An alternative solution is to employ a window function. The autocorrelation metric (6) of the preamble is a feasible window [8]. Therefore, we take the product of the window (6) and the proposed timing metric (18) as the modified timing metric for timing offset estimate, that is, the timing offset can be estimated from

$$\hat{\varepsilon} = \arg\max_d \left(M(d) \cdot M_1(d) \right). \tag{22}$$

After the timing synchronization, the starting point of the received signal can be determined. In the next section, we will discuss the frequency synchronization when the proposed timing synchronization method is employed in OFDM digital receiver.

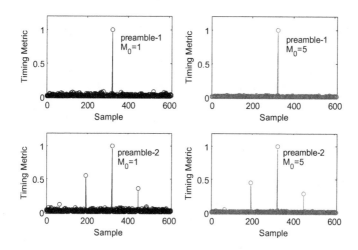

Fig. 2. The timing metric for the proposed method

3.3 Corresponding Frequency Estimation

Similar to the frequency synchronization in [7], the frequency offset v is divided into fractional part q and integer part ξ, $v = q + \xi$. In order to estimate the frequency offset, the preamble containing two identical halves (5) is exploited. The fractional frequency offset estimation is given as

$$\hat{\xi} = \frac{1}{\pi} angle \left(\sum_{n=0}^{N/2-1} r^*(\varepsilon + n) \cdot r(\varepsilon + n + N/2) \right). \tag{23}$$

After compensating the fractional frequency offset of the preamble signal, we multiply the received preamble by $a_1^*(n)$ in (5) as

$$r_1(n) = r(n) \cdot \exp\left[-j\left(2\pi\hat{\xi}n/N\right)\right] \cdot a_1^*(n). \tag{24}$$

The integer frequency offset is then estimated by

$$\hat{q} = \arg\max_{q} \left(\Gamma(q)\right), \tag{25}$$

where

$$\Gamma(q) = \left| \sum_{n=0}^{N-1} r_1(n) \cdot e^{-j2\pi qn/N} \right|^2 , \quad q = -\frac{N}{2}, \cdots, \frac{N}{2}. \tag{26}$$

Therefore, the total frequency offset estimate can be estimated as

$$\hat{v} = \hat{q} + \hat{\xi}. \tag{27}$$

As shown in (26), the range of the frequency offset estimation method in (27) is $\pm N/2$. The accuracy of different timing estimation methods (9), (14) and (22) will affect the succeeding frequency estimation in (23)–(27). In the following section, the performance of timing synchronization will be studied, whereas that of frequency estimation still keeps with the results in [7].

4   Simulation Results and Discussions

In this section, we will investigate the performance of the proposed timing estimation method in terms of bias and MSE by simulations. The results are further compared with those of the conventional methods.

These estimators work in the OFDM system with QPSK modulation, $N_u = 180$ active subcarriers, $N = 256$ size of IFFT/FFT, and $N_g = 32$ samples of cyclic prefix. The OFDM system bandwidth is 3 MHz, and the subcarrier spacing is 15 kHz. The channel conditions are described in the following. The Rayleigh fading channel has an exponential power delay profile given by $A_i = e^{-(i/3)}$. In our simulation, the corresponding time delays of a multipath channel with 6 taps are set as [0 0.333 0.667 1.0 1.333 1.667] μs. A normalized carrier frequency offset of $v = 1.2$ is considered, and 10000 simulation runs are applied.

Figure 3 shows the estimation means of the timing offset, which reflect the bias of timing estimates. It can be found that both of the proposed method (22) and Ren's method (14) have smaller bias values than those of Schmidl's method. Furthermore, adopting different M_0 values for the differential correlations, the proposed method (22) achieves approximate estimation bias values. For example, for $M_0 = 3, 8$, as given in Fig. 3, the proposed method (22) achieves much smaller bias values than those of Ren's methods for SNR < 10 dB.

Figure 4 illustrates the MSE performance of timing offset estimation. We find that both of the proposed method and Ren's method outperform Schmidl's method. And the proposed method will achieve better MSE performance for increasing M_0 especially at low SNRs. What's more, for $M_0 \geq 2$, the MSEs of the proposed method are smaller than those of Ren's method for SNR < 10 dB, whereas their MSEs get tight for SNR $\geq$ 10 dB.

It is noted that the performance improvements become tiny for $M_0 \geq 3$, whereas the computational complexity will grow continuously. Therefore, in practical OFDM system, the parameter M_0 should make a compromise between estimation performance and implementation complexity and $M_0 = 1, 2, 3$ are recommended here.

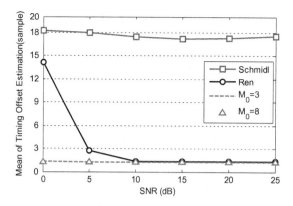

Fig. 3. Mean of timing offset estimate in a Rayleigh fading channel

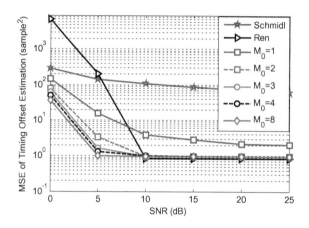

Fig. 4. MSE of timing offset estimate in a Rayleigh fading channel

5   Conclusions

In this paper, we firstly describe an OFDM system and then analyze the advantages and weaknesses of conventional timing synchronization methods. Then, we propose a novel timing synchronization method in the time-domain for OFDM systems with a single-symbol preamble. The proposed timing synchronization method has an impulse-shaped timing metric and outperforms the conventional methods in multi-path fading channels by averaging the M_0 differential correlations of a conjugate structure preamble. Furthermore, performances of the proposed estimator are evaluated by simulations in terms of bias and MSE. This work may be helpful for designing OFDM wireless transceivers.

References

1. ETSI: Framing Structure, Channel Coding and Modulation for Satellite Services to Handheld Devices (SH) Below 3 GHz, ETSI TS 102 583, European Telecommunication Standards Institute
2. Kelley, P.: Overview of the DVB-SH specifications. Int. J. Satell. Commun. Netw. **27**(4/5), 198–214 (2009)
3. IEEE 802 LAN/MAN standards committee: wireless LAN medium access control (MAC) and physical layer (PHY) specifications, higher speed physical layer extension in the 5 GHz band. IEEE Stand. 802.11 (1999)
4. Abdzadeh-Ziabari, H., Shayesteh, M.G.: Robust timing and frequency synchronization for OFDM systems. IEEE Trans. Veh. Technol. **60**(8), 3646–3656 (2011)
5. Schmidl, T., Cox, D.C.: Robust frequency and timing synchronization for OFDM. IEEE Trans. Commun. **45**(12), 1613–1621 (1997)
6. Minn, H., Bhargava, V.K., Letaief, K.B.: A robust timing and frequency synchronization for OFDM systems. IEEE Trans. Wirel. Commun. **2**(4), 822–839 (2003)
7. Ren, G.L., Chang, Y., Zhang, H., Zhang, H.: Synchronization methods based on a new constant envelope preamble for OFDM systems. IEEE Trans. Broadcast. **51**(1), 139–143 (2005)
8. Awoseyila, A.B., Kasparis, C., Evans, B.G.: Robust time-domain timing and frequency synchronization for OFDM systems. IEEE Trans. Consum. Electron. **55**(2), 391–399 (2009)
9. Cosovic, I., Brandes, S., Schnell, M.: Subcarrier weighting: a method for sidelobe suppression in OFDM systems. IEEE Commun. Lett. **10**(6), 44–446 (2006)
10. Ma, Y.S., Zhou, S.W., Yan, C.X.: Design and implementation of OFDM transmitter in UAV TT&C system. J. Telemetry Tracking Command **36**(4), 45–50 (2015)
11. Kang, Y., Kim, S., Ahn, D., Lee, H.: Timing estimation for OFDM systems by using a correlation sequence of preamble. IEEE Trans. Consum. Electron. **54**(4), 1600–1608 (2008)

A Printed Wideband MIMO Antenna for Mobile and Portable Communication Devices

Chan H. See[1,2], Elmahdi Elkazmi[2,3], Khalid G. Samarah[2,4], Majid Al Khambashi[2,3], Ammar Ali[2], Raed A. Abd-Alhameed[2(✉)], Neil J. McEwan[5], and Peter S. Excell[6]

[1] School of Engineering, University of Bolton, Deane Road, Lancashire, Bolton, BL3 5AB, UK
[2] School of Electrical Engineering and Computer Science, University of Bradford, Richmond Road, West Yorkshire, Bradford, BD7 1DP, UK
{chsee2,raaabd}@bradford.ac.uk
[3] The Higher Institute of Electronics, Bani Walid, Libya
[4] Mutah University, Al-Karak 61710, Jordon
[5] AlZahra College for Women, Muscat, Oman
[6] School of Computing and Communication Technology, Glyndwr University, Mold Road, Wrexham LL11 2AW, Wales, UK

Abstract. A printed crescent-shaped monopole MIMO antenna is presented for handheld wireless communication devices. The mutual coupling between the two antenna elements can be minimised by implementing a I-shaped common radiator. Both the simulated and measured results agree that the antenna covers the operating frequency band from 1.6 to 2.8 GHz with the return loss and isolation better than 10 dB and 14 dB respectively. To further verifying the MIMO characteristic including far-field, gain, radiation efficiency, channel capacity loss and envelope correlation, the results confirm that the antenna can operate effectively in a rich multipath environment.

Keywords: MIMO · Mutual coupling · Channel capacity loss · Envelope correlation

1 Introduction

The increasing use of smart phones and tablet computers has resulted in substantial teletraffic growth. Clearly, the current capacity and bandwidth of the existing wireless communication systems are insufficient to meet the needs of next generation mobile requirements. In anticipation of this, new high data-rate wireless systems, such as HSPA (3.5G), LTE (4G), WiFi, WiMAX and UWB, have been developed [1–3]. It is expected that future mobile devices will be required to inter-operate seamlessly with all the existing second and third generation mobile standards as well as with these new enhanced standards. To help cater for this unabated demand for high data-rate traffic resulting from ever-growing mobile applications, these existing wireless standards require a multiple-input-multiple-output (MIMO) antenna system.

© Institute for Computer Sciences, Social Informatics and Telecommunications Engineering 2015
P. Pillai et al. (Eds.): WiSATS 2015, LNICST 154, pp. 239–248, 2015.
DOI: 10.1007/978-3-319-25479-1_18

It is well known that the use of multiple antennas on both the transmitting and receiving ends of a communication system promises to deliver robust communication links with enhanced data-rate, without increasing the power and spectrum requirement, when operating in a multi-user rich scattering environment [2, 4, 5]. Nevertheless, implementing the MIMO antenna technology in a handheld device remains a challenge for antenna engineers due to the limited space within an existing commercial mobile chassis. MIMO antennas have to satisfy all the performance indicators of a single-element antenna and also require a good port-to-port isolation between the closely placed antenna elements.

Over the last five years, much research effort has been invested in conceiving solutions capable of minimizing the mutual coupling between MIMO antenna elements. These solutions can be classified into three categories, i.e. narrow band [6, 8–17], wideband [18–21] and dual band [22–25]. In the narrow band applications, MIMO antennas were developed to operate with the following wireless standards, LTE 700–800 MHz [6–8], UMTS [9], WLAN 2.4 GHz [10–13], 5.2 GHz [14] and 5.8 GHz [15, 16]. To suppress the mutual coupling between the antenna elements, these antennas have adopted several innovative methods including ferrite [6] or magneto-dielectric substrates [7], the insertion of parasitic elements between the radiators [8, 9], introducing resonators between the antenna elements [10], connecting the elements via a neutralization line [11], etching a series of slits in the ground plane [12], utilizing lumped elements [13] or inductive coils [14], heavily slotting the ground plane [15], placing the antenna elements in an orthogonal formation [16], and adopting varactor diode and lumped elements [17]. These solutions achieve a good port-to-port isolation ranging from 10 dB to 40 dB, and reasonable inter-element distances from 0.03 λ_0 to 0.17 λ_0.

By compromising between the bandwidth, mutual coupling level and inter-element spacing distance of a MIMO antenna, some wideband isolation methods were proposed in [18–21] to enable multiband operation. In [18], it was found that by separating the two antenna elements by 0.25 λ_0, mutual coupling as good as −11 dB can be attained across the frequency band from 1.65 to 2.5 GHz. Other work in [19] demonstrates that by placing the antenna elements orthogonally, the inter-port isolation level can be enhanced to 20 dB with an element spacing of 0.1λ_0, covering an operating band from 1.63 to 2.05 GHz. Moreover, the author in [20] suggests that using inverted L-parasitic monopoles offers a wideband inter-port isolation of better than 14.8 dB from 1.85 to 2.170 GHz. In addition, the work in [21] proposes placing branched neutralization lines between the antenna elements which can suppress the mutual coupling to better than −17 dB over a wide frequency band, from 2.4 to 4.2 GHz.

Apart from narrow band and wideband inter-port isolation methods, some dual band methods have also been introduced in [22–25]. Interestingly, it was noticed that good dual band port-to-port isolation can be realized by using either a modified version of single band methods or combinations of them. In [22], the authors show that inserting both a T-shaped and a dual inverted L-shaped branch in the ground plane can provide a good isolation of 13 dB and 18 dB over the UMTS and WLAN 2.4 GHz bands respectively. Furthermore, to suppress the mutual coupling of two antenna elements in two distinct sub-bands, i.e. WLAN 2.4 GHz and WiMAX 3.5 GHz, the authors in [23] recommend that implementing a T-shaped slot in the ground plane and a folded L-slot in the radiator can effectively reduce mutual coupling to 19.2 dB and 22.8 dB over the two bands correspondingly. On the other hand, authors in [24] also show that, by carefully adjusting the lengths of proposed parasitic elements, isolation better than 15 dB and 22 dB can be obtained over the two frequency bands, i.e. WLAN 2.4 GHz and 5.8 GHz. However, these methods only offer operation in narrow band wireless standards. To overcome this limitation, a dual wideband MIMO antenna was studied in [25]. In this work, two U-shaped slots were etched in the ground plane to act as a decoupling network. Both the outer and inner slots improve the inter-port isolation to better than 15 dB and 20 dB over the two wide sub-bands, i.e. 1.5–2.8 GHz and 4.7 to 8.5 GHz, respectively.

In summary, it is found that most of these antennas only provide either single or dual narrow band operation [15–17, 22–24], or wideband operation [18–21] but do not fully cover many existing mobile standards, except for the work in [25]. To address this limitation, this paper presents a MIMO/diversity monopole antenna for GSM1800/1900 (1710–1880 MHz, 1850–1990 MHz), UMTS2000 (1920–2170 MHz), LTE2300 (2300–2400 MHz), LTE2600 (2500–2690 MHz), WiFi (2400–2485 MHz) and WiMAX (2500–2690 MHz) applications. By inserting an **I** - shaped metal strip on a defected ground plane which is located underneath the two antenna elements, an inter-port isolation as good as 14 dB can be realized with return loss better than 10 dB across all the operating frequency bands.

2   Antenna Design Concept and Geometry

Figure 1 illustrates the configuration of the proposed MIMO/diversity antenna. It consists of two metallic layers where the top layer has two radiating elements and the bottom layer is a modified defected ground plane. Both layers are printed on a Duroid 5870 substrate material with a thickness of 0.79 mm, dielectric constant (ε_r) of 2.33 and a loss tangent of 0.0012. The single element of the radiator is very similar to the one in [21, 26], except that the ends of the radiator have straight extensions. The crescent-shaped radiator is fed by an 18 x 1 mm microstrip line. This line has lower and upper sections of length 11.25 mm and 6.75 mm respectively. The lower section which lies over the ground plane has a characteristic impedance of 83 Ω and is required for matching to 50 Ω at the input port. The dimensions of the ground plane are 71.5 x 50 mm^2 with a defected area of 26.8 x 12 mm^2 in its top edge, this being used to improve the impedance matching at the input port. The overall dimensions of the proposed antenna are 90 × 50 × 0.8 mm which is suitable for application in a typical handheld device.

The design procedure of this antenna begins with optimizing a single element crescent-shaped radiator on a corner of a 50 x 90 mm ground plane. By manipulating the radii r1 and r2 of the geometry parameters (Fig. 1), the lower and upper resonant frequencies can be adjusted to establish a wide impedance bandwidth [21]. In order to improve the impedance matching for return loss to better than 10 dB over the band, the length of the crescent-shaped radiator's straight extensions (d) was further optimized and a defected area of 26.8 × 11.25 mm^2 was introduced in the ground plane. Once the targeted performance of a single element antenna was realized, two identical copies of the radiators were symmetrically placed on the two corners of the ground plane. These two radiators are separated by the largest possible distance of 24 mm which is equivalent to 0.128 λ_0 (where λ_0 is the free space wavelength) at 1.6 GHz (i.e. the lowest working frequency) for the optimal mutual coupling suppression. The final design stage was to insert an **I** -shaped strip on the ground plane of the antenna and to find its geometric parameters for best performance in terms of reflection coefficient and mutual coupling.

Confirming the effectiveness of the **I** -shaped strip, Fig. 2 shows the S-parameters of the proposed antenna with and without the strip. In the $|S_{11}|$ plot, without the strip, the impedance bandwidth covers 1.55 GHz to 2.5 GHz band, thus setting $|S11|$ as better than -10 dB. However, the insertion of the strip results in a wider impedance bandwidth which accommodates the band from 1.6 to 2.8 GHz with two clearly distinct resonant frequencies, i.e. 1.7 GHz and 2.4 GHz. Examining the $|S_{21}|$ curves, the mutual coupling has been improved by -5 dB from -9 to -14 dB over the entire band comparing the cases with and without the **I** -shape. Thus the **I** -shaped structure is acting as an impedance bandwidth enhancement and inter-element decoupling network.

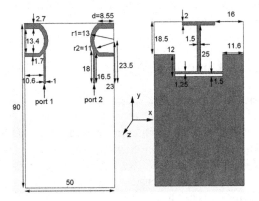

Fig. 1. The geometry of proposed printed planar MIMO antenna

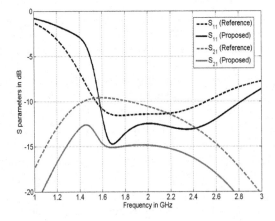

Fig. 2. Simulated S-parameters of the proposed antenna with and without the I-shaped structure.

To further understand the contribution of the **I** - shaped structure to the performance, the surface current distributions of the antenna with and without this structure are also illustrated in Fig. 3. In this study, two frequencies of 1.7 and 2.5 GHz were selected, representing lower and upper points in the desired frequency band. As can be clearly noticed, when port 1 is excited and port 2 is terminated in 50 Ω without the **I** - shaped structure, a strong induced current appears on the port 2 antenna element. This can be attributed to the presence of a near field coupling current and a shared common ground plane current when the two antennas are closely placed. However, when the structure is introduced on the ground plane, it is found that the surface current is trapped on this structure and this attenuates the induced current on the port 2 of the antenna elements. This tends to decouple the currents on the port 2 radiator and hence improves the inter-port isolation between the antenna elements.

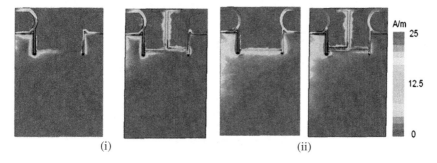

(i)   (ii)

Fig. 3. Contour plot surface current distributions with and without the **I** - shape at (i) 1.7 GHz, and (ii) 2.5 GHz. Port 1 (left) is excited and port 2 (right) is terminated in 50Ω.

3   Results and Discussion

A prototype of the proposed antenna was fabricated as shown in Fig. 4 and tested in order to validate the simulated results. Figure 5 compares the computed and experimental reflection |S$_{11}$| and mutual coupling |S$_{21}$| coefficients of the antenna. Observing the |S$_{11}$| plots, these exhibit an impedance bandwidth of 1.2 GHz, from 1.6 to 2.8 GHz for the criterion of |S$_{11}$| better than −10 dB, which is equivalent to 54.5 %. This wide bandwidth enables the antenna to meet the requirements of the GSM1800/1900, UMTS2000, LET2300, LT2600, WiFi and WiMAX frequency bands. Further scrutinizing the mutual coupling |S$_{21}$| curves, it is notable that inter-element mutual coupling is better than −14 dB over the entire operating frequency band. Both simulated and measured results are in satisfactory agreement, and minor discrepancies in these results can be attributed to fabrication tolerances, SMA connector effects and some uncertainty in the electrical properties of the substrate material.

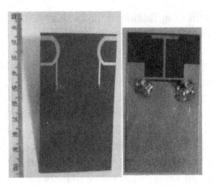

Fig. 4. Practical prototype of the proposed antenna

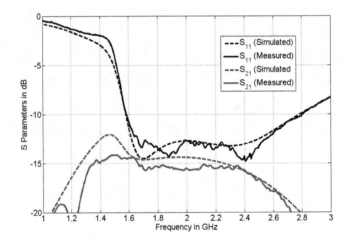

Fig. 5. S parameters of the proposed antenna

To check the MIMO/diversity performance of the proposed antenna, the two key performance indicators, i.e. envelope correlation coefficient (ECC) and channel capacity loss, will be considered. It is well known that ECC can be computed by using 3D far-field data [4, 5] and S-parameters [11, 21, 25] methods, but as the far-field method is very time consuming and involves complex integral calculations of radiation pattern data, the S-parameters method, as described in [11, 21, 25], was adopted. The simulated and measured ECC of the MIMO antenna are presented in Fig. 6(a), showing that computed and experimental results are in good agreement. ECC values for the antenna are less than 0.006 over the targeted operating frequency band. These results are comparable to the published results in [11, 21, 25]. The simplified channel capacity loss of a 2×2 MIMO system can be evaluated by using the following equation, given in [11, 21]:

$$C_{loss} = - \log_2 \det(\psi^R) \tag{1}$$

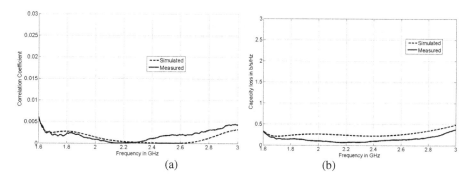

Fig. 6. Simulated and measured MIMO characteristics of the proposed antenna, (a) correlation coefficient, (b) capacity loss

where ψ^R is the receiving antenna correlation matrix: $\psi^R = \begin{bmatrix} \rho_{11} & \rho_{12} \\ \rho_{21} & \rho_{22} \end{bmatrix}$,

with $\rho_{ii} = 1 - \left(|S_{ii}|^2 + |S_{ij}|^2 \right)$, and $\rho_{ij} = - \left(S_{ii}^* S_{ij} + S_{ji}^* S_{jj} \right)$, for i, j = 1 or 2.

Using the above formulas, simulated and measured channel capacity loss of the MIMO antenna can be estimated as less than 0.3 b/s/Hz, which is acceptable for practical MIMO systems [11, 25]. Small discrepancies between the results can be attributed to fabrication errors and feed cable effects.

Figure 7 shows the simulated and measured peak gains of the MIMO antenna over the 1.6 to 2.8 GHz frequency band. Here the peak gain is about 1.7–2.5 dBi and 1.2–2.3 dBi for the computed and experimental results respectively over the band. The worst disagreement between the simulation and measurement is about 0.5 dBi at the lowest operating frequency, i.e. 1.6 GHz. To demonstrate the consistency of the radiation pattern over all designated operating frequency bands, simulated and measured radiation patterns of the proposed antenna are shown in Fig. 8. Since the structure of the antenna element differs substantially from conventional designs, it is difficult to predict its radiation pattern behavior other than by simulation. In the measurements, three pattern cuts (i.e. x-z, y-z and x-y planes) were taken at two representative operating frequencies, i.e. 1.8 and 2.4 GHz. The presented results show that the radiation patterns are consistent at these two frequencies, and the computed and measured radiation patterns are seen to be in acceptable agreement. Some small discrepancies between the simulated and measured results can be attributed to the physical feeding arrangements. It should be highlighted that the cross-polarization ratio of this antenna is acceptable, as the issue of polarization purity is not critical for this antenna's use in portable devices.

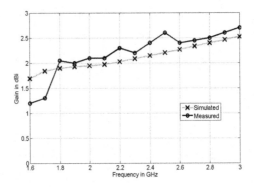

Fig. 7. Measured and simulated peak gains of the proposed antenna.

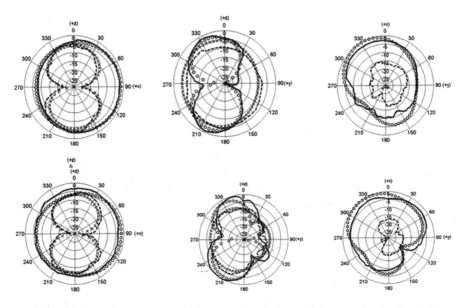

Fig. 8. Simulated and measured normalized radiation patterns of the proposed antenna for three planes (left: x-z plane, centre: y-z plane and right: x-y plane) at (a) 1.8 GHz and (b) 2.4 GHz 'xxxx' simulated cross-polarization 'oooo' simulated co-polarization '------' measured cross-polarization '———' measured co-polarization

4 Conclusion

A broadband printed MIMO monopole antenna which is suitable for commercial mobile/wireless handheld devices was developed. It offers a wide operating frequency band from 1.6 GHz to 2.8 GHz corresponding to a bandwidth of 1.2 GHz. To achieve a good isolation between the two antenna elements, an **I** - shaped strip was introduced in a defected ground plane to effectively enhance the bandwidth and suppress the mutual coupling. By implementing this, inter-port isolation as good as 14 dB and return loss better than 10 dB across the designated frequency band was realized. This diversity antenna, with envelope dimensions of 50 x 90 x 0.8 mm^3, exhibits sufficient impedance bandwidth, suitable radiation characteristics, adequate gains, and low correlation coefficient and channel capacity loss for GSM1800/1900, UMTS2000, LTE 2300, LTE2600, WiFi (2.4 GHz) and WiMAX (2.5 GHz) applications.

References

1. Hanzo, L., Haas, L., Imre, S., O'Brien, D., Rupp, M., Gyongyosi, L.: Wireless myths, realities, and futures: from 3G/4G to optical and quantum wireless. Proc. IEEE **100**, 1853–1888 (2012)
2. Hanzo, L., El-Hajjar, M., Alamri, O.: Near-capacity wireless transceivers and cooperative communications in the MIMO era: evolution of standards, waveform design, and future perspectives. Proc. IEEE **99**(8), 1343–1385 (2011)
3. Ying, L.: Antennas in cellular phones for mobile communications. Proc. IEEE **101**, 2286–2296 (2012)
4. Shin, H., Lee, J.H.: Capacity of multiple-antenna fading channels: spatial fading correlation, double scattering, and keyhole. IEEE Trans. Inform. Theory **49**, 2636–2647 (2003)
5. Wallace, J., Jensen, M., Swindlehurst, A., Jeffs, B.: Experimental characterization of the MIMO wireless channel: data acquisition and analysis. IEEE Trans. Wirel. Commun. **2**, 335–343 (2003)
6. Shin, Y.S., Park, S.O.: A monopole antenna with a magneto-dielectric material and its MIMO applications for 700 MHz LTE-band. Microw. Opt. Technol. Lett. **52**, 603–606 (2010)
7. Lee, J., Hong, Y.-K., Bae, S., Abo, G.S., Seong, W.-M., Kim, G.-H.: Miniature long-term evolution (LTE) MIMO ferrite antenna. IEEE Antennas Wirel. Propag. Lett. **10**, 2364–2367 (2011)
8. Sharawi, M.S., Iqbal, S.S., Faouri, Y.S.: An 800 MHz 2 X 1 compact MIMO antenna system for LTE handsets. IEEE Trans. Antennas Propag. **59**, 3128–3131 (2011)
9. Li, Z., Du, Z., Takahashi, M., Saito, K., Ito, K.: Reducing mutual coupling of MIMO antennas with parasitic elements for mobile terminals. IEEE Trans. Antennas Propag. **60**, 473–481 (2012)
10. Minz, L., Garg, R.: Reduction of mutual coupling between closely spaced PIFAs. Electron. Lett. **46**, 392–394 (2010)
11. Su, S.-W., Lee, C.T., Chang, F.-S.: Printed MIMO-antenna system using neutralization-line technique for wireless USB-Dongle applications. IEEE Trans. Antennas Propag. **60**, 456–463 (2012)
12. Li, H., Xiong, J., He, S.: A compact planar MIMO antenna system of four elements with similar radiation characteristic and isolation structure. IEEE Antennas Wirel. Propag. Lett. **8**, 1107–1110 (2009)
13. Chen, S.-C., Wang, Y.-S., Chung, S.-J.: A decoupling technique for increasing the port isolation between two strongly coupled antennas. IEEE Trans. Antennas Propag. **56**, 3650–3658 (2008)
14. Krairiksh, M., Keowsawat, P., Phongcharoenpanich, C.: Two-probe excited circular ring antenna for MIMO application. Prog. Electromagnetics Res. **97**, 417–431 (2009)
15. OuYang, J., Yang, F., Wang, Z.M.: Reducing mutual coupling of closely spaced microstrip MIMO antennas for WLAN application. IEEE Antennas Wirel. Propag. Lett. **10**, 310–313 (2011)

16. Mallahzadeh, A.R., Es'haghi, S., Alipour, A.: Design of an E shaped MIMO antenna using IWO algorithm for wireless application at 5.8 GHz. Prog. Electromagnetics Res. **97**, 417–431 (2009)

17. Tang, X., Mouthaan, K., Coetzee, J.C.: Tunable decoupling and matching network for diversity enhancement of closely spaced antennas. IEEE Antennas Wirel. Propag. Lett. **11**, 268–271 (2012)

18. Zhou, X., Li, R., Jin, G., Tentzeris, M.M.: A compact broadband MIMO antenna for mobile handset applications. Microw. Opt. Technol. Lett. **53**, 2773–2776 (2011)

19. Zhang, S., Zetterberg, P., He, S.: Printed MIMO antenna system of four closely-spaced elements with large bandwidth and high isolation. Electron. Lett. **46**, 1052–1053 (2010)

20. Kang, G., Du, Z., Gong, K.: Compact broadband printed slot-monopole-hybrid diversity antenna for mobile terminal. IEEE Antennas Wirel. Propag. Lett. **10**, 159–162 (2011)

21. See, C.H., Abd-Alhameed, R.A., Abidin, Z.Z., McEwan, N.J., Excell, P.S.: Wideband printed MIMO/Diversity monopole antenna for WiFi/WiMAX applications. IEEE Trans. Antennas Propag. **60**, 2028–2035 (2012)

22. Ding, Y., Du, Z., Gong, K., Feng, Z.: A novel dual-band printed diversity antenna for mobile terminals. IEEE Trans. Antenna Propag. **55**, 2088–2096 (2007)

23. Zhang, S., Lau, B.K., Tan, Y., Ying, Z., He, S.: Mutual coupling reduction of two PIFA with a T-shape slot impedance transformer for MIMO mobile terminals. IEEE Trans. Antennas Propag. **60**, 1521–1531 (2012)

24. Addaci, R., Diallo, A., Luxey, C., Thuc, P.L., Staraj, R.: Dual-Band wlan diversity antenna-system with high port-to-port isolation. IEEE Antennas Wireless Propag. Lett. **11**, 244–247 (2012)

25. Zhou, X., Quan, X., Li, R.: A dual-broadband MIMO antenna system for GSM/UMTS/LTE and WLAN handsets. IEEE Antennas Wirel. Propag. Lett. **11**, 514–551 (2012)

26. See, C.H., Abd-Alhameed, R.A., Zhou, D., Lee, T.H., Excell, P.S.: A crescent-shaped multiband planar monopole antenna for mobile wireless applications. IEEE Antenna Wirel. Propag. Lett. **9**, 152–155 (2010)

A New Approach for Implementing QO-STBC Over OFDM

Yousef A.S. Dama[1], Hassan Migdadi[2], Wafa Shuaieb[2],
Elmahdi Elkazmi[1,3], Eshtiwi A. Abdulmula[1,4],
Raed A. Abd-Alhameed[1(✉)], Walaa Hammoudeh[1],
and Ahmed Masri[1]

[1] Telecommunication Engineering Department,
An-Najah National University, Nablus, Palestine
yasdama@najah.edu, r.a.a.abd@bradford.ac.uk
[2] Electrical Engineering and Computer Science,
University of Bradford, Bradford BD7 1DP, UK
[3] The Higher Institute of Electronics, Bani Walid, Libya
[4] Higher Institute for Comprehensive Careers, Tarhuna, Libya

Abstract. A new approach for implementing QO-STBC and DHSTBC over OFDM for four, eight and sixteen transmitter antennas is presented, which eliminates interference from the detection matrix and improves performance by increasing the diversity order on the transmitter side. The proposed code promotes diversity gain in comparison with the STBC scheme, and also reduces Inter Symbol Interference.

Keywords: MIMO-OFDM system · Quasi-Orthogonal space time block coding (QO-STBC) over OFDM · Full rate, full diversity order · Eigenvector · Diagonalized hadamard space time code (DHSTBC) over OFDM

1 Introduction

Single-Input Single-Output (SISO) communication systems have a single antenna at both the transmitter and the receiver, with resulting limitations in capacity. To increase the capacity of SISO systems, large bandwidths and high transmit power would be required. Alternatively, MIMO systems could give improvements without the need to increase the transmission power or the bandwidth, also decreasing the error rates in comparison with single-antenna systems [1].

High data-rate wireless systems with very small symbol periods usually face unacceptable Inter-Symbol Interference (ISI) originating from multipath propagation and the resulting delay spread. Orthogonal Frequency Division Multiplexing (OFDM) is a multicarrier-based technique for mitigating ISI whose spectral efficiency improves capacity. [2]

The structure of a MIMO-OFDM system is described in Fig. 1

In 2013, Dama et al. proposed a new approach for Quasi-Orthogonal Space-Time Block Coding (QO-STBC), which eliminated interference from the detection matrix, thus improving the diversity gain compared with the conventional QO-STBC scheme

© Institute for Computer Sciences, Social Informatics and Telecommunications Engineering 2015
P. Pillai et al. (Eds.): WiSATS 2015, LNICST 154, pp. 249–259, 2015.
DOI: 10.1007/978-3-319-25479-1_19

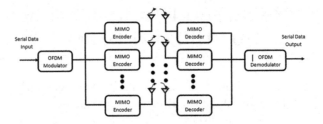

Fig. 1. MIMO-OFDM block diagram

[3]. The method was then extended to Diagonalized Hadamard Space-Time Block Coding (DHSTBC), to provide full rate diversity. These approaches were implemented for MIMO systems with three and four transmitter antennas [3–5].

In the present paper, QO-STBC and DHSTBC are implemented for OFDM systems using four, eight and sixteen transmitter antennas.

2   Quasi-Orthogonal Space Time Block Coding (QO-STBC)

2.1   QO-STBC with Four Transmit Antennas

In quasi-orthogonal coding, the columns of the transmission matrix are divided into groups. Columns within each group are not orthogonal to each other but those from different groups are mutually orthogonal [6]. Pairs of transmitted symbols can be decoded independently, but there is some loss of diversity in QOSTBC due to coupling terms between the estimated symbols [7].

For four symbols x_1, x_2, x_3 and x_{41}, the encoding matrix X_{ABBA} is formed from two (2×2) Alamouti code matrices X_{12} and X_{34}:

$$X_{12} = \begin{bmatrix} x_1 & x_2 \\ -x_2^* & x_1^* \end{bmatrix} X_{34} = \begin{bmatrix} x_3 & x_4 \\ -x_4^* & x_3^* \end{bmatrix} \qquad (1)$$

And so

$$X_{ABBA} = \begin{bmatrix} X_{12} & X_{34} \\ X_{34} & X_{12} \end{bmatrix} \qquad (2)$$

The equivalent virtual channel matrix H_v can be written as:

$$H_v = \begin{bmatrix} h_1 & h_2 & h_3 & h_4 \\ h_2^* & -h_1^* & h_4^* & -h_3^* \\ h_3 & h_4 & h_1 & h_2 \\ h_4^* & -h_3^* & h_2^* & -h_1^* \end{bmatrix} \qquad (3)$$

Considering a linear system of the form

$$Y = HX + n \tag{4}$$

A simple method to decode QO-STBC over OFDM is by applying the maximum ratio combining (MRC) technique: the received vector Y is multiplied by H_v^H thus:

$$X = H_v^H Y = H_v^H . H_v X_{ABBA} + H_v^H n$$
$$= D_4 X_{ABBA} + H_v^H n \tag{5}$$

where $D_4 = H_v^H H_v$ is a non-diagonal detection matrix, H_v^H is the Hermitian of H_v and n is the noise vector of AWGN channel.

$$D_4 = \begin{bmatrix} \alpha & 0 & \beta & 0 \\ 0 & \alpha & 0 & \beta \\ \beta & 0 & \alpha & 0 \\ 0 & \beta & 0 & \alpha \end{bmatrix} \tag{6}$$

The diagonal elements α and β in Eq. 6 α represent the channel gain and the interference from other signals respectively, and they are defined as follows,

$$\alpha = |h_1|^2 + |h_2|^2 + |h_3|^2 + |h_4|^2$$
$$\beta = h_1^* h_3 + h_2 h_4^* + h_3^* h_1 + h_4 h_2^* \tag{7}$$

Since the interference terms β will cause performance degradation, more complex decoding methods have been introduced to estimate $\hat{X}$ [3, 4].

The solution of the eigenvalue problem of the detection matrix D_4 can be written as,

$$D_4 V_{QO-STBC} - V_{QO-STBC} D_{QO-STBC} = 0 \tag{8}$$

where $D_{QO-STBC}$ and $V_{QO-STBC}$ are the eigenvectors and eigenvalues of D_4 respectively,

$$D_{4QO-STBC} = \begin{bmatrix} \alpha + \beta & 0 & 0 & 0 \\ 0 & \alpha + \beta & 0 & 0 \\ 0 & 0 & \alpha - \beta & 0 \\ 0 & 0 & 0 & \alpha - \beta \end{bmatrix} \tag{9}$$

$$V_{4QO-STBC} = \begin{bmatrix} 1 & 0 & -1 & 0 \\ 0 & 1 & 0 & -1 \\ 1 & 0 & 1 & 0 \\ 0 & 1 & 0 & 1 \end{bmatrix} \tag{10}$$

From this basis, the channel matrix for four transmit antennas can be defined as:

$$H_{4QO-STBC} = H_v V_{4QO-STBC} \tag{11}$$

where $H_{4QO-STBC}$ is given by:

$$H_{4QO-STBC} = \begin{bmatrix} h_1 + h_3 & h_2 + h_4 & h_3 - h_1 & h_4 - h_2 \\ h_2^* + h_4^* & -h_1^* - h_3^* & h_4^* - h_2^* & h_1^* - h_3^* \\ h_1 + h_3 & h_2 + h_4 & h_1 - h_3 & h_2 - h_4 \\ h_2^* + h_4^* & -h_1^* - h_3^* & h_2^* - h_4^* & h_3^* - h_1^* \end{bmatrix} \tag{12}$$

$H_{4QO-STBC}^H \cdot H_{4QO-STBC}$ is a diagonal matrix which can achieve simple linear decoding, because of the orthogonal characteristics of the channel matrix $H_{4QO-STBC}$

The encoding matrix $X_{4QO-STBC}$ corresponding to the channel matrix $H_{4QO-STBC}$ can be derived as follows:

$$X_{4QO-STBC} = \begin{bmatrix} x_1 - x_3 & x_2 - x_4 & x_1 + x_3 & x_2 + x_4 \\ x_4^* - x_2^* & -x_3^* + x_1^* & -x_4^* - x_2^* & x_3^* + x_1^* \\ x_1 + x_3 & x_2 + x_4 & x_3 - x_1 & x_4 - x_2 \\ -x_4^* - x_2^* & x_3^* + x_1^* & x_4^* - x_2^* & -x_3^* + x_1^* \end{bmatrix} \tag{13}$$

2.2   QO-STBC for Eight and Sixteen Transmit Antennas

In the case of four transmitter antennas the diagonal terms α are the channel gains and off-diagonal terms β represent interference. With eight and sixteen transmitter antennas α_8, α_{16} are the channel gains described in Eqs. (14) and (15) respectively, and β_8, γ_8, σ_8, β_{16}, γ_{16}, σ_{16}, ω_{16}, ζ_{16}, η_{16} and φ_{16} are the interference from neighboring signals

Using the same methodology as in section the channel gains and the interference terms for eight transmitter antennas can be written as follows,

$$\begin{aligned} \alpha_8 &= \alpha + |h_5|^2 + |h_6|^2 + |h_7|^2 + |h_8|^2 \\ \beta_8 &= \beta + h_5^* h_7 + h_6 h_8^* + h_7^* h_5 + h_8 h_6^* \\ \gamma_8 &= h_1^* h_5 + h_2 h_6^* + h_3^* h_7 + h_4 h_8^* + h_5^* h_1 + h_6 h_2^* + h_7^* h_3 + h_8 h_4^* \\ \sigma_8 &= h_1^* h_7 + h_2 h_8^* + h_3^* h_5 + h_4 h_6^* + h_5^* h_3 + h_6 h_4^* + h_7^* h_1 + h_8 h_2^* \end{aligned} \tag{14}$$

In the same way the channel gains and the interference terms are derived for sixteen transmitter antennas:

$$\begin{aligned} \alpha_{16} &= \alpha_8 + |h_9|^2 + |h_{10}|^2 + |h_{11}|^2 + |h_{12}|^2 + |h_{13}|^2 + |h_{14}|^2 + |h_{15}|^2 + |h_{16}|^2 \\ \beta_{16} &= \beta_8 + h_9^* h_{11} + h_{10} h_{12}^* + h_{11}^* h_9 + h_{12} h_{10}^* + h_{13}^* h_{15} + + h_{14} h_{16}^* + h_{15}^* h_{13} + h_{16} h_{14}^* \\ \gamma_{16} &= \gamma + h_9^* h_{13} + h_{10} h_{14}^* + h_{11}^* h_{15} + h_{12} h_{16}^* + h_{13}^* h_9 + h_{10}^* + h_{15}^* h_{11} + h_{16} h_{12}^* \\ \sigma_{16} &= \sigma + h_9^* h_{15} + h_{10} h_{16}^* + h_{11}^* h_{13} + h_{12} h_{14}^* + h_{13}^* h_{11} + h_{14} h_{12}^* + h_{15}^* h_9 + h_{16} h_{10}^* \end{aligned}$$

$$\omega_{16} = h_1^*h_9 + h_2h_{10}^* + h_3^*h_{11} + h_4h_{12}^* + h_5^*h_{13} + h_6h_{14}^* + h_7^*h_{15} + h_8h_{16}^* + h_9^*h_1 + h_{10}h_2^*$$
$$+ h_{11}^*h_3 + h_{12}h_{14}^* + h_{13}^*h_{11} + h_{14}h_{12}^* + h_{15}^*h_9 + h_{16}h_{10}^*$$

$$\zeta_{16} = h_1^*h_{11} + h_2h_{12}^* + h_3^*h_9 + h_4h_{10}^* + h_5^*h_{15} + h_6h_{16}^* + h_7^*h_{13} + h_8h_{14}^* + h_9^*h_3 + h_{10}h_4^*$$
$$+ h_{11}^*h_1 + h_{12}h_2^* + h_{13}^*h_7 + h_{14}h_8^* + h_{15}^*h_5 + h_{16}h_6^*$$

$$\eta_{16} = h_1^*h_{13} + h_2h_{14}^* + h_3^*h_{15} + h_4h_{16}^* + h_5^*h_9 + h_6h_{10}^* + h_{13}^*h_1 + h_{14}h_2^* + h_{15}^*h_3 + h_{16}h_4^*$$

$$\varphi_{16} = h_1^*h_{15} + h_2h_{16}^* + h_3^*h_{13} + h_4h_{14}^* + h_5^*h_{11} + h_6h_{12}^* + h_7^*h_9 + h_8h_{10}^* + h_9^*h_7 + h_{10}h_8^*$$
$$+ h_{11}^*h_5 + h_{12}h_6^* + h_{13}^*h_3 + h_{14}h_4^* + h_{15}^*h_1 + h_{16}h_2^*$$

$$(15)$$

The eigenvalues matrix $D_{8QO-STBC}$ and the corresponding eigenvectors $V_{8QO-STBC}$ for eight transmitter antennas are given by Eqs. (16) and (17).

$$D_{8QO-STBC} = \begin{bmatrix} \beta+\alpha-\sigma-\gamma & 0 & 0 & 0 & 0 & 0 & 0 \\ 0 & \beta+\alpha-\sigma-\gamma & 0 & 0 & 0 & 0 & 0 \\ 0 & 0 & -\beta-\alpha+\sigma+\gamma & 0 & 0 & 0 & 0 \\ 0 & 0 & 0 & -\beta-\alpha+\sigma+\gamma & 0 & 0 & 0 \\ 0 & 0 & 0 & 0 & -\beta+\alpha+\sigma-\gamma & 0 & 0 \\ 0 & 0 & 0 & 0 & 0 & -\beta+\alpha+\sigma-\gamma & 0 \\ 0 & 0 & 0 & 0 & 0 & 0 & \beta+\alpha+\sigma+\gamma \\ 0 & 0 & 0 & 0 & 0 & 0 & 0 & \beta+\alpha+\sigma+\gamma \end{bmatrix}$$

$$(16)$$

$$V_{8QO-STBC} = \begin{bmatrix} -1 & 0 & -1 & 0 & 1 & 0 & 1 & 0 \\ 0 & -1 & 0 & -1 & 0 & -1 & 0 & 1 \\ -1 & 0 & 1 & 0 & -1 & 0 & 1 & 0 \\ 0 & -1 & 0 & 1 & 0 & 1 & 0 & 1 \\ 1 & 0 & -1 & 0 & -1 & 0 & 1 & 0 \\ 0 & 1 & 0 & -1 & 0 & 1 & 0 & 1 \\ 1 & 0 & 1 & 0 & 1 & 0 & 1 & 0 \\ 0 & 1 & 0 & 1 & 0 & -1 & 0 & 1 \end{bmatrix} \qquad (17)$$

From Eq. (18), the new channel matrix is derived based on the virtual channel matrix as shown in Eq. (19).

$$H_{8QO-STBC} = H_{v8}V_{8QO-STBC} \qquad (18)$$

Where,

$$H_{v8} = \begin{bmatrix} h_1 & h_2 & h_3 & h_4 & h_5 & h_6 & h_7 & h_8 \\ h_2^* & -h_1^* & h_4^* & -h_3^* & h_6^* & -h_5^* & h_8^* & -h_7^* \\ h_3 & h_4 & h_1 & h_2 & h_7 & h_8 & h_5 & h_6 \\ h_4^* & -h_3^* & h_2^* & -h_1^* & h_8^* & -h_7^* & h_6^* & -h_5^* \\ h_5 & h_6 & h_7 & h_8 & h_1 & h_2 & h_3 & h_4 \\ h_6^* & -h_5^* & h_8^* & -h_7^* & h_2^* & -h_1^* & h_4^* & -h_3^* \\ h_7 & h_8 & h_5 & h_6 & h_3 & h_4 & h_1 & h_2 \\ h_8^* & -h_7^* & h_6^* & -h_5^* & h_4^* & -h_3^* & h_2^* & -h_1^* \end{bmatrix} \qquad (19)$$

Then the encoding matrix $X_{8QO-STBC}$ is derived corresponding to the channel matrix $H_{8QO-STBC}$ as in Eqs. (20) and (21)

$H_{8QO-STBC} =$

$$
\begin{bmatrix}
-h_1-h_3+h_5+h_7-h_2-h_4+h_6+h_8 & h_1+h_3+h_5+h_7 & h_2+h_4+h_6+h_8 & h_1-h_3+h_5-h_7 & h_2-h_4+h_6-h_8 & h_2-h_4-h_6+h_8 & -h_1+h_3+h_5-h_7 \\
-h_2^*-h_4^*+h_6^*+h_8^* & h_1^*+h_3^*-h_5^*-h_7^* & h_2^*+h_4^*+h_6^*+h_8^* & -h_1^*-h_3^*-h_5^*-h_7^* & h_2^*-h_4^*+h_6^*-h_8^* & -h_1^*+h_3^*-h_5^*+h_7^* & -h_1^*+h_3^*+h_5^*-h_7^*-h_2^*+h_4^*+h_6^*-h_8^* \\
-h_1-h_3+h_5+h_7-h_2-h_4+h_6+h_8 & h_1+h_3+h_5+h_7 & h_2+h_4+h_6+h_8 & h_3-h_1+h_7-h_5 & h_4-h_2+h_8-h_6 & -h_3+h_1+h_7-h_5 \\
-h_2^*-h_4^*+h_6^*+h_8^* & h_1^*+h_3^*-h_5^*-h_7^* & h_2^*+h_4^*+h_6^*+h_8^* & -h_1^*-h_3^*-h_5^*-h_7^* & h_2^*-h_4^*+h_6^*-h_8^* & -h_3^*+h_1^*-h_7^*+h_5^*-h_3^*+h_1^*+h_7^*-h_5^*-h_4^*+h_2^*+h_8^*-h_6^* \\
-h_5-h_7+h_1+h_3-h_6-h_8+h_2+h_4 & h_1+h_3+h_5+h_7 & h_2+h_4+h_6+h_8 & h_1-h_3+h_5-h_7 & h_2-h_4+h_6-h_8 & h_4-h_2-h_8+h_6 \\
-h_6^*-h_8^*+h_2^*+h_4^* & h_5^*+h_7^*-h_1^*-h_3^* & h_2^*+h_4^*+h_6^*+h_8^* & -h_1^*-h_3^*-h_5^*-h_7^* & h_2^*-h_4^*+h_6^*-h_8^* & -h_1^*+h_3^*-h_5^*+h_7^*+h_1^*+h_7^*-h_5^*+h_2^*+h_8^*-h_6^* \\
-h_5-h+h_1+h_3-h_6-h_8+h_2+h_4 & h_1+h_3+h_5+h_7 & h_2+h_4+h_6+h_8 & h_3-h_1+h_7-h_5 & h_4-h_2+h_8-h_6 & h_2-h_4-h_6+h_8-h_1+h_3+h_5-h_7 \\
-h_6^*-h_8^*+h_2^*+h_4^* & h_5^*+h_7^*-h_1^*-h_3^* & h_2^*+h_4^*+h_6^*+h_8^* & -h_1^*-h_3^*-h_5^*-h_7^* & h_2^*-h_4^*+h_6^*-h_8^* & -h_3^*+h_1^*+h_7^*-h_5^*-h_1^*+h_3^*+h_5^*-h_7^*-h_2^*+h_4^*+h_6^*-h_8^*
\end{bmatrix}
$$

(20)

$X_{8QO-STBC} =$

$$
\begin{bmatrix}
-x_1-x_3+x_5+x_7 & -x_2-x_4-x_6+x_8 & -x_1+x_3-x_5+x_7 & -x_2+x_4+x_6+x_8 & x_1-x_3-x_5+x_7 & x_2-x_4+x_6+x_8 & x_1+x_3+x_5+x_7 & x_2+x_4-x_6+x_8 \\
x_2^*+x_4^*+x_6^*-x_8^* & -x_1^*-x_3^*+x_5^*+x_7^* & x_2^*-x_4^*-x_6^*-x_8^* & -x_1^*+x_3^*-x_5^*+x_7^* & x_2^*-x_3^*-x_5^*+x_7^* & -x_2^*-x_4^*+x_6^*-x_8^* & x_1^*+x_3^*+x_5^*+x_7^* \\
-x_1+x_3-x_5+x_7 & -x_2-x_4+x_6+x_8 & -x_1-x_3+x_5+x_7 & -x_2-x_4-x_6+x_8 & x_1+x_3+x_5+x_7 & x_2+x_4-x_6+x_8 & x_1-x_3-x_5+x_7 & x_2-x_4+x_6+x_8 \\
x_2^*-x_4^*-x_6^*-x_8^* & -x_1^*+x_3^*-x_5^*+x_7^* & x_2^*+x_4^*+x_6^*-x_8^* & -x_1^*-x_3^*+x_5^*+x_7^* & x_2^*+x_4^*-x_6^*-x_8^* & x_1^*-x_3^*-x_5^*+x_7^* \\
x_1-x_3-x_5+x_7 & x_2-x_4+x_6+x_8 & x_1+x_3+x_5+x_7 & x_2+x_4-x_6+x_8 & -x_1-x_3+x_5+x_7 & -x_2-x_4-x_6+x_8 & -x_1+x_3-x_5+x_7-x_2+x_4+x_6+x_8 \\
-x_2^*+x_4^*-x_6^*-x_8^* & x_1^*-x_3^*-x_5^*+x_7^* & -x_2^*-x_4^*+x_6^*-x_8^* & x_1^*+x_3^*+x_5^*+x_7^* & x_2^*+x_4^*+x_6^*-x_8^* & -x_1^*-x_3^*+x_5^*+x_7^* & x_2^*-x_4^*-x_6^*-x_8^* & -x_1^*+x_3^*-x_5^*+x_7^* \\
x_1+x_3+x_5+x_7 & x_2+x_4-x_6+x_8 & x_1-x_3-x_5+x_7 & x_2-x_4+x_6+x_8 & -x_1+x_3-x_5+x_7 & -x_2+x_4+x_6+x_8 & -x_1-x_3+x_5+x_7-x_2-x_4-x_6+x_8 \\
-x_2^*-x_4^*+x_6^*-x_8^* & x_1^*+x_3^*+x_5^*+x_7^* & -x_2^*+x_4^*-x_6^*-x_8^* & x_1^*-x_3^*-x_5^*+x_7^* & x_2^*-x_4^*-x_6^*-x_8^* & -x_1^*+x_3^*-x_5^*+x_7^* & x_2^*+x_4^*+x_6^*-x_8^* & -x_1^*-x_3^*+x_5^*+x_7^*
\end{bmatrix}
$$

(21)

Similarly, the detection matrix for sixteen transmitters can be derived to eliminate the interference terms. The resultant channel model and coding matrices result in an interference-free detection matrix.

3   DHSTBC for Multiple Transmit Antennas

In this section a full-rate full-diversity order Diagonalized Hadamard Space-Time Code (DHSTBC) over OFDM for 4, 8 and 16 transmitter antennas is implemented. The detection matrix generated, $D = X.X^H$, is a diagonal matrix [5, 8]. The generated codes provide full rate and full diversity when the number of the receiver antennas are at least equal to the number of transmitter antennas, the code matrices for DHSTBC are Hadamard matrices of size $N = 2^n$ where $n \geq 1$.

Let $s_1, s_2, \ldots\ldots, s_N$ be the transmitted symbols. These symbols are sorted to form the cyclic matrix S_8 as in Eqs. (22) to (24) as follows,

$$
\begin{aligned}
S_{12} &= \begin{bmatrix} s_1 & s_2 \\ s_2 & s_1 \end{bmatrix} & S_{34} &= \begin{bmatrix} s_3 & s_4 \\ s_4 & s_3 \end{bmatrix} \\
S_{56} &= \begin{bmatrix} s_5 & s_6 \\ s_6 & s_5 \end{bmatrix} & S_{78} &= \begin{bmatrix} s_7 & s_8 \\ s_8 & s_7 \end{bmatrix} \\
S_4 &= \begin{bmatrix} S_{12} & S_{34} \\ S_{34} & S_{12} \end{bmatrix} & S_5 &= \begin{bmatrix} S_{56} & S_{78} \\ S_{78} & S_{56} \end{bmatrix}
\end{aligned}
$$

(22)

The transmitted matrix is,

$$
S_8 = \begin{bmatrix} S_4 & S_5 \\ S_5 & S_4 \end{bmatrix}
$$

(23)

The same procedure is applied to form the transmitted matrix S_{16}, as follows,

$$S_9 = \begin{bmatrix} S_{9-10} & S_{11-12} \\ S_{11-12} & S_{9-10} \end{bmatrix} \quad S_{10} = \begin{bmatrix} S_{13-14} & S_{15-16} \\ S_{15-16} & S_{13-14} \end{bmatrix} \tag{24}$$

$$S_{11} = \begin{bmatrix} S_9 & S_{10} \\ S_{10} & S_9 \end{bmatrix} \quad S_{16} = \begin{bmatrix} S_8 & S_{11} \\ S_{11} & S_8 \end{bmatrix} \tag{25}$$

The Hadamard matrices of order four, eight and sixteen which are used to form the new channel matrix are given in Eqs. (26, 27 and 28) respectively:

$$H_4 = \begin{bmatrix} 1 & 1 & 1 & 1 \\ 1 & -1 & 1 & -1 \\ 1 & 1 & -1 & -1 \\ 1 & -1 & -1 & 1 \end{bmatrix} \tag{26}$$

$$H_8 = \begin{bmatrix} 1 & 1 & 1 & 1 & 1 & 1 & 1 & 1 \\ 1 & -1 & 1 & -1 & 1 & -1 & 1 & -1 \\ 1 & 1 & -1 & -1 & 1 & 1 & -1 & -1 \\ 1 & -1 & -1 & 1 & 1 & -1 & -1 & 1 \\ 1 & 1 & 1 & 1 & -1 & -1 & -1 & -1 \\ 1 & -1 & 1 & -1 & -1 & 1 & -1 & 1 \\ 1 & 1 & -1 & -1 & -1 & -1 & 1 & 1 \\ 1 & -1 & -1 & 1 & -1 & 1 & 1 & -1 \end{bmatrix} \tag{27}$$

$$H_{16} = \begin{bmatrix} 1 & 1 & 1 & 1 & 1 & 1 & 1 & 1 & 1 & 1 & 1 & 1 & 1 & 1 & 1 & 1 \\ 1 & -1 & 1 & -1 & 1 & -1 & 1 & -1 & 1 & -1 & 1 & -1 & 1 & -1 & 1 & -1 \\ 1 & 1 & -1 & -1 & 1 & 1 & -1 & -1 & 1 & 1 & -1 & -1 & 1 & 1 & -1 & -1 \\ 1 & -1 & -1 & 1 & 1 & -1 & -1 & 1 & 1 & -1 & -1 & 1 & 1 & -1 & -1 & 1 \\ 1 & 1 & 1 & 1 & -1 & -1 & -1 & -1 & 1 & 1 & 1 & 1 & -1 & -1 & -1 & -1 \\ 1 & -1 & 1 & -1 & -1 & 1 & -1 & 1 & 1 & -1 & 1 & -1 & -1 & 1 & -1 & 1 \\ 1 & 1 & -1 & -1 & -1 & -1 & 1 & 1 & 1 & 1 & -1 & -1 & -1 & -1 & 1 & 1 \\ 1 & -1 & -1 & 1 & -1 & 1 & 1 & -1 & 1 & -1 & -1 & 1 & -1 & 1 & 1 & -1 \\ 1 & 1 & 1 & 1 & 1 & 1 & 1 & 1 & -1 & -1 & -1 & -1 & -1 & -1 & -1 & -1 \\ 1 & -1 & 1 & -1 & 1 & -1 & 1 & -1 & -1 & 1 & -1 & 1 & -1 & 1 & -1 & 1 \\ 1 & 1 & -1 & -1 & 1 & 1 & -1 & -1 & -1 & -1 & 1 & 1 & -1 & -1 & 1 & 1 \\ 1 & -1 & -1 & 1 & 1 & -1 & -1 & 1 & -1 & 1 & 1 & -1 & -1 & 1 & 1 & -1 \\ 1 & 1 & 1 & 1 & -1 & -1 & -1 & -1 & -1 & -1 & -1 & -1 & 1 & 1 & 1 & 1 \\ 1 & -1 & 1 & -1 & -1 & 1 & -1 & 1 & -1 & 1 & -1 & 1 & 1 & -1 & 1 & -1 \\ 1 & 1 & -1 & -1 & -1 & -1 & 1 & 1 & -1 & -1 & 1 & 1 & 1 & 1 & -1 & -1 \\ 1 & -1 & -1 & 1 & -1 & 1 & 1 & -1 & -1 & 1 & 1 & -1 & 1 & -1 & -1 & 1 \end{bmatrix} \tag{28}$$

The resultant encoding matrix X. for 4, 8 and 16 transmitter antennas is a DHSTBC over OFDM and hence, the overall expression is given by,

$$X = H.S \tag{29}$$

The encoding matrix for four transmitter antennas can be generated using Eq. (30) as,

$$X_4 = \begin{bmatrix} s_1 + s_2 + s_3 + s_4 & s_1 + s_2 + s_3 + s_4 & s_1 + s_2 + s_3 + s_4 & s_1 + s_2 + s_3 + s_4 \\ s_1 - s_2 + s_3 - s_4 & s_2 - s_1 + s_4 - s_3 & s_1 - s_2 + s_3 - s_4 & s_2 - s_1 + s_4 - s_3 \\ s_1 + s_2 - s_3 - s_4 & s_1 + s_2 - s_3 - s_4 & s_3 + s_4 - s_1 - s_2 & s_3 + s_4 - s_1 - s_2 \\ s_1 - s_2 - s_3 + s_4 & s_2 - s_1 - s_4 + s_3 & s_2 - s_1 - s_4 + s_3 & s_1 - s_2 - s_3 + s\_4 \end{bmatrix} \tag{30}$$

The Same Procedure Can Be Followed to Generate the Encoding Matrices for 8 and 16 Transmit Antennas

$$X_4.X_4^H = \begin{bmatrix} 4(s_1 + s_2 + s_3 + s_4) & 0 & 0 & 0 \\ 0 & 4(s_1 - s_2 + s_3 - s_4) & 0 & 0 \\ 0 & 0 & 4(s_1 + s_2 - s_3 - s_4) & 0 \\ 0 & 0 & 0 & 4(s_1 - s_2 - s_3 + s_4) \end{bmatrix}$$
$$\tag{31}$$

One can notice that the detection $X_4 X_4^H$ is diagonal matrix where the interference terms have been eliminated which can achieve simple linear decoding as the shown in Eq. (31).

4 Simulation and Results

The performance of QO-STBC and DHSTBC over OFDM was evaluated over Rayleigh fading channel using MATLAB. The signals were modulated using 16-QAM, and the total transmit power was divided equally among the number of transmitter antennas. The fading was assumed to be constant over four, eight and sixteen consecutive symbol periods for four, eight and sixteen transmitter antennas respectively and the channel was known at the receiver. Finally the results of these methods were compared with STBC results, using the same data and channel parameters. Table 1 shows the simulation parameters.

Table 1. OFDM Simulation Parameter

Parameter	Specifications
Carrier frequency (MHz)	5.8
Sample frequency (MHz)	40
Bandwidth (MHz)	40
FFT size	128
Cyclic prefix ratio	0.25
Constellation	16-QAM
Data subcarrier/Pilots	108/6
Virtual carrier	14

Figure 2 shows the BER performance of QO-STBC over OFDM for four, eight and sixteen transmit antennas. The best BER is achieved by using sixteen transmitter antennas, since this gives the largest diversity order.

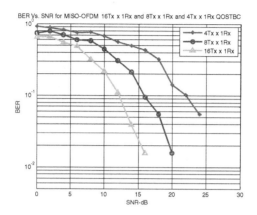

Fig. 2. BER performance of QO-STBC over OFDM for Four, Eight and Sixteen Transmitter Antennas.

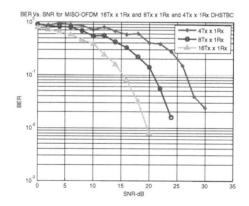

Fig. 3. BER performance of DHSTBC over OFDM for Four, Eight and Sixteen Transmitter Antennas

Figure 3 shows BER performance of DHSTBC over OFDM for four, eight and sixteen transmitter antennas. Again the best BER performance is achieved using sixteen transmitter antennas.

Next we compare the BER performances of QO-STBC, DHSTBC and the conventional STBC method with the same number of transmitter antennas. It's noticeable in Figs. 4, 5 and 6 that proposed DHSTBC achieves the best performance, and proposed QO-STBC outstrips conventional STBC.

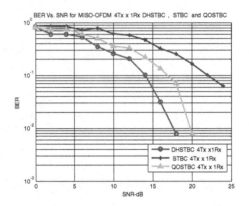

Fig. 4. BER performance of STBC, QO-STBC and DHSTBC over OFDM for Four Transmitter Antennas.

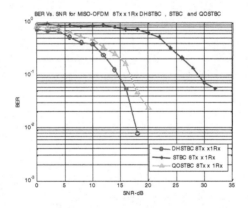

Fig. 5. BER performance of STBC, QO-STBC and DHSTBC over OFDM for Eight Transmitter Antennas

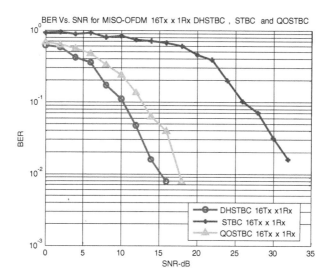

Fig. 6. BER performance of STBC, QO-STBC and DHSTBC over OFDM for Sixteen Transmitter Antennas.

5 Conclusions

New methods for QO-STBC and DHSTBC over OFDM for four, eight and sixteen transmitter antenna were implemented by deriving the orthogonal channel matrix that results in simple decoding scheme. The performance of QO-STBC and DHSTBC over OFDM was evaluated by varying the number of transmitter antennas and tested with different modulation schemes. When these compared with real STBC it shows a better performance.

References

1. Li, Y., Sollenberger, N.R.: Adaptive antenna arrays for OFDM systems with co-channel interference. IEEE Trans. Commun. **47**(2), 217–229 (1999)
2. Chiueh, T.-.D., Tsai, P.-.Y., Lai, I.-.W.: Baseband Receiver Design For Wireless MIMO-OFDM Communications, vol. 360, p. 127. Wiley-IEEE Press, New York (2012)
3. Dama, Y.A.S., Abd-Alhameed, R.A., Ghazaany, T.S., Zhu, S.: A new approach for OSTBC and QOSTBC. Int. J. Comput. Appl. **67**(6), 45–48 (2013)
4. Dama, Y.A.S., Abd-Alhameed, R.A., Jones, S.M.R., Migdadi,H.S.O., Excell, P.S.: A new approach to quasi-orthogonal space-time block coding applied to quadruple MIMO transmit antennas. In: 4th International Conference on Internet Technologies and Applications (2011)
5. Anoh, K.O.O., Dama, Y.A.S., Abd-Alhameed, R.A.A., Jones, S.M.R.: A simplified improvement on the design of QO-STBC based on hadamard matrices. Int. J. Commun. Netw. Syst. Sci. **7**, 37–42 (2014)
6. Jafarkhani, H.: A quasi-orthogonal space-time block code. IEEE Trans. Commun. **49**, 1–4 (2001)
7. Sharma, N., Papadias, C.B.: Improved quasi-orthogonal codes through constellation rotation. IEEE Trans. Commun. **51**, 332–335 (2003)
8. Seberry, J., Yamada, M.: Hadamard Matrices, Sequences and Block Designs Contemporary Design Theory: A Collection of Surveys. Wiley, New York (1992)

Special Session on Network Coding for Satellite Communication Systems

Network Coding for Multicast Communications over Satellite Networks

Esua Kinyuy Jaff[✉], Misfa Susanto, Muhammadu Ali, Prashant Pillai, and Yim Fun Hu

Faculty of Engineering and Informatics, University of Bradford, Bradford, UK
{e.k.jaff,m.susanto,m.ali70,p.pillai,
y.f.hu}@bradford.ac.uk

Abstract. Random packet errors and erasures are common in satellite communications. These types of packet losses could become significant in mobile satellite scenarios like satellite-based aeronautical communications where mobility at very high speeds is a routine. The current adaptive coding and modulation (ACM) schemes used in new satellite systems like the DVB-RCS2 might offer some solutions to the problems posed by random packet errors but very little or no solution to the problems of packet erasures where packets are completely lost in transmission. The use of the current ACM schemes to combat packet losses in a high random packet errors and erasures environment like the satellite-based aeronautical communications will result in very low throughput. Network coding (NC) has proved to significantly improve throughput and thus saves bandwidth resources in such an environment. This paper focuses on establishing how in random linear network coding (RLNC) the satellite bandwidth utilization is affected by changing values of the generation size, rate of packet loss and number of receivers in a satellite-based aeronautical reliable IP multicast communication. From the simulation results, it shows that the bandwidth utilization generally increases with increasing generation size, rate of packet loss and number of receivers.

Keywords: Aeronautical communications networks · IP multicast · Network coding · Satellite networks

1 Introduction

Satellites with their large geographical coverage and the ability to reach large satellite terminal populations present an unrivaled platform for group communication. For satellite broadband service providers to satisfy these large terminal populations and also meet the service level agreements, the efficient utilization of the available satellite bandwidth capacity becomes essential. In order to increase the available satellite capacity and high data rates, new satellite systems are now designed to support multiple spot beams (frequency reuse) and operate at high frequency bands (e.g., the Ka-band). Satellites operating at Ka-band (and higher) will likely experience an increase in random packet

© Institute for Computer Sciences, Social Informatics and Telecommunications Engineering 2015
P. Pillai et al. (Eds.): WiSATS 2015, LNICST 154, pp. 263–271, 2015.
DOI: 10.1007/978-3-319-25479-1_20

errors and erasures, thus reducing throughput in the satellite network as a whole and wasting bandwidth resources especially in reliable communication.

Nowadays, most satellite terminals have return channel capabilities. These return channel satellite terminals (RCSTs) [1] in remote locations can therefore send back acknowledgments, their channel conditions, etc. to the Network Control Centre (NCC) [1] or satellite gateway for retransmission in reliable communication and for efficient utilization of the channel respectively. In satellite communications, efficient utilization of the allocated bandwidth resources is crucial to both the satellite service providers and the customers if they are to maximize their profits and get best value for money respectively. For reliable multicast communication, one of the most intriguing challenge when designing a satellite-based content distribution platform is the efficient bandwidth management scheme. Despite the use of negative acknowledgments (NACKs) in reliable multicasting in order to reduce bandwidth overhead, the lossy nature of satellite channels and the potentially large satellite terminal populations imply that a huge volume of NACKs can still be generated. This bandwidth management challenge can become more acute in mobile satellite scenarios like the satellite-based aeronautical communication which is more likely to witness higher random packet errors and erasures due to mobility at high speeds.

Although the introduction of adaptive coding and modulation (ACM) in some satellite systems like the DVB-RCS2 [2] has proved to increase throughput and save some bandwidth in satellite networks, ACM offers no solution to the problems presented by erasures. Network coding (NC) has been shown to effectively solve the problems of erasure channels and random packets errors which are common in satellite networks [3] as well as increasing throughput in the network by communicating more information with fewer packet transmissions. For content distribution using reliable IP multicasting in satellite-based aeronautical communication, NC can save considerable amount of satellite bandwidth resources in scenarios where many aircrafts at different locations with varying channel conditions (under one satellite footprint) are subscribed to the same multicast group. The transmission of a few redundant coded multicast packets in such a scenario could reduce or even prevent the NACKs from being generated as the redundant coded packets can compensate for various original packets lost. This will not only lead to an increase in throughput and bandwidth conservation but also, will reduce the time required to successfully transmit a certain number of packets over the satellite considering the long satellite propagation delay.

Using Random Linear Network Coding (RLNC) [4], this paper investigates the impact on bandwidth utilization of varying generation size, number of receivers and rate of packet loss on content distribution using reliable IP multicasting in a satellite-based aeronautical communications. Unlike in most existing NC schemes over satellite networks which require the satellite terminals to be multi-homed in order to exploit the multipath scenario, the proposed scheme here is designed for a RCST with only one satellite interface. The focus here is to determine how the changing values of the above stated parameters affect the bandwidth usage i.e., number of transmissions required for all group members (aircrafts) to receive all the multicast packets transmitted over the lossy satellite network.

2   Literature Review of Network Coding over Satellite

Recently, some research works on NC over satellite networks have been published in open literature. In [5], the authors examined the feasibility of applying NC on different types of satellite network architectures. For transparent (bent-pipe) satellites, the authors proposed that the Analogue Network Coding (ANC) be implemented on-board the satellite. Here, two satellite terminals that want to exchange data via satellite use the same time-slot. ANC is performed at the physical layer on-board the satellite on the signals transmitted by the two terminals using the same time-slot. During the second time-slot, the satellite then transmit the coded signal to the two terminals which will individually use their own transmission and signal cancellation techniques to recover the signal from the other terminal. The advantage of using ANC is that only 2 time slots are used to transmit and also receive data by the two terminals instead of 4, implying savings to satellite bandwidth resources and an increase in network throughput. The ANC scheme faces two major drawbacks: signal saturation and interference caused by mixing signals from the same uplink beam before NC is performed and the requirement for the two signals to arrive at the satellite at the same instant. In reality, a time shift always exist between two signals even when transmitted at same time. For regenerative satellites, the authors in [5] proposed the use of XOR NC where the mixing of the two data streams is done at bit level on-board the satellite and the coded data broadcasted to the two terminals. According to [5], the advantage of using XOR NC in regenerative satellites is that the downlink capacity required by the two terminals is reduced by half.

The authors in [3, 6, 7] proposed the use of RLNC in multipath scenarios in a satellite network. In a satellite-based reliable multicast scenario in [3], RLNC is implemented at layer 2 in the satellite Hub for all traffic destined for the satellite network. The multipath scenario is created here by making use of gap-fillers which relay transmission from the satellite to the group members (subscribers). Each subscriber can therefore receive coded packets directly from the satellite (line-of sight) and also indirectly through gap-fillers. So, the redundancy of receiving two copies of each coded packet (one from each path) will compensate for any loses in any of the two paths. This compared with the traditional NACK-based reliable multicast, showed a remarkable improvement in terms of throughput. The cost of gap-fillers and the requirement for the receivers to be multi-homed are some of the weaknesses of this proposal. In [6], multipath scenario is created by making use of the overlapping area of 2 beams as each terminal located here can receive transmissions through both beams. This is also a satellite-based reliable multicast communication where RLNC is implemented at layer 2 in the satellite gateway. With NC according to [6], the multicast receivers within the overlapping area (where the erasure rate is generally higher due to weak signal strength at beam edge), witnessed a 25 % increase in throughput compared to the scheme with no NC. The authors in [7] proposed how RLNC can be used to support soft-handovers by mobile multi-homed satellite terminals in reliable unicast communication. Once the mobile terminal enters the overlapping area of two beams, the satellite starts transmitting coded packets through both beams for the roaming terminal. The advantages here are: increase in throughput due redundant packets provided by multipath which compensates for lost packets, thus preventing any retransmissions over the satellite; no specific coordination between codes at physical layer is required and load-balancing between two beams for terminals located with the overlapping area. In all proposed multipath-based NC schemes [3, 6, 7], one main common drawback is that they offer no solution to single-interface satellite receivers which cannot benefit from the multipath scenario.

In all the proposed NC schemes described above, none has examined how the throughput increases or bandwidth conservation due to NC implementation is affected by varying the generation size, rate of packet loss and number of receivers (i.e., in multicast scenarios). This paper seeks to study how changing the values of these parameter will affect the satellite bandwidth saved by implementing NC.

3 Proposed Network Architecture

Figure 1 shows the network architecture for IP multicast application in satellite-based aeronautical communications. The multicast source (content delivery) is located in the terrestrial network while the receivers are aircrafts, each equipped with a RCST for satellite communication. The satellite has a transparent payload (bent-pipe). The satellite gateway (GW) or its local network is assumed to have a multicast enabled router.

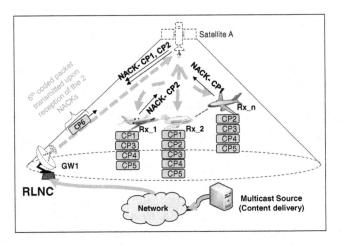

Fig. 1. Reliable IP multicast in satellite-based aeronautical communications with RLNC

Aircrafts wishing to join any multicast group send their Internet Group Management Protocol (IGMP) [8] or Multicast Listener Discovery (MLD) [9] to the GW. Upon reception of IGMP/MLD, the GW subscribes to the multicast group on behalf of the aircrafts. So, when GW1 (Fig. 1) receives multicast packets from the source, it performs RLNC on a number of original packets according to the set generation size (i.e., number of original packets mixed together) to produce a coded packet (CP) i.e.,

$$CP1 = a_{11}P1 + a_{12}P2 + a_{13}P3 + \ldots\ldots a_{1n}Pn. \tag{1}$$

Where n = generation size, a_{11}, a_{12}, a_{13}, a_{1n} are randomly chosen coefficients from a finite Galois field. The coded packets are then forwarded to the aircrafts according to their subscription. The minimum number of coded packets required at the receiver to correctly retrieve the original packets from the coded packets is equal to the number of original packets contained in each coded packet.

With content delivery, reliability is the key parameter for quality of service consideration unlike in real-time applications where delay and jitter are the main parameters. So, reliable IP multicast where a NACK is sent back to the multicast router (located at GW) for each packet lost is used for content delivery. Due to the different location and therefore channel conditions of the aircrafts, different packets are likely to be lost by each aircraft. Figure 1 shows a RLNC example of generation size 5. As illustrated in Fig. 1, out of the 5 coded packets transmitted by GW1, aircrafts Rx_1 lost packet *CP2*, Rx_n packet *CP1* and Rx_2 received all the 5 coded packets transmitted. Aircrafts Rx_1 and Rx_n then generate 2 NACKs and sent to GW1 as shown in Fig. 1. Upon reception of the 2 NACKs, GW1 then transmits the 6th coded packet (CP6) to the multicast group. If *CP6* is successfully received by Rx_1 and Rx_n, then all the multicast receivers in Fig. 1 will therefore be able to retrieve the 5 original packets contained in the received coded packets. It should be noted that all the coded packets received for each generation by the aircrafts are linearly independent packets but contain exactly the same original packets. Figure 2 shows the minimum number of transmissions over the satellite air interface required for the 3 aircrafts to receive all the original packets for the example describe above (Fig. 1). It is the minimum number of transmissions required because if any of the NACKs (or the 6th coded packet) are lost then the total number of transmissions required will definitely increase.

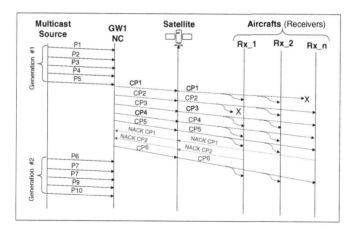

Fig. 2. RLNC example for satellite–based multicast scenario

4   Bandwidth Utilization in RLNC with Changing Generation Size, Number of Receivers and Rate of Packets Loss

In a reliable IP multicast scenario like the one illustrated in Fig. 1 above, the impact on satellite bandwidth utilization of changing generation size, number of receivers and rate of packet loss in RLNC is investigated. The amount satellite bandwidth utilization is measured here in terms of the number of transmissions over the satellite required for all the receivers to successfully retrieve all the original packets from the received coded packets. To carry out this investigation, the number of receivers are set to 20, 50 and 100, and the generation size 5 and 10. For the generation size of 5 the minimum number of data transmissions over the satellite required for all 20, 50, and 100 aircrafts to successfully retrieve all the original packets from the coded packets are separately measured for the rate of packet loss of 10 %–50 %. This process is repeated for generation size of 10.

5   Simulation Results and Analysis

Using Network Simulator 3 (NS3), the scenario illustrated in Fig. 1 was simulated with the generation size, number of receivers and rate of packet loss set as described in Sect. 4. The minimum number of data transmissions over the satellite required for all aircrafts to successfully retrieve all the original packets from the coded packets were measured. This was done for all settings of the generation size, number of receivers and rate of packet loss described in Sect. 4.

Figure 3 shows how the generation size and the number of receivers affect the number of data transmissions required for all receivers (aircrafts) to successfully retrieve the original packets contained in the received coded packets.

Figure 3 shows that for a generation size of 5 and rate of packet loss of 10 %, there is an average increase of about 10.0 % and 20.0 % in the number of data transmissions required when the number of receivers increases from 20 to 50 and from 20 to 100 respectively. One of the main reasons why the minimum number of data transmissions required increases with increasing number of receivers is that there is a higher probability for the transmitted packets to be lost when the number of receivers is higher than when it is smaller. From the percentage increase in the minimum number of data transmissions required, it is clear that this increase is not directionally proportional to the increase in the number of receivers.

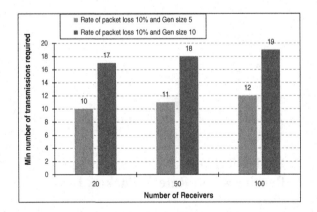

Fig. 3. Impact of varying generation size and number of receivers on minimum number of data transmissions required

For 20, 50 and 100 receivers at a constant rate of packet loss of 10 % in Fig. 3, the minimum number of data transmissions required increases on the average by 70.0 %, 63.6 % and 58.3 % respectively when the generation size is increased from 5 to 10. The general increase in the minimum number of data transmissions required when the generation size increases is mainly because the minimum number of data transmissions required must be at least equal to the generation size.

With the generation size of 10 and rate of packet loss 10 %, the average increase in the minimum number of data transmissions required when the number of receivers increases from 20 to 50 and from 20 to 100 is about 5.9 % and 11.8 % respectively. Compared to those with the generation size of 5, these increases are significantly smaller (i.e., almost half). This makes sense since combining more original packets to produce one coded packet implies one coded packet can compensate for many different lost packets which could have required more than one coded packet to compensate for.

Figure 4 shows how the minimum number of data transmissions over the satellite required for all aircrafts to successfully retrieve the original packets contained in the received coded packets is affected by varying rate of packet loss at a constant generation size for 20, 50 and 100 receivers. From Fig. 4, it can be seen that the minimum number of data transmissions required generally increases as the rate of packet loss increases for each set of receivers. Although the relationship between the minimum number of data transmissions required and rate of packet loss is not linear, at a constant generation size of 5, the average increase in the minimum number of data transmissions required for 20, 50 and 100 receivers as the rate of packet loss changes from 10 % to 50 % is 58.00 %, 58.18 % and 73.33 % respectively. This trend is expected since an increase in rate of packet loss implies an increase in the minimum number of coded packets transmissions required in order to compensate for the high loss.

Also, Fig. 4 shows that the minimum number of data transmissions required increases with both increasing rate of packet loss and number of receivers. This is mainly due to the fact that at a constant generation size, increasing the number of receivers increases the probability of different packets being lost. If many different packets are lost, then more transmissions will have be made to compensate for them since the generation size is constant or fixed.

Figure 5 shows a similar scenario to Fig. 4 except for the fact that the generation size here is now set to 10. The general trend here is similar to that in Fig. 4. Similarly to Fig. 4, at a constant generation size of 10, the average increase in the minimum number of data transmissions required for 20, 50 and 100 receivers as the rate of packet loss changes from 10 % to 50 % is 42.35 %, 50.00 % and 52.63 % respectively. Comparing these values to those in Fig. 4 shows that the average increases in the minimum number of data transmissions required for 20, 50 and 100 receivers in Fig. 5 are generally lower compared to those in Fig. 4. This is expected due to the fact that increasing the generation size will increase the probability of each coded packet compensating many more different packets lost, thus reducing the minimum number of data transmissions required.

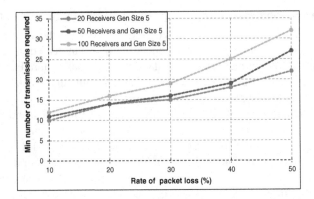

Fig. 4. Effects of rate of packet loss on minimum number of data transmissions required - at generation 5

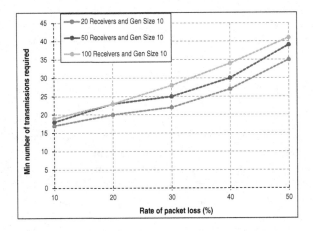

Fig. 5. Effects of rate of packet loss on minimum number of data transmissions required - at generation 10

6   Conclusion

This paper presents a detailed account of how RLNC could be implement in an IP multicast scenario over satellite-based aeronautical communications. The paper sets out to investigate the effects in RLNC of varying generation size, rate of packet loss and number of multicast receivers on the bandwidth utilization of the allocated satellite bandwidth resources.

From the investigation, it was discovered that for a constant:

- Generation size and rate of packet loss, the satellite bandwidth utilization (minimum number of data transmissions required) increases generally as the number of multicast receivers increases.

- Rate of packet loss, the satellite bandwidth utilization increases with both increasing generation size and number of receivers.
- Generation size, the satellite bandwidth utilization increases with both increasing rate of packet lost and number of receivers

References

1. Digital Video Broadcasting (DVB); Interaction channel for satellite distribution systems, ETSI EN 301 790 (2009)
2. Digital Video Broadcasting (DVB); Second Generation DVB Interactive Satellite System (DVB-RCS2); Part 2: Lower Layers for Satellite standards, ETSI EN 301 545-2 (2012)
3. Vieira, F., Barros, J.: Network Coding Multicast in Satellite Networks. In: 2009 Next Generation Internet Networks, NGI 2009, pp. 1–6 (2009)
4. Esmaeilzadeh, M., Aboutorab, N., Sadeghi, P.: Joint optimization of throughput and packet drop rate for delay sensitive applications in TDD satellite network coded systems. IEEE Trans. Commun. **62**, 676–690 (2014)
5. Vieira, F., Shintre, S., Barros, J.: How feasible is network coding in current satellite systems? In: 2010 5th Advanced Satellite Multimedia Systems Conference (ASMS) and The 11th Signal Processing for Space Communications Workshop (SPSC), pp. 31–37 (2010)
6. Alegre, R., Gheorghiu, S., Alagha, N., Vazquez Castro, M.A.: Multicasting optimization methods for multi-beam satellite systems using network coding. In: 29th AIAA International Communications Satellite Systems Conference (ICSSC-2011), ed. American Institute of Aeronautics and Astronautics (2011)
7. Vieira, F., Lucani, D.E., Alagha, N.: Load-aware soft-handovers for multibeam satellites: a network coding perspective. In: 2012 6th Advanced Satellite Multimedia Systems Conference (ASMS) and 12th Signal Processing for Space Communications Workshop (SPSC), pp. 189–196 (2012)
8. Cain, B., Deering, S., Kouvelas, I., Fenner, B., Thyagarajan, A.: Internet Group Management Protocol, Version 3, IETF RFC 3376 (2002)
9. Vida, R., Costa, L.: Multicast Listener Discovery Version 2 (MLDv2) for IPv6, IETF RFC 3810 (2004)

Network Coding over SATCOM: Lessons Learned

Jason Cloud$^{(\boxtimes)}$ and Muriel Médard

Massachusetts Institute of Technology, Cambridge, MA 02139, USA
{jcloud,medard}@mit.edu

Abstract. Satellite networks provide unique challenges that can restrict users' quality of service. For example, high packet erasure rates and large latencies can cause significant disruptions to applications such as video streaming or voice-over-IP. Network coding is one promising technique that has been shown to help improve performance, especially in these environments. However, implementing any form of network code can be challenging. This paper will use an example of a generation-based network code and a sliding-window network code to help highlight the benefits and drawbacks of using one over the other. In-order packet delivery delay, as well as network efficiency, will be used as metrics to help differentiate between the two approaches. Furthermore, lessoned learned during the course of our research will be provided in an attempt to help the reader understand when and where network coding provides its benefits.

Keywords: Intra-session network coding · Implementation concerns · Satellite Networks · In-order delivery delay · Lessons learned

1 Introduction

Space-based packet data networks are becoming a necessity in everyday life, especially when considering world-wide Internet connectivity. It is estimated that over half of the world's population still does not have access to broadband Internet due to a variety of factors including a lack of infrastructure and low affordability, especially in rural areas and developing countries [1]. To overcome these barriers, a number of companies such as SpaceX, Google, and FaceBook have recently launched projects that incorporate some form of space-based or high altitude data packet network. However, significant challenges such as large latencies, high packet erasure rates, and legacy protocols (e.g., TCP) can seriously degrade performance and inhibit the user's quality of service. One promising approach to help in these challenged environments is network coding. This paper will investigate some of the gains that network coding provides, as well as outline some of the lessons learned from our research.

Space-based networks have a number of unique characteristics that challenge high quality of service applications. Large packet latencies and relatively high packet erasure rates can negatively impact existing protocols. Fading due to

© Institute for Computer Sciences, Social Informatics and Telecommunications Engineering 2015
P. Pillai et al. (Eds.): WiSATS 2015, LNICST 154, pp. 272–285, 2015.
DOI: 10.1007/978-3-319-25479-1_21

scintillation or other atmospheric effects are more pronounced than in terrestrial networks. The high cost in terms of both deployment and bandwidth make efficient communication a requirement. Finally, the broadcast nature of satellite networks create unique challenges that are non-existent in terrestrial networks. While existing physical and data link layer techniques help improve performance in these conditions, we will show that coding above these layers can also provide performance gains.

Various forms of network coding can be used with great benefits in space-based networks. In general, these can be characterized into two broad categories: inter-session network coding, and intra-session network coding. Figure 1 provides a simple example of both. Inter-session network coding combines information flows together to improve the network capacity. A summary of the various methods that can be used for satellite communications is provided by Vieira *et al.* [2]. Intra-session network coding, on the other hand, is used to add redundancy into a single information flow. Adding this redundancy has shown that file transfer times can be decreased for both multicast [3] and unicast [4,5] sessions.

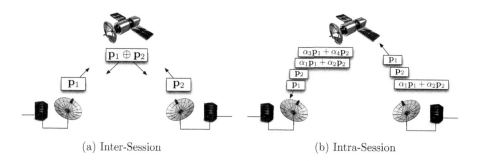

(a) Inter-Session   (b) Intra-Session

Fig. 1. Examples of inter-session (a) and intra-session (b) network coding. This paper focuses completely on intra-session network coding.

While there are merits to both techniques, our focus will be on intra-session network coding techniques that help achieve the following goals: provide consistent performance for protocols not designed for space systems; decrease delay for real-time or near real-time data streams; efficiently use any network resources that are available; and reduce packet erasure rates due to correlated losses. A generation-based approach [5,6] and a sliding-window approach [7] will be used to help highlight the potential gains, design choices, and implementation decisions that need to be taken into account. Several performance metrics including the in-order delivery delay, efficiency, and upper layer packet erasure rates will be used to help differentiate between the approaches.

The remainder of the paper is organized as follows. Section 2 will provide details on the coding algorithms considered. Section 3 provides information about the assumed network model and evaluates the performance of these coding algorithms when used for both reliable and unreliable data streams. Section 4 discusses various considerations that need to be taken into account when implementing network coding into real systems. Finally, conclusions are summarized in Sect. 5.

2 Network Coding over Packet Streams

Network coding has been shown to dramatically improve network performance; however, implementing it can be a challenge. In order to develop practical coding techniques, random linear network coding (RLNC) [8] has been used by a large number of coding schemes because of its simplicity and effectiveness in most network scenarios. While both practical inter and intra-session techniques have been proposed, we are primarily interested in the latter due to the inherent limitations of existing satellite communication networks (i.e., typical satellite communication networks employ a bent-pipe architecture or have very limited on-orbit processing power). Assume that we want to send a file consisting of information packets $\mathbf{p}_i$, $i \in \mathcal{P}$, where $\mathcal{P}$ is the set of information packet indexes (i.e., the file has size $|\mathcal{P}|$ packets). Within these intra-session packet streams, RLNC can be used to add redundancy by treating each $\mathbf{p}_i$ as a vector in some finite field $\mathbb{F}_{2^q}$. Random coefficients $\alpha_{ij} \in \mathbb{F}_{2^q}$ are chosen, and linear combinations of the form $\mathbf{c}_i = \sum_{j \in \mathcal{P}} \alpha_{ij} \mathbf{p_j}$ are generated. These coded packets are then inserted at strategic locations to help overcome packet losses in lossy networks.

Management of the coding windows for these intra-session network coding schemes generally fall within the following two categories: fixed-length/generation-based schemes, or variable/sliding window based schemes. Fixed-length or generation-based schemes first partition information packets into blocks, or generations, $G_i = \left\{ \mathbf{p}_{(i-1)k+1}, \ldots, \mathbf{p}_{\min(ik, |\mathcal{P}|)} \right\}$ for $i = [1, \lceil |\mathcal{P}|/k \rceil]$ and generation size $k \geq 1$. Coded packets are then produced based on the information packets contained within each individual generation. As a result, coded packets consisting of linear combinations of packets in generation G_i cannot be used to help decode generation $G_j, i \neq j$. Alternatively, sliding window schemes do not impose this restriction. Instead, information packets are dynamically included or excluded from linear combinations based on various performance requirements.

Examples of both schemes are provided in Fig. 2. Columns within the figure represent information packets that need to be sent, rows represent the time when a specific packet is transmitted, and the elements of the matrix indicate the composition of the transmitted packet. For example, packet $\mathbf{p}_1$ is transmitted in time-slot 1, while coded packet $\mathbf{c}_5 = \sum_{i=1}^{4} \alpha_i \mathbf{p}_i$ is transmitted in time-slot 5. The double-arrows on the right of each matrix indicate when an information packet is delivered, in-order, to an upper-layer application, and the red crosses mark lost packets.

Each approach has its benefits and drawbacks. It is easy from a coding perspective to implement the generation-based coding scheme, and these schemes achieve capacity when $k \to \infty$. However, partitioning packets into generations adds artificial restrictions on the code's capability to recover from losses, and may not be as efficient as sliding window schemes. Furthermore, generation-based

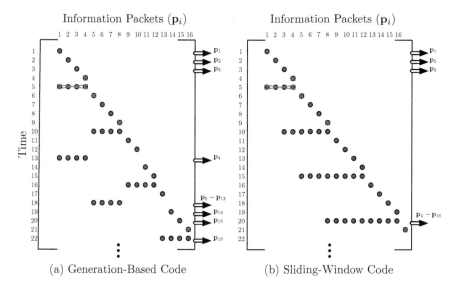

Fig. 2. Examples of generation-based and sliding-window network coding schemes. It is important to note that the generation-based coding scheme requires feedback and retransmissions to ensure reliable delivery while the sliding-window coding scheme only requires feedback to help slide the coding window.

schemes can increase the complexity of the feedback process, especially for reliable data transfers. Sliding-window schemes, on the other hand, can outperform generation-based schemes in terms of efficiency and delay. Unfortunately, coding window management can be difficult and these schemes typically cannot guarantee a decoding event occurs before the termination of a session. In addition, the size of the coding window maybe much larger than the generation-based schemes leading to increased decoding complexity and communication overhead.

The examples shown in Fig. 2 will be used throughout the remainder of this paper in order to provide some intuition into the trade-offs of using one type of coding approach over the other. Algorithm 1 describes the packet generation policy for the generation-based scheme shown in Fig. 2(a), while Algorithm 2 describes the policy for the sliding-window coding scheme shown in Fig. 2(b). Each algorithm uses a systematic approach where information packets $\mathbf{p}_i, i \in \mathcal{P}$, are first sent uncoded and redundancy is added to help correct packet erasures by inserting coded packets into the packet stream. We will assume that the amount of redundancy added to the packet stream is defined by $R \geq 1$ (e.g., the code rate is $c = 1/R$).

It is important to note that feedback is not addressed in these algorithms. In general, feedback is necessary to accurately estimate the network packet erasure rate. Furthermore, feedback maybe required to ensure reliable delivery in some instances. For the generation-based scheme, the server may need to know the number of received degrees of freedom from each transmitted generation. This

Algorithm 1: Generation-based coding algorithm [6]	**Algorithm 2:** Sliding window coding algorithm [7]
for each $j \in \left[1, \lceil \frac{|\mathcal{P}|}{k} \rceil \right]$ **do** $\quad w_l \leftarrow (j-1)\,k + 1$ $\quad w_u \leftarrow \min(jk, |\mathcal{P}|)$ $\quad$ **for each** $i \in [w_l, w_u]$ **do** $\quad\quad$ Transmit $\boldsymbol{p}_i$ $\quad$ **for each** $m \in [1, k\,(R-1)]$ **do** $\quad\quad$ Transmit $\quad\quad c_{j,m} = \sum_{i=w_l}^{w_u} \alpha_{i,j,m} \boldsymbol{p}_i$	Initialize $k = 1$, $u = 1$, and $n = \frac{R}{R-1}$ **for each** $k \in \mathcal{P}$ **do** $\quad$ **if** $u < n$ **then** $\quad\quad$ Transmit packet $\boldsymbol{p}_k$ $\quad\quad u \leftarrow u + 1$ $\quad$ **else** $\quad\quad$ Transmit $\boldsymbol{c}_k = \sum_{i=1}^{k} \alpha_{k,i} \boldsymbol{p}_i$ $\quad\quad u \leftarrow 1$

feedback can be used by the server to retransmit additional degrees of freedom if a particular generation cannot be decoded. Details are provided in [6]. In the sliding window scheme, knowledge of the number of received degrees of freedom may not be necessary [7]; but feedback can be used to help slide the coding window or facilitate decode events if there are delay constraints.

3   Network Coding Performance for Packet Streams

As we mentioned in the previous section, we will compare the performance of two types of intra-session network coding schemes (see Algorithms 1 and 2) for both a reliable data stream (e.g., a TCP session) and an unreliable data stream (e.g., a UDP session). The metrics used to evaluate both coding schemes will depend slightly on the type of data stream; however, the following definitions will be used throughout this section.

Definition 1. *The in-order delivery delay D is the difference between the time an information packet is first transmitted and the time that the same packet is delivered, in-order.*

Definition 2. *The efficiency η of a coding scheme is defined as the total number of degrees of freedom (i.e., the total number of information packets) that need to transfered divided by the actual number of packets (both uncoded and coded) received by the sink.*

Both of these metrics are particularly important for satellite communication systems. In the case of reliable data streams, large propagation delays can compound the effects of packet losses by creating considerable backlogs and in-order delivery delays. For large file transfers or non-time sensitive applications, this may not be an issue. However, a large number of time-sensitive applications (e.g., non-real-time video streaming) use TCP. Lost packets can result in very large resequencing delays that can seriously degrade the quality of user experience. Network coding is particularly useful in these situations to help recover from packet losses without excessive retransmissions. Furthermore, bandwidth is expensive for these systems. Any coding scheme that promises to provide a specified quality of service needs to be efficient.

The remainder of this section will provide an outline of the network model and examine the performance of the two coding schemes presented above. The two metrics defined earlier will be used in addition to any additional metrics that are important for the specific type of data stream.

3.1   Network Model

We will assume a time-slotted model where each time-slot has a duration t_s equal to the time it takes to transmit a single packet. The network propagation delays will be taken into account by defining $t_p = {RTT}/2$ where RTT is the round-trip time. As a reminder, we will assume that the amount of redundancy added ($R \geq 1/{1-\epsilon}$ given that ϵ is the packet erasure probability) defines the code rate $c = 1/R$. For the generation-based scheme, c is equal to the generation size divided by the number of degrees of freedom transmitted for that generation (i.e., $c = k/{Rk}$ where k is the generation size). In the case of the sliding window scheme, c is dependent on the number of consecutively transmitted information packets (i.e., $c = {n-1}/n$ where $n = R/{R-1}$ is the number of packets between each inserted coded packet).

The satellite channel will be modeled using a simple Gilbert channel with transition probability matrix

$$P = \begin{bmatrix} 1 - \gamma & \gamma \\ \beta & 1 - \beta \end{bmatrix} \tag{1}$$

where γ is the probability of transitioning from the "good" state (which has a packet erasure rate equal to zero) to the "bad" state (which has a packet erasure rate equal to one) and β is the probability of transitioning from the "bad" state to the "good" one. The steady-state distribution of the "bad" state $\pi_B = \gamma/{\gamma+\beta}$ and the expected number of packet erasures in a row $\mathbb{E}[L] = 1/\beta$ will be used as the primary parameters for determining the transition probabilities of the channel model. It should be noted that this model does not necessarily reflect the effects of fading due to scintillation or rain, which generally have a duration equal to hundreds of milliseconds to hours. Instead, the model is intended to help model the cases where the SNR is such that the performance of the underlying physical layer code is degraded; but the situation does not warrant the need to change to a more robust modulation/coding scheme.

Lesson Learned: Network coding is not a cure-all solution. It cannot mitigate the effects of deep fades with very large durations.

3.2   Reliable Data Stream Performance

Reliable data delivery is a fundamental requirement for some applications. This section will focus on the performance of both a generation-based and a sliding-window coding scheme by looking at the following metrics: the ability of the scheme to provide 100 % reliability, the in-order delivery delay, and the coding schemes' efficiency. Furthermore, the performance of an idealized version of selective-repeat ARQ will be provided to highlight the gains network coding can provide in satellite communications systems.

Before proceeding, feedback maybe necessary to ensure reliability. With regard to the two example coding schemes presented here, the generation-based scheme requires feedback while the sliding-window scheme does not. Algorithm 1 can be modified to include this feedback with only a few changes. Assume that delayed feedback contains information regarding the success or failure of a specific generation being decoded by the client. If a decoding failure occurs, the server can then produce and send additional coded packets from that generation to overcome the failure. On the other hand, the construction of the sliding-window scheme outlined in Algorithm 2 has been shown in [7] to provide a finite in-order delivery delay with probability one. Therefore, our results will assume that no feedback is available when using this scheme even though feedback may actually increase the algorithm's performance.

A detailed analysis of the in-order delivery delay and the efficiency for the generation-based scheme ($\mathbb{E}[D_G]$ and η_G respectively) is provided in [6], while the same is provided in [7] for the sliding-window scheme ($\mathbb{E}[D_S]$ and η_S respectively). The analysis of the generation-based scheme shows that $\mathbb{E}[D_G]$ and the delay's variance σ_G^2 are dependent on both the generation size k and the amount of added redundancy R. For a given R that is large enough and independent and identically distributed (i.i.d.) packet losses, $\mathbb{E}[D_G]$ is convex with respect to k and has a global minimum. Determining this minimum, $\mathbb{E}[D_G^*] = \arg\min_k \mathbb{E}[D_G]$, is difficult due to the lack of a closed form expression; however it can be found numerically. The following results will only show $\mathbb{E}[D_G^*]$ for a given R since the behavior of $\mathbb{E}[D_G]$ and σ_G^2 as a function of k is provided in [6]. The analysis of the sliding-window scheme's in-order delivery delay shows that $\mathbb{E}[D_S]$ is only dependent on R since there is no concept of generation or block size. Therefore, a simple renewal process can be defined and a lower-bound for the expected in-order delay can be derived. While the efficiency of this scheme is not explicitly given in [7], it can easily be shown that the efficiency is $\eta_S = 1/R(1-\epsilon)$ for i.i.d. packet losses that occur with probability ϵ. Regardless of this existing analysis, the in-order delivery delay and efficiency used below for both the generation-based and sliding-window coding schemes are found using simulations developed in Matlab.

Figures 3 and 4 show $\mathbb{E}[D]$ and η respectively for both coding schemes as a function of R. Furthermore, each sub-figure shows the impact correlated losses have on the schemes' performance where $\mathbb{E}[L]$ is the expected number of packet losses that occur in a row. For uncorrelated losses (e.g., $\mathbb{E}[L] = 1$), both coding schemes provide an in-order delivery delay that is superior to the idealized version of selective repeat ARQ. This performance gain becomes less pronounced as $\mathbb{E}[L]$ increases. In fact, the sliding-window coding scheme performs worse than ARQ for small R when $\mathbb{E}[L] = 8$. The cause of this is due to the lack of feedback, which can help overcome the large number of erasures if it is implemented correctly. Regardless, Fig. 3 shows that coding can help in the cases where losses are correlated; although the gains come with a cost in terms of efficiency.

> Lesson Learned: While feedback is necessary for estimating the channel/network state, it also aids in decreasing in-order delivery delay.

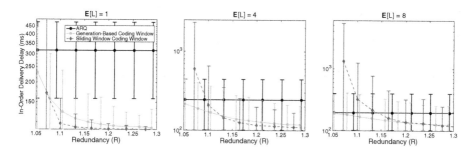

Fig. 3. In-order packet delay ($\mathbb{E}[D]$) as a function of the redundancy (R) where $RTT = 200$ ms, $t_s = 1.2$ ms, and $\pi_B = 0.05$.

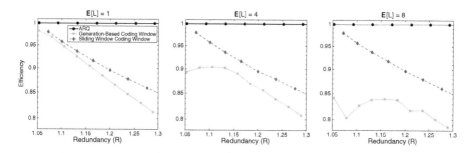

Fig. 4. Efficiency (η) as a function of the redundancy (R) where $RTT = 200$ ms, $t_s = 1.2$ ms, and $\pi_B = 0.05$.

Decreasing $\mathbb{E}[D]$ results in decreased η, which can be observed in Fig. 4. The figure shows that the sliding-window coding scheme is more efficient than the generation-based scheme. There are two major contributors to this behavior. First, code construction has a major impact on efficiency. Since coding occurs over more information packets in the sliding-window scheme, coded packets can help recover from packet erasures that occur over a larger span of time (i.e., multiple generations if we compare it with the generation-based scheme). Second, the decrease in the generation-based scheme's efficiency, as well as the non-decreasing behavior of η_G, for $\mathbb{E}[L] > 1$, is an indication that retransmissions are necessary to provide reliability. In fact, the generation-based scheme almost always requires retransmissions to be made when $\mathbb{E}[L] = 8$. This behavior helps illustrate that artificially restricting the coding window's size can have negative impacts and may not be the appropriate strategy in certain circumstances.

> Lesson Learned: Generation-based coding schemes perform poorly when packet losses are correlated due to the limited number of packets that are used to form a coded packet.

3.3   Unreliable Data Stream Performance

Data streams such as real-time voice and video do not necessarily require 100 % reliability. However, decreasing the underlying packet erasure rates may still drastically improve upper layer quality of service. Recent work in this area has shown that network coding is one tool that can help improve performance [9, 10]. This section will compare both the generation-based and sliding-window coding schemes with respect to the upper-layer packet erasure probabilities and the expected in-order delivery delays.

The generation-based coding scheme shown in Algorithm 1, where feedback is only necessary to identify the packet erasure rate, is ideally suited to the case where there is a delay constraint and packet delivery is not guaranteed. Packets within each generation are delivered in-order until the first packet loss is encountered. Once the entire generation has been received, the client attempts to decode it. If the generation cannot be decoded, only the successfully received information packets are delivered. If the generation can be decoded, every information packet contained in the generation is delivered in-order.

Modifying the sliding-window coding scheme shown in Algorithm 2 for unreliable data streams is somewhat difficult. If a delay constraint exists, the coding window cannot be arbitrary changed to accommodate these constraints. For example, assume that a lost information packet $\mathbf{p}_i$ is no longer necessary due to its delivery time exceeding some specified value. One approach would be to move the left side of the coding window to the right so that $\mathbf{p}_i$ is no longer used in the generation of future coded packets (i.e., $\mathbf{c}_j = \sum_{k=i+1}^{j} \alpha_{j,k} \mathbf{p}_k$). In order for these new coded packets to be useful, the decoder must discard any coded packet containing $\mathbf{p}_i$ that it has already received. Not only does this decrease the efficiency of the coding scheme, but it also potentially increases the delay for subsequent packets $\mathbf{p}_j, i < j$. As a result, we will assume that Algorithm 2 is left unchanged in this scenario.

> Lesson Learned: Great care must be taken when modifying a sliding-window coding schemes' coding window when trying to meet a delay constraint. Not doing so properly can lead to decreased efficiency and increased in-order delivery delay for subsequent packets.

Figure 5 and 6 show the expected upper-layer packet erasure rate (PER) and expected in-order delivery delay $\mathbb{E}\left[D\right]$ respectively for both the generation-based (GB) and sliding-window (SW) coding schemes. Three values of the expected number of packet losses in a row $\mathbb{E}[L]$ and two levels of efficiency η (indicated by the values shown in parentheses) are provided. Due to the sliding-window coding scheme's construction, the PER and $\mathbb{E}[D_S]$ are constant with respect to k.

These figures illustrate some of the trade-offs that need to be taken into account when selecting the appropriate code. First, the larger the generation size in the generation-based scheme, the better the error performance. This is expected since you are essentially averaging losses over more packets. However, the cost is increased latency. Second, correlated losses can have a significant impact on the performance of the generation-based code. This is a result of partitioning information packets into generations, which places artificial constraints the ability of the code to recover from packet losses. The sliding-window scheme has no such constraints. On the other hand, the redundancy inserted into the packet stream must be enough to ensure that any delay constraints are satisfied. For example, Fig. 6 shows that $\mathbb{E}[D_S]$ and σ_S can be very large if your goal is to be highly efficient (e.g., $\eta_S \approx 0.97$). In order to match the delay of the generation-based code, a significant amount of redundancy must be added to the packet stream.

> Lesson Learned: Decreasing the efficiency of sliding-window coding schemes is necessary to outperform generation-based schemes in terms of in-order delivery delay.

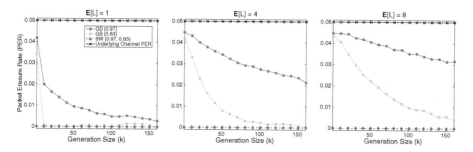

Fig. 5. Upper layer packet erasure rate (PER) as a function of the generation-based coding scheme's generation size (k) where $RTT = 200$ ms, $t_s = 1.2$ ms, and $\pi_B = 0.05$. The values shown within the parentheses for each item in the legend indicate the efficiency η.

4   Implementation Considerations

Implementing any type of network coding scheme presents its own challenges. Sections 2 and 3 highlighted just a few of them. However, there are a number of items that also affect how we code, especially in satellite networks. While we

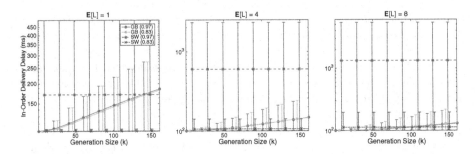

Fig. 6. In-order delivery delay $\mathbb{E}[D]$ as a function of the generation-based coding scheme's generation size (k) where $RTT = 200$ ms, $t_s = 1.2$ ms, and $\pi_B = 0.05$. The error bars show two standard deviations above and below the mean. The values shown within the parentheses for each item in the legend indicate the efficiency η.

cannot address everything, we do provide a brief discussion on some of the items that we believe are important.

The first major consideration is where to perform the coding and decoding operations. Ideally, redundancy should be added at any point in the network where packet losses occur. This includes locations such as queues or links where the physical layer cannot provide 100 % reliability. Furthermore, the amount of added redundancy should only be enough to help recover from losses that occur between network nodes that can code. This can be motivated by the simple example shown in Fig. 7 where a source S wants to transmit $|\mathcal{P}|$ packets to the destination D. However, these packets must travel over a tandem network where each link $i \in \{1, 2, 3\}$ has an i.i.d. packet erasure probability ϵ_i. If end-to-end coding is used, $|\mathcal{P}| \left(\prod_i (1 - \epsilon_i)^{-1} - 1 \right)$ coded packets must be generated at S and transmitted through the network. This results in an inefficient use of links closer to the source than would be necessary if redundancy is included into the packet stream at each node $R_i, i \in 1, 2$.

Lesson Learned: Coding at intermediate nodes, rather than coding end-to-end increases overall network efficiency.

$$\underset{\substack{\eta_1^{\bar{E}} = 0.72 \\ \eta_1^{E} = 1}}{\boxed{S} \overset{\epsilon_1 = 0}{\text{———}}} \underset{\substack{\eta_2^{\bar{E}} = 0.9 \\ \eta_2^{E} = 1}}{\boxed{R_1} \overset{\epsilon_2 = 0.2}{\text{———}}} \underset{\substack{\eta_3^{\bar{E}} = 1 \\ \eta_3^{E} = 1}}{\boxed{R_2} \overset{\epsilon_3 = 0.1}{\text{———}}} \boxed{D}$$

Fig. 7. A simple example showing that coding within the network is more efficient than end-to-end coding. η_i^j is the efficiency on link $i \in 1, 2, 3$ when coding is performed end-to-end ($j = \bar{E}$) or at each intermediate network node ($j = E$).

This simple fact can have major implications for satellite networks since bandwidth is limited and very expensive. As a result, coding should be performed at each satellite gateway or performance enhancing proxy (PEP) at a minimum; and if possible, at each hop in the satellite network. While coding should be performed as often as possible, network codes do not need to be decoded at each hop. This is also extremely beneficial in satellite networks since you can essentially shift a large portion of the required processing to the satellite gateway or end client. In other words, coded packets can be generated at multiple points within the network while only needing to decode once at the client or satellite network gateway. In the example provided in Fig. 7, coding can take place at S, R_1, and R_2; however, only D needs to decode.

Lesson Learned: Decoding only needs to be performed once regardless of the number of times coding occurs within the network.

The second consideration that needs to be taken into account is how to communicate the coding coefficients α_i used to the decoder. For generation-based coding schemes where k is typically small, one can simply insert each coding coefficient into the header, which would require qk bits assuming each $\alpha_i \in \mathbb{F}_{2^q}$. Coding within the network only needs to modify the existing coefficients and does not increase the size of the coding coefficient vector. Of course, other approaches that require less than qk bits such as [11] or [12] can be used to decrease overhead.

Communicating the coefficients efficiently for sliding-window schemes is more challenging since the coding windows can be quite large. Existing methods typically use a pseudo-random number generator and communicate only the seed. This seed is then used by the decoder to generate the coefficients used to create each coded packet. Unfortunately, this does not scale well when coding occurs at intermediate network nodes. As an example, assume that an intermediate node's coding window contains multiple coded packets that were generated by previous nodes. When the node generates a new coded packet, it must communicate the seed used to generate the packet; in addition to all of the seeds for each of the coded packets contained within its coding window. If the coding window and the number of coded packets contained within the window are large, the amount of overhead required to reproduce the coefficients can far exceed the payload size.

Lesson Learned: The overhead required to communicate coding coefficents for sliding-window based schemes can be significant if not done correctly.

Finally, congestion control and file size can potentially dictate the coding approach used. Regardless of the type of data stream, some form of congestion control is typically needed at either the client/server or at the satellite network gateway. Common congestion control algorithms can cause bursts of packets, or packet trains, while they are ramping up to fully utilize the network. This behavior is even more pronounced when considering TCP flows over satellite networks. In these situations, it maybe preferable to use a coding scheme that provides a high probability of delivering every packet within a burst without needing retransmissions or waiting for the next packet burst to arrive. For example, a generation-based coding scheme can be used for small congestion window sizes and a sliding-window scheme can be used for large ones.

In a similar fashion, the coding strategy can also significantly impact the overall throughput for some file sizes. For example, consider a small file that can be transmitted using less than a single bandwidth-delay product worth of packets. A generation-based coding scheme, or a mixture of the generation-based and sliding-window schemes, should be used so that the the probability of decoding the file after the first transmission attempt is made very large. While this may impact the efficiency of the network, it can have major benefits for the user's quality of service or experience.

> Lesson Learned: Congestion control and the length of the data stream may affect the network coding strategy.

5   Conclusion

Intra-session network coding is a promising technique that can help improve application layer performance in challenging space-based data packet networks. However, implementing it can be problematic if done incorrectly. This paper used two common examples of intra-session network codes to show the benefits and drawbacks of one over the other. The first example used was a generation-based network code and the second a sliding-window based network code. While generation-based network codes are easier to implement, sliding-window network codes can provide improved performance in terms of in-order delivery delay and efficiency. This is especially the case when reliability is required. However, generation-based network codes are able to provide strict delay guarantees and improved upper layer packet erasure rates with little impact to the overall network efficiency when reliability is not a constraint. On the other hand, implementation considerations typically limit the performance of sliding-window network codes in these environments.

Lessons learned, as well as other implementation tips, were provided in addition to the above comparison. Some of the more important lessons learned include the facts that restricting the size of the coding window in any way limits the network code's performance gains; and feedback is useful for not only estimating the channel/network state information, but it also can be used to decrease delay. Both of these are apparent when considering the effects correlated packet losses have on the delay for reliable data streams. Various implementation considerations were also highlighted. These include where coding and decoding within the network should occur, how congestion control affects the way we code, and the challenges regarding the communication of RLNC coefficients between the source and sink. While properly implementing network coding in real networks can be difficult, we hope that our lessons learned will aid in the deployment of network codes in future satellite communication systems.

References

1. Sprague, K., Grijpink, F., Manyika, J., Moodley, L., Chappuis, B., Pattabiraman, K., Bughin, J.: Offline and falling behind: barriers to internet adoption. McKinsey & Company,Technical Report (2014)
2. Vieira, F., Shintre, S., Barros, J.: How feasible is network coding in current satellite systems? In: 5th Advanced Satellite Multimedia Systems Conference (ASMA) and the 11th Signal Processing for Space Communications Workshop (SPSC), pp. 31–37. IEEE Press, New York (2010)
3. Rezaee, A., Zeger, L., Médard, M.: Speeding multicast by acknowledgment reduction technique (SMART). In: IEEE Global Telecommunications Conference (GLOBECOM), pp. 1–6. IEEE Press, New York (2011)
4. Lucani, D.E., Stojanovic, M., Médard, M.: Random linear network coding for time division duplexing: when to stop talking and start listening. In: IEEE INFOCOM, pp. 1800–1808. IEEE Press, New York (2009)
5. Lucani, D.E., Médard, M., Stojanovic, M.: Systematic network coding for time-division duplexing. In: 2010 IEEE International Symposium on Information Theory Proceedings (ISIT), pp. 2403–2407. IEEE Press, New York (2010)
6. Cloud, J., Leith, D.J., Médard, M.: A coded generalization of selective repeat ARQ. In: IEEE INFOCOM, pp. 1–9. IEEE Press, New York (2015)
7. Karzand, M., Leith, D.J.: Low delay random linear coding over a stream. In: 52nd Annual Allerton Conference on Communication, Control, and Computing (Allerton), pp. 521–528. IEEE Press, New York (2014)
8. Ho, T., Médard, M., Koetter, R., Karger, D.R., Effros, M., Shi, J., Leong, B.: A random linear network coding approach to multicast. IEEE Trans. Inf. Theory **52**(10), 4413–4430 (2006)
9. Teerapittayanon, S., Fouli, K., Médard, M., Montpetit, M.-J., Shi, X., Seskar, I., Gosain, A.: Network coding as a WiMAX link reliability mechanism. In: Bellalta, B., Vinel, A., Jonsson, M., Barcelo, J., Maslennikov, R., Chatzimisios, P., Malone, D. (eds.) MACOM 2012. LNCS, vol. 7642, pp. 1–12. Springer, Heidelberg (2012)
10. Adams, D.C., Du, J., Médard, M., Yu, C.C.: Delay constrained throughput-reliability tradeoff in network-coded wireless systems. In: IEEE Global Communications Conference (GLOBECOM), pp. 1590–1595. IEEE Press, New York (2014)
11. Lucani D.E., Pedersen, M.V., Heide, J., Fitzek, F.H.P.: Fulcrum network codes: a code for fluid allocation of complexity. In: CoRR, Cornell University Library, New York (2014). abs/1404.6620
12. Thomos, N., Frossard, P.: Toward one symbol network coding vectors. IEEE Commun. Lett. **16**(11), 1860–1863 (2012)

Network Coding Applications to High Bit-Rate Satellite Networks

Giovanni Giambene[1(✉)], Muhammad Muhammad[1], Doanh Kim Luong[1],
Manlio Bacco[2], Alberto Gotta[2], Nedo Celandroni[2], Esua Kinyuy Jaff[3],
Misfa Susanto[3], Yim Fun Hu[3], Prashant Pillai[3], Muhammad Ali[3],
and Tomaso de Cola[4]

[1] CNIT - University of Siena, Via Roma, 56, 53100 Siena, Italy
giambene@unisi.it
[2] CNR - ISTI, Pisa Research Area, Via G. Moruzzi, 1, 56124 Pisa, Italy
[3] University of Bradford, West Yorkshire, Bradford, BD7 1DP, UK
[4] DLR, Institute of Communications and Navigation, 82234 Wessling, Germany

Abstract. Satellite networks are expected to support multimedia traffic flows, offering high capacity with QoS guarantees. However, system efficiency is often impaired by packet losses due to erasure channel effects. Reconfigurable and adaptive air interfaces are possible solutions to alleviate some of these issues. On the other hand, network coding is a promising technique to improve satellite network performance. This position paper reports on potential applications of network coding to satellite networks. Surveys and preliminary numerical results are provided on network coding applications to different exemplary satellite scenarios. Specifically, the adoption of Random Linear Network Coding (RLNC) is considered in three cases, namely, multicast transmissions, handover for multi-homed aircraft mobile terminals, and multipath TCP-based applications. OSI layers on which the implementation of networking coding would potentially yield benefits are also recommended.

Keywords: Satellite networks · Network coding · Multipath communications · OSI layers · Robustness and resiliency

1 Introduction

Satellite networks are expected to satisfy stringent Quality of Service (QoS) requirements for broadband service delivery. Towards this end, new solutions that exploit the benefits of multipath transmissions and Network Coding (NC) to minimise packet losses are being explored. The main idea of NC is to allow nodes in the network to perform coding operations at the packet level. The application of network coding to communication networks is relatively recent, dating back to year 2000 [1]. Since then, NC has shown great potentials in correcting random packet errors and errors introduced by malicious nodes, making it a powerful tool to achieve efficiency and reliability with many potential areas of application to satellite networks.

© Institute for Computer Sciences, Social Informatics and Telecommunications Engineering 2015
P. Pillai et al. (Eds.): WiSATS 2015, LNICST 154, pp. 286–300, 2015.
DOI: 10.1007/978-3-319-25479-1_22

This paper aims to identify compatibility issues and best approaches as well as to investigate pros and cons on the application of NC to a satellite network at different OSI layers, taking multipath capabilities of satellite user terminals into account; our main focus is on the adoption of Random Linear Network Coding (RLNC), ranging from integrated satellite-terrestrial networks to multicast networks [2]. In particular, the combination of multipath connectivity and NC are analysed for the following broad-scope scenarios:

– Multicast transmissions with satellite/terrestrial component and erasure channels in the presence of mobile nodes and Complementary Ground Component (CGC);
– Mobile multicast for satellite-based aeronautical applications;
– Multipath TCP-based connections with simultaneous use of multiple paths.

Taking into account the work carried out in the Network Coding Research Group (NWCRG) of the Internet Research Task Force (IRTF) [3], this paper contains a preliminary study carried out within the "Network Coding Applications in Satellite Communication Networks" working group of the ESA funded project Satellite Network of Experts (SatNEx) IV [4].

2 A Cooperative Scenario in a Vehicular Land Mobile Satellite Environment

2.1 Fundamental Concepts

IP multicasting is a key networking technique for reaching a large number of users with a single transmit operation. The most notable application of this technique is the use of satellites for distributing audio/video contents due to the inherent broadcast nature of satellites and their large coverage area. With DVB-SH (Digital Video Broadcasting - Satellite Handheld) devices [5], the satellite version of DVB-H (Digital Video Broadcasting - Handheld) [6] for both handheld and in car retrofit devices, many mobile/vehicular applications can also benefit from satellite broadcast networks for reaching a large number of customers. As far as mobile/vehicular applications are concerned, satellite transmissions can be impaired by a number of factors such as the presence of buildings and obstacles in cities. To overcome this, the use of terrestrial gap-fillers [7], also known as CGC in the DVB-SH standard, has been proposed [8]. Gap-fillers act as repeaters, extending the satellite coverage in areas where the satellite signal degrades because of the presence of obstacles. In the future, the concept of ITS (Intelligent Transportation Systems), together with a plethora of new services for customers, will foster the use of Road Side Units (RSUs) that will provide a CGC system to allow short range communications with vehicles. They are the ideal complement to existing communication infrastructures to provide high mobility support in large networks. Here, the paradigms of V2V (Vehicle-to-Vehicle), V2I (Vehicle-to-Infrastructure), I2V and, more generally, V2X arise.

Enabling Technologies. As far as V2X is concerned, IEEE 802.11p [9] is the de facto standard for terrestrial wireless communications. It is an approved amendment and enhancement to the IEEE 802.11 standard to support ITS applications for the Wireless Access in Vehicular Environments (WAVE), which was published in 2010. This standard includes data exchange between moving vehicles and between vehicles and RSUs. WAVE is in the roadmap of many ITS projects, where the satellite component may play a role as complementary network without further significant investments for setting up a consistent coverage.

Land Mobile Satellite Channel Models. Experimental Land Mobile Satellite (LMS) propagation data have been processed in [10, 11], among others, in order to characterise the channel behaviour under narrow-band transmission conditions for different environments, degrees of shadowing and elevation angles. In the Lutz's model [10], a two-state channel model was proposed: a good state under Rician fading and a bad state with Rayleigh/Lognormal fading. In the Fontan's model [11], a three-state channel model is described, accounting for Line of Sight (LOS), moderate shadow, and deep shadow conditions.

Mobility Models. The mobility model plays an important role in establishing the effectiveness of cooperation between the satellite and terrestrial segments. Data multicast by the satellite are exchanged between the mobile nodes via 802.11p, thus filling data holes that mobile nodes may experience because of signal losses. The mobility pattern in a city is different than the one outside a city. This difference may alter the effectiveness of a V2V data distribution model, thus raising the aforementioned question about the use of RSUs. In cities, vehicle clusters are frequent formed, for instance, during traffic jams or vehicles approaching traffic lights. Clearly, a model based on real collected traces, able to capture at the same time sparse and clustered network partitions, can help in simulating a scenario close to the reality. In [12], the authors proposed a mobility model, named Heterogeneous Random Walk, which is able to capture the presence of a cluster as well as isolated nodes, and the correlation between the speed and the clustering factor. Intuitively speaking, a cluster implies that nodes are of slow speed due to the large number of vehicles moving temporary together in the same direction. The slow speed of nodes inside a cluster facilitates V2V data transmissions and the short distance between them increases the probability of correct data transmissions. The cluster formation process is desirable for improving the effectiveness of a V2V data distribution scenario. Mobile nodes can also be isolated nodes outside a cluster, spreading out on the network (roads system) and moving at higher speeds. This mobility model is a better representation of the real situation than the use of a pure city section mobility model.

2.2 Cooperative Scenario in a Vehicular Land Mobile Satellite System

The scenario under investigation is depicted in Fig. 1. A transparent satellite multicasts data from a single source to multiple terrestrial nodes, including RSU units. Data packets are coded together using NC [13], applied before the network level, generating N packets out of K source packets, $N > K$. Data sent through the satellite may not be correctly received by mobile nodes because of fading and shadowing effects [10, 11]. The RSUs, equipped with a DVB-SH and an 802.11p interface, cooperate in propagating the information received on the DVB-SH interface, retransmitting it without modification on the 802.11p interface. The mobile nodes receive data via DVB-SH and also retransmit them via 802.11p, increasing the probability that closer nodes can fill possible data holes. Finally, each RSU is assumed sufficiently apart to possibly experience different satellite channel statistics. RSUs are connected via a terrestrial link, which makes the terrestrial segment as robust as possible.

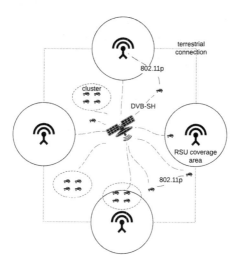

Fig. 1. Cooperative scenario for a vehicular land mobile satellite system with spatial diversity

The node mobility model is taken from [12] and described in Sect. 2.1. A cluster here can be modelled as a sort of mobile super-node, in which nodes share the maximum available fraction of received data. Four different cases are possible: (i) a set of nodes inside a cluster and inside the coverage area of an RSU; (ii) a set of nodes inside a cluster but far away from an RSU; (iii) a single node not inside a cluster but in the coverage area of an RSU; (iv) finally, a single node not part of a cluster and far away from an RSU.

Case (i) is the most favourable one: the nodes can take advantage of both situations, i.e., to be inside a cluster and close to an RSU, while case (iv) represents the worst situation: a node is isolated and can only rely on the satellite channel.

NC helps in protecting transmitted data coding different packets together at the source and also allows for recoding at intermediate nodes, for instance to deal with different channel statistics.

In a multicast scenario, the absence of a feedback channel makes it impossible for the source to know if data have been correctly received. Large redundancy can help in reducing losses but, on the other hand, it reduces the channel goodput because a fraction of the channel capacity is merely used for error correction. Thus, a trade-off must be identified between scenario requirements and channel utilization. The use of RLNC codes in a multicast scenario has been analysed in depth in [14]. It allows for a decentralized architecture (i.e., no need for network codes planned or known by a central authority), while keeping a high level of robustness. In [15], the authors dealt with the use of several communication links, for example IEEE 802.11, IEEE 802.16 and a satellite link for communication between a fixed station and mobile nodes. NC techniques are used to code together data and fully exploit the available links, achieving significant QoS improvement even in the presence of large losses on a link. In [16], multiple sources cooperate to reach a single receiver via satellite (ON/OFF channel model). Sources are supposed to be able to exchange packets among them; therefore, each source sends coded combinations of packets (RLNC) to the receiver via satellite. The different sources were spaced apart, introducing spatial diversity when transmitting to the satellite. The different geographical positions helped in reducing the system outage even in conditions of deep fading produced by the randomness of the surrounding environment. It was shown that RLNC was an effective strategy to counteract random losses in communication channels at the expenses of channel capacity: a trade-off must be identified, taking bandwidth requirements and channel statistics into account. The large performance gain shown in real satellite scenarios [17, 18] proved that the benefits of NC far exceed its shortcoming introduced by the delay in collecting at least K packets for coding in the source buffer plus the coding/decoding delay.

3   Network Coding for Mobile Multicast in a Satellite-Based Aeronautical Scenario

In satellite-based aeronautical communications, IP multicasting remains the most bandwidth efficient technology for group communication. It can further take advantage of NC to minimize the effects of random packet errors and erasures that frequently occur in satellite communications, especially in a mobile environment. During handover in a multi-beam satellite scenario where the mobile multicast receiver is in the overlapping area of two satellite beams, the benefits of NC can be even more significant as the overlapping area is always at the beam edge, which is prone to random packet errors and erasures due to the weak signal strength.

There are typically three types of handover for satellite communications, namely, beam handover, gateway handover and inter-satellite systems handover. Gateway handover entails beam handover and inter-satellite systems handover entails gateway handover. Both gateway handover and inter-satellite system handover require handover at the IP layer, while beam handover of the same satellite system is carried out at the link layer. This paper concentrates only on gateway handover in order to investigate the effect of NC on the IP layer.

3.1   Network Coding and IP Multicast Receiver Mobility in Satellite-Based Aeronautical Communications

Figures 2 and 3 present the gateway handover scenario considered in this paper. The footprint of each satellite is divided into two gateway (GW) beams (GW_B1 and GW_B2) where each GW beam represents a separate IP network. The IP multicast source is located in the terrestrial network and receivers are aircrafts equipped with a return channel satellite terminal, for example, a DVB Return Channel Satellite Terminals (RCSTs) [19]. As the reception of every single multicast packet is essential, then there is the need of feedback/acknowledgement channels from receivers (aircrafts). Different IP multicast receiver mobility support schemes do exist today, but here Home Subscription (HS)-based and Remote Subscription (RS)-based approaches are considered [20].

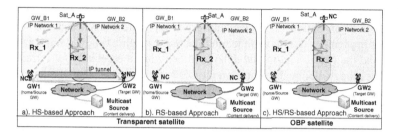

Fig. 2. NC and IP multicast receiver mobility support at gateway handover

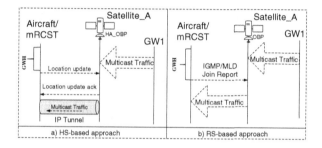

Fig. 3. Multicast reception signalling after gateway handover

3.2 Gateway Handover

Transparent (Bent-Pipe) Payload. In this scenario, intra-flow Systematic Random Linear Network Coding (S-RLNC) [21] is implemented at the satellite air interface of each GW. S-RLNC here implies that for transparent payloads the GWs will first transmit the original packets to the satellite and then the coded packets of the already transmitted packets. The transmission of the coded packets could be due to requests from some receivers or to pre-emptive measures to prevent receivers from generating retransmission requests since the satellite channel is prone to random packet errors and erasures. If there are no packet losses, then any redundant transmission is a waste of satellite resources. But if packets losses occur in a reliable IP multicast scenario over satellite where the receiving satellite terminals' population could be very large, then pre-emptive transmissions of a few coded packets could compensate for the lost packets, thus saving satellite resources by preventing terminals from generating and transmitting NACKs. The gain in throughput is proportional to the number of receivers that suffer from packets loss and also to the NC generation size. In Fig. 2, the multicast receiver aircraft RX_2 entering the overlapping area between GW_B1 and GW_B2 will signal a handover for a specified target beam.

On-Board Processing (OBP) Payload. Figure 3 presents the gateway handover scenario with OBP satellites for both HS-based and RS-based approaches. With a layer 3 regenerative OBP payload, the satellite can join the multicast groups on behalf of all multicast receivers within the satellite footprint and can also replicate multicast packets. This also gives an option of implementing NC on-board the satellite. If intra-flow S-RLNC is implemented on-board the satellite, more satellite bandwidth resources will be saved and the packet end-to-end delay will be reduced especially for retransmitted packets since they will now be sent from the OBP satellite. Since the OBP satellite acts as a multicast router to receivers in both IP networks 1 and 2, as shown in Fig. 2c, the path taken by the multicast traffic before and after the gateway handover remains the same in both HS-based and RS-based approaches. Figure 3a and b show the signalling required to receive multicast traffic after the gateway handover. For the HS-based approach, when aircraft RX_2 completes the gateway handover, it will register its newly acquired Care-of-Address (CoA) to its Home Agent (HA) located at the OBP satellite [22]. The HA now intercepts and tunnels to aircraft RX_2 the traffic from all multicast groups that aircraft RX_2 is a member of as shown in Fig. 3a. For the RS-based approach, after gateway handover is completed, aircraft RX_2 simply uses its CoA to re-subscribe to all the multicast groups that it belonged to before gateway handover, as shown in Fig. 3b. Therefore, the multicast router in the OBP satellite adds aircraft RX_2 to the list of downstream receivers in GW_B2.

3.3 Performance Evaluation

Suppose that n aircrafts in the overlapping beam area are subscribed to receive IP multicast traffic from the multicast source (see Figs. 2 and 3) and that the packet loss rate is $R_L\%$. Assuming that there are no packet losses in the terrestrial/wired network and that Negative ACKnowledgement (NACK) is used to request for any lost packet to ensure reliability, then the total expected number of transmissions (multicast packet + NACKs) required per multicast session, $E[N_{T/S}]$, for all n aircrafts to receive E_s packets successfully over the overlapping area of the two beams with and without networking coding for transparent and OPB satellites is as follows:

Without Network Coding:
For transparent satellites:

$$E\left[N_{T/S}\right] = \left\lceil \frac{\psi_s h_{GW\_A}}{1 - R_L/100}(1 + n) \right\rceil \qquad \text{for } R_L > 0 \tag{1}$$

For OBP satellites:

$$E\left[N_{T/S}\right] = \left\lceil \frac{\psi_s}{1 - R_L/100}\left(h_{GW\_A} + nh_{A\_s}\right) \right\rceil \qquad \text{for } R_L > 0 \tag{2}$$

where $E[N_{T/S}]$ = expected value of $N_{T/S}$, h_{GW_A} = number of hops between GW and aircraft via satellite; h_{A_S} = number of hops between aircraft and satellite; $\lceil x \rceil$ = ceiling function of x; E_s = average multicast session length in number of packets; n = total number of receivers (aircrafts).

With intra-flow S-RLNC, after sending K original packets, NC is performed on copies of the K original packets (generation size) to produce each coded packet that is transmitted as redundant packet. The number of redundant coded packets transmitted depends on the packet loss rate R_L. With S-RLNC, it assumed that the original packets received are used to decode each redundant coded packet received.

With Network Coding:

If Ψ_s/K is the number of coded packets produced from one multicast session, then the total number of transmissions (original + coded packets) required for all n aircrafts to receive all packets in one multicast session $N_{T/S}^{NC}$ is given by the sum of the number of transmissions of original packets plus redundant coded packets.

For transparent satellites:

$$E\left[N_{T/S}^{NC}\right] = \left\lceil \left(\frac{\Psi_s\, h_{GW\_A}}{1 - R_L/100}\right) + \left(\frac{h_{GW\_A}\left(\Psi_s/K\right)}{1 - R_L/100}\right)\right\rceil \qquad \text{for } R_L > 0 \qquad (3)$$

$$E\left[N_{T/S}^{NC}\right] = \left\lceil \frac{\Psi_s\, h_{GW\_A}}{1 - R_L/100}\left(1 + \frac{1}{K}\right)\right\rceil \qquad \text{for } R_L > 0 \qquad (4)$$

For satellites with OBP:

$$E\left[N_{T/S}^{NC}\right] = \left\lceil \left(\frac{\Psi_s\, h_{GW\_A}}{1 - R_L/100}\right) + \left(\frac{h_{A\_S}\left(\Psi_s/K\right)}{1 - R_L/100}\right)\right\rceil \qquad \text{for } R_L > 0 \qquad (5)$$

$$E\left[N_{T/S}^{NC}\right] = \left\lceil \frac{\Psi_s}{1 - R_L/100} + \left(h_{GW\_A} + \frac{h_{A\_S}}{K}\right)\right\rceil \qquad \text{for } R_L > 0 \qquad (6)$$

where $E\left[N_{T/S}^{NC}\right]$ is the expected value of $N_{T/S}^{NC}$.

Numerical Results. The following parameters are used for numerical results: $n = 50$, $R_L = 20\,\%$, $E_s = 10$, $K = 5$, $h_{GW\_A} = 2$, $h_{A\_S} = 1$. From Fig. 4, it can be seen that the total number of transmissions required with NC is 97.6 % less compared with that without NC for a transparent satellite payload. Similarly, for OBP payload, the total number of transmissions required with NC is 95.8 % less compared with that without NC.

Figure 5 shows the effect of packet loss rate in the overlapping area on the total number of transmissions required for all receivers to receive all transmitted multicast packets for transparent/OBP satellite payloads with and without NC. It can be seen that the difference in the number of transmissions required for transparent satellite payload with and without NC is huge throughout the whole range of R_L. Similar huge differences are seen for OBP satellite payload with and without NC.

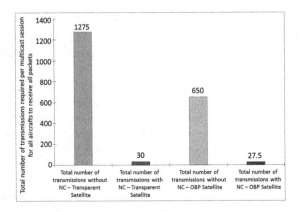

Fig. 4. Total number of transmissions with and without NC for transparent and OBP satellites

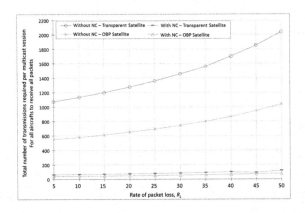

Fig. 5. Effects of packet loss rate on the total number of transmissions with and without NC

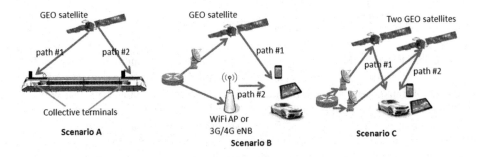

Fig. 6. Envisaged MSS scenarios for multipath TCP-based connections

4 Multipath TCP-Based Scenario

Mobile Satellite Systems (MSSs) can provide communication services in areas where a terrestrial cellular infrastructure is not available. This section considers mobile users affected by ON/OFF (Markov) channel due to their movement and the presence of obstacles. The focus is on MSS scenarios where an end user can connect via two paths simultaneously using two transceivers having either the same air interface but different carriers or different air interfaces connecting to different wireless systems. In particular, the following subcases are considered as depicted in Fig. 7:

- **Scenario A**: A train with a collective terminal and two antennas in the front and at the back of the train;
- **Scenario B**: A multi-Radio Access Technology (multi-RAT) system where the mobile terminals can use different air interfaces (hybrid system), such as: satellite, WiFi, and 3G/4G;
- **Scenario C**: A satellite diversity case, where two GEO satellites are adopted to reach the mobile user.

In Scenario A, the train employs two antennas to receive the traffic and a collective terminal is used to exploit the data received from both paths. The collective terminal can connect local users on the train by means of an onboard WiFi system. In Scenario B, mobile terminals are expected to simultaneously use multiple air interfaces. The presence of different paths with different propagation delays and packet loss conditions may be a critical issue. This asymmetry could cause the receiver buffer to fill up while waiting for the recovery of lost packets on the slowest path. Finally, Scenario C considers a complex mobile device (or collective terminal) that uses two antennas and two independent transceivers to connect and to simultaneously synchronise with two GEO satellites. This allows path diversity and can better cope with user mobility and the occurrence of occasional path disconnections due to obstacles. In all these scenarios, the mobile user receives from two independent paths affected by independent ON/OFF channel behaviours on the wireless segment. Multipath protocols allow the exploitation of the inherent path diversity; these protocols are considered here in combination with NC solutions. Note that each multipath scenario above has its own unique characteristics. For instance, the two paths of Scenario B are characterised by different air interface conditions and different propagation delays.

The ON/OFF Markov channel model is characterised as follows: the mean ON (OFF) time is T_{ON} (T_{OFF}). During the OFF phase, packet transmissions are affected by erasures according to probability p, while in the ON phase all packet transmissions are considered to be received correctly. The ON and OFF state probabilities are:

$$P_{ON} = \frac{T_{ON}}{T_{ON} + T_{OFF}} \quad , \quad P_{OFF} = \frac{T_{OFF}}{T_{ON} + T_{OFF}} \tag{7}$$

The values of T_{ON} and T_{OFF} are on the order of seconds for MSSs and can be determined according to [10] considering user speed and S and L (2 GHz) bands.

RLNC is adopted here as it seems to offer a simple and robust solution; each coded packet generated from a block of packets is just another packet that can contribute to fulfil the degrees of freedom needed at the decoder. Other NC codes such as Raptor codes [23] could be more complex to implement even though the decoding complexity is linear with the block size K, $O(K)$, while RLNC have complexity $O(K^3)$. Hence, RLNC requires to keep a small-enough block size K for an efficient decoding; the encoding block size could also be differentiated from path to path in case of different channel conditions on the two paths (asymmetry).

This Section investigates the combination of MP-TCP at transport layer with NC of the RLNC type implemented as shim layer. Note that in order to achieve the maximum transparency at end-hosts (both servers and end-users), MP-TCP and RLNC are not implemented end-to-end, but inside the network between two transport-layer Perform-ance Enhancing Proxies (PEPs), at an intermediate router and at a collective terminal (user side) with a feedback loop between them. If the receiving PEP is unable to decode a block due to a loss of degrees of freedom, it is then possible to ask the PEP-sender to transmit further encoded packets to recover the losses. The PEP implements a TCP split approach, performing a 'conversion' from TCP to MP-TCP that is used inside the satel-lite network [24]: each TCP flow is divided into two subflows that exploit two inde-pendent paths (see Fig. 6). Each subflow is protected by RLNC that is implemented as a shim layer below the transport layer (i.e., MP-TCP/TCP subflow/NC) to recover packet losses due to channel effects. If IP packets are end-to-end encrypted with IPsec before they are sent via the PEP-based satellite network, it would be impossible to perform PEP functions at the intermediate node, since we could not access IP packet payload data (TCP header). However, IPsec could be applied between the two PEPs.

In our scenario, the two paths experience independent ON/OFF channel behaviours. Hence, even if one path is affected by losses, the other path may experience good condi-tions, thus allowing path diversity with error recovering capabilities that can be exploited by NC.

There are other techniques proposed in the literature to combine TCP-based trans-port-layer protocols with NC, as those in [15, 25, 26]. In particular, the authors of [25] introduced TCP/NC where TCP is combined with NC at a shim layer between transport and network layers. A TCP/NC source transmits random linear combinations of all packets in a coding (sliding) window that is related to the congestion window. The receiver acknowledges every degree of freedom (i.e., a new encoded packet that provides new information). Another TCP version that includes NC is called Coded TCP (CTCP) [26]. The CTCP sender divides the data stream into blocks with a fixed number of packets; then, linear combinations are generated for the packets of each block. CTCP estimates the packet loss rate and adaptively computes the number of necessary coded packets to be transmitted. Finally, MPTCP/NC adopts two layers of network coding [15]. The first NC layer is applied before packets are injected into a TCP subflow; this layer does not add redundancy, but is useful so that packets of both subflows can be combined together for NC decoding purposes. Instead, the second NC layer, based on TCP/NC (subflow level), introduces redundancy for protecting subflows from packet losses.

In this project, we will apply MP-TCP to our scenarios in Fig. 6, where encoded packets (RLNC) of one path are sent on the other path to improve robustness to packet losses in the case of mobile users (path-coding diversity). This technique is called Path-

Based Network Coding MP-TCP (PBNC-MP-TCP). The encoding scheme can be applied both as intra-flow NC at the PEP (i.e., using multiple encoders, one for each subflow) or as inter-flow coding (i.e., using a common buffer and one encoder at the intermediate PEP). The number of redundancy packets generated for each subflow depends on the channel conditions experienced on the path and can be determined according to a cross-layer approach to maximize transport layer goodput.

In our scenario, we adopt the S-RLNC analysis proposed in [27] to study the successful decoding probability P_{suc} of an encoded block, taking the behaviour of the ON/OFF channel into account (coded blocks can pick the channel in OFF or ON state according to the respective state probabilities). If the transmission of the coded block occurs while the channel is in the OFF state, packets are subject to an erasure rate p. If the transmission occurs in the ON state, packets are received successfully. This approach is possible because the satellite channel has a much slower behaviour than the transmission time of a coded block: the transmission of a coded block just samples the satellite channel in ON or OFF states according to the corresponding probabilities of the ON/OFF Markov chain. In what follows, K is the information block size, N is the size of the encoded block, $N - K$ is the number of redundancy packets, $\delta \in [0, 1, ..., N - K]$ is the number of redundancy packets sent on the secondary path, q is the field size of the Galois field used for NC. The following formulas (8)–(10) are adapted from [27] to account for δ packets out of $N - K$ redundancy packets sent on the secondary path and experiencing an ideal lossless channel. This assumption is made to emphasize the impact of using a secondary path.

$$P_{suc} = \left[\sum_{r=K-\delta}^{N-\delta} \binom{N-\delta}{r} (1-p)^r p^{N-\delta-r} f_k(r+\delta, N) \right] P_{OFF} + f_K(N,N) P_{ON} \qquad (8)$$

where $f_K(r + \delta, N)$ is determined as follows:

$$f_K(r+\delta,N) = \frac{\binom{N-K}{r+\delta-K} + \sum_{h=h_{min}}^{K-1} \binom{K}{h} \binom{N-K}{r+\delta-h} \prod_{j=0}^{K-h-1} (1-q^{h+j-r-\delta})}{\binom{N}{r+\delta}} \qquad (9)$$

$$h_{min} = \max \{0, r+\delta-N+K\} \qquad (10)$$

Hence, the block decoding failure P_{fail} can be obtained as $P_{fail} = 1 - P_{suc}$. Figure 7 shows the performance of S-RLNC ($K = 5$ packets is the encoding block size and $N = 9$ is the coded block size) for different field sizes q for an ON-OFF channel (primary path) with $T_{ON} = 4$ s, $T_{OFF} = 2$ s, and packet erasure rate in the OFF state $p = 0.5$. The $N - K = 4$ redundancy packets can be sent in part ($\delta \leq N - K$) on a secondary path.

Note that without network coding the average packet loss rate is equal to $p \times P_{OFF} \approx 0.17$ that would practically cause TCP goodput to drop to zero. Hence, NC can significantly reduce the packet loss corresponding to P_{fail} as shown in the graph, thus allowing a better TCP behaviour. In particular, the case where all redundancy packets are sent on the secondary path permits to reduce the packet decoding failure with respect

G. Giambene et al.

to the case where all redundancy packets are sent on the main path (classical, MPTCP/NC-like case). For instance, with field size $q = 256$, this reduction is by a factor about equal to 15 that can roughly correspond to a TCP goodput increase of four times. Hence, we expect that our diversity approach for sending redundancy packets can provide a positive impact on the PBNC-MP-TCP technique proposed.

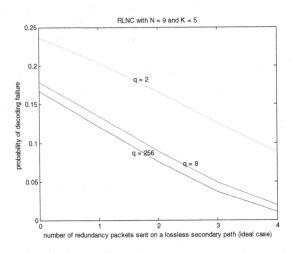

Fig. 7. RLNC performance for different field sizes and different number of redundancy packets sent on an ideal path without losses

5 Conclusion and Future Directions

NC has many potentialities to improve the performance of satellite networks. This paper surveys some of these opportunities identifying key scenarios and presenting preliminary results. The future activity will concern with the implementation of simulators dealing with the described scenarios to provide numerical quantitative evidences to the considerations made in this paper.

Acknowledgments. This work has been partially funded by ESA within the framework of the SatNex IV project. The view expressed herein can in no way be taken to reflect the official opinion of ESA. A special thank is to Dr. Nader Alagha of the ESA Technical Centre (ESTEC) for his useful comments and suggestions.

References

1. Bassoli, R., et al.: Network codingtheory: a survey. IEEE Commun. Surv. Tutor. **15**(4), 1950–1978 (2013)
2. Vieira, F., Shintre, S., Barros, J.: How feasible is network coding in current satellite systems? In: Proceedings of the 5th ASMS and the 11th SPSC, Cagliari, Italy, September 2010
3. NWCRG home page with. https://irtf.org/nwcrg

4. ESA SatNex IV project home page with. http://www.satnex4.org/
5. European Telecommunications Standards Institute, ETSI EN 302 304 (2004–11), Digital Video Broadcasting (DVB); Transmission System for Handheld Terminals (DVB-H)
6. European Telecommunications Standards Institute, ETSI TS 102 584 V1.2.1, Digital Video Broadcasting (DVB); DVB-SH Implementation Guidelines Issue 2, January 2011
7. Evans, B., et al.: Integration of satellite and terrestrial systems in future multimedia communications. IEEE Wirel. Commun. **12**(5), 72–80 (2005)
8. Cocco, G., Alagha, N., Ibars, C.: Network-coded cooperative extension of link level FEC in DVB-SH. In: Proceedings of the 29th AIAA International Communications Satellite Systems Conference 2011, Nara Japan (2011). http://arc.aiaa.org/doi/abs/10.2514/6.2011-8034
9. Jiang, D., Delgrossi, L.: IEEE 802.11p: towards an international standard for wireless access in vehicular environments. In: Proceedings of the Vehicular Technology Conference 2008. Spring, Korea, Seoul, May 2008
10. Lutz, E., Cygan, D., Dippold, M., Dolainsky, F., Papke, W.: The land mobile satellite communications channel-recording, Statistics, and channel model. IEEE Trans. Veh. Technol. **40**(2), 368–375 (1991)
11. Fontain, F.P., et al.: Statistical modeling of the LMS channel. IEEE Trans. Veh. Technol. **50**(6), 1549–1567 (2001)
12. Pirkowski, M., et al.: On clustering phenomenon in mobile partitioned networks. In: Proceedings of the 1st ACM SIGMOBILE Workshop on Mobility Models, pp. 1–8 (2008)
13. Ahlswede, R., Cai, N., Li, S.-Y.R., Yeung, R.W.: Network information flow. IEEE Trans. Inf. Theory **46**(4), 1204–1216 (2000)
14. Ho, T., et al.: A random linear network coding approach to multicast. IEEE Trans. Inf. Theory **52**(10), 4413–4430 (2006)
15. Cloud, J., et al.: Multi-path TCP with network coding for mobile devices in heterogeneous networks. In: Proceedings of the 78th IEEE Vehicular Technology Conference 2013 - Fall, pp. 1–5 (2013)
16. Alegre-Godoy, R., Vazquez-Castro, M.A.: Spatial diversity with network coding for on/off satellite channels. IEEE Commun. Lett. **17**(8), 1612–1615 (2013)
17. Vieira, F., Barros, J.: Network coding multicast in satellite networks. In: Proceedings of Next Generation Internet Networks 2009 (NGI 2009), pp. 1–6 (2009)
18. Alegre, R., Steluta, G., Alagha, N., Vazquez-Castro, M. A.: Multicasting optimization methods for multi-beam satellite systems using network coding. In: Proceedings of the 29th AIAA International Communications Satellite Systems Conference (2011)
19. European Telecommunications Standards Institute, ETSI TR 102 768 (2009–04) Digital Video Broadcasting (DVB); interaction channel for Satellite Distribution Systems; Guidelines for the use of EN 301 790 in Mobile Scenarios, April 2009
20. Romdhani, I., et al.: IP mobile multicast: challenges and solutions. IEEE Commun. Surv. Tutor. **6**, 18–41 (2004)
21. Lucani, D.E., Medard, M., Stojanovic, M.: On coding for delay-network coding for time-division duplexing. IEEE Trans. Inf. Theory **58**, 2330–2348 (2012)
22. IRIS Space Router Key Capabilities and Benefits, Cisco Systems, Inc. and Astrium White Paper (2011)
23. Shokrollahi, A.: Raptor codes. IEEE/ACM Trans. Netw. **14**, 2551–2567 (2006)
24. Muhammad, M., Berioli, M., Colade, T.:A simulation study of network-coding- enhanced PEP for TCP flows in GEO satellite networks. In: Proceedings of ICC, Sydney, Australia, June 2014
25. Sundararajan, J.K., et al.: Network coding meets TCP: theory and implementation. Proc. IEEE **99**(3), 490–512 (2011)

26. Kim, M., Cloud, J., Parandeh Gheibi, A., Urbina, L., Fouli, K., Leith, D., Medard, M.: Network coded TCP (CTCP) (2013). http://arxiv.org/pdf/1212.2291.pdf
27. Jones, A. L., Chatzigeorgiou, I., Tassi, A.: Binary systematic network coding for progressive packet decoding. In: Proceedings of IEEE ICC 2015 - Communication Theory Symposium, 8–12 June 2015, London, UK (2015)

Can Network Coding Mitigate TCP-induced Queue Oscillation on Narrowband Satellite Links?

Ulrich Speidel[1]([✉]), Lei Qian[1], 'Etuate Cocker[1], Péter Vingelmann[2],
Janus Heide[2], and Muriel Médard[3]

[1] Department of Computer Science, The University of Auckland,
Auckland, New Zealand
{ulrich,lqia012,ecoc005}@cs.auckland.ac.nz
[2] Steinwurf ApS, Aalborg, Denmark
{peter,janus}@steinwurf.com
[3] EECS, Massachusetts Institute of Technology, Cambridge, MA, USA
medard@mit.edu

Abstract. Satellite-based Internet links often feature link bandwidths significantly below those of the ground networks on either side. This represents a considerable bottleneck for traffic between those networks. Excess traffic banks up at IP queues at the satellite gateways, which can prevent conventional TCP connections from reaching a transmission rate equilibrium. This well-known effect, known as *queue oscillation* can leave the satellite link severely underutilised, with a corresponding impact on the goodput of TCP connections across the link. Key to queue oscillation are sustained packet losses from queue overflow at the satellite gateway that the TCP senders cannot detect quickly due to the long satellite latency. Network-coded TCP (TCP/NC) can hide packet loss from TCP senders in such cases, allowing them to reach equilibrium. This paper reports on three scenarios in the Pacific with two geostationary and one medium earth orbit connection. We show by simulation and circumstantial evidence that queue oscillation is common, and demonstrate that tunneling TCP over network coding allows higher link utilisation.

Keywords: Queue oscillation · TCP · Network coding · Satellite links

1 Introduction

A large number of locations around the world rely on satellite links as their only feasible means of connecting to the Internet. Traditionally, this connectivity used to be supplied by geostationary (GEO) satellites, although in the last few years, a medium earth orbit (MEO) satellite operator has entered the market [1]. In practice, this means a connection with long round trip times (typ. >500 ms for GEO, or >125 ms for MEO) and generally expensive and hence often narrow bandwidth. Even though the advent of MEO service has brought some relief

© Institute for Computer Sciences, Social Informatics and Telecommunications Engineering 2015
P. Pillai et al. (Eds.): WiSATS 2015, LNICST 154, pp. 301–314, 2015.
DOI: 10.1007/978-3-319-25479-1_23

in this respect, many small and medium-sized communities cannot afford bandwidths comparable to those that connect to the satellite gateways at either end of the link.

The resulting bottleneck presents a significant challenge to the Internet's staple transport protocol, TCP [2], leading to both bandwidth *underutilisation* and slow or even stalling data transfers. These problems have been discussed in the literature for many years [3–5], leading to the development of TCP variants for connections involving wideband satellite links [6,7]. However, these do not necessarily work well in the common narrowband scenarios, where the link is shared with conventional TCP varieties and widespread deployment of satellite-friendly TCP variants is not a feasible option.

This paper presents two main results: Firstly, we show by simulation that native TCP behaviour of multiple parallel TCP flows suffices to produce the effects observed in three satellite-connected Pacific Island locations. Secondly, we show by experiment that network coding of some TCP flows into these locations can result in better goodput in many cases.

The next section discusses the principle behind queue oscillation. Section 3 considers how to model a satellite link and the traffic across it in a simulation, followed by a selection of simulation results in Sect. 4. We then discuss the network coding scheme used in our live experiment across the three actual satellite links in Sect. 5, and provide some experimental evidence on its effectiveness in Sect. 6.

2   TCP-induced Queue Oscillation

The Transmission Control Protocol [2] handles the bulk of Internet traffic. TCP relies on a family of complex flow and congestion control algorithms aimed at achieving reliability, high goodput rates and fair bandwidth sharing among multiple TCP flows. A common feature of all these algorithms is their use of acknowledgment packets (ACK) as feedback to the TCP sender. The sender infers from the ACKs whether packets have been lost or delayed, and how much data it may entrust to the channel with acknowledgments outstanding. Under ideal circumstances, this results in links that carry a maximum of goodput and a minimum of retransmissions.

Roughly speaking, a TCP sender starts data transmission at a slow packet rate to elicit ACKs from the receiver. As long as the receiver returns ACKs and indicates readiness to receive more data, the sender progressively increases the packet rate. This continues either until the receiver throttles the sender through its advertised window in the ACKs, or until packet loss in either direction disrupts the ACK sequence arriving at the sender. In the case of packet loss (missing ACKs), the sender responds with a more-or-less aggressive reduction in packet rate.

Packet loss can occur in the physical layer (interference, noise, fading, etc.) but also in network equipment (routers, switches, modems, ...) unable to buffer sufficient incoming packets in their input queue for dispatch on outgoing links, dropping excess packets arriving for a full queue. In this paper, we argue that the latter effect suffices to explain the low performance reported across many satellite links.

In the case of satellite links, the satellite bandwidth is generally much smaller than that of the network infrastructure it connects to (typically fibre optic networks). Viewed from both ends of a TCP connection across the satellite, the sat link thus represents a bottleneck. The queue at which traffic banks up when the bottleneck gets congested is the transmission queue at the satellite ground station. We note here as an aside that the bottleneck bandwidth also tends to be the most expensive, meaning that the size of the bottleneck is generally severely constrained by economic factors rather than traffic demand dimensioning.

TCP's congestion control algorithms are of course designed to cope with bottlenecks, albeit those for which the algorithm can obtain quasi-realtime information. However, our satellite transmission queue is both the location at which congestion events are most likely to occur, and the location at which congestion relief through sender response must become effective. Information about packet drops generally propagates in the form of "non-events" via the receiver to the sender: A data packet dropped at the satellite queue subsequently fails to arrive at the receiver, which therefore does not emit an ACK for the packet. The ACK then doesn't return via the satellite link, and only once it is overdue at the sender, the sender will take corrective action by lowering its packet rate. The lower packet rate in turn does not become effective at the queue until the data held back by the sender eases the overflow situation at the queue.

That is, event and response are separated by a whole round-trip-time (RTT), which is particularly long in satellite links. In other words: The TCP sender on a satellite link always works with severely out-of-date congestion information. On links carrying multiple TCP flows, there is an additional problem: Multiple senders are all independently afflicted by the same issue. Sensing capacity on the link (=senders receive regular ACKs for packets that passed through the bottleneck some time ago), the senders *all* more or less simultaneously increase their congestion window and hence their sending rate. As the packet arrival rate at the queue is the sum of the (latency time-shifted) sender rates, this can considerably accelerate the rate at which the queue fills. Now note that this acceleration continues even after the queue overflows – it only ceases once the senders learn about the packet drops through the missing ACKs, and the resulting back-off only has an effect at the queue once the packets sent at the fast rates have all arrived there.

Similarly, when the senders detect packet loss and back off in response, the effect on the queue arrival rate is in effect multiplied by the number of flows involved: The flood now slows to a trickle. It remains a trickle until the senders receive ACKs again, one RTT later. This often allows the queue to drain completely, leaving the link idle. As the ACKs return again, the cycle may repeat.

This effect is known as *queue oscillation* and is well documented in the literature, not just for satellite links. However, satellite links are prime candidates for queue oscillation due to their long latencies and the severe bandwidth bottleneck they often represent.

In principle, we may distinguish three scenarios for a satellite link with multiple-sender TCP traffic:

- **Pre-oscillation:** Low traffic demand; existing TCP flows do not fill the bandwidth of the link and the queue does not overflow. Characteristic for this scenario are low packet loss and low link utilisation.
- **Oscillation:** The queue oscillates between entirely empty and overflow and reaches both extremes in the order of around 2–3 RTT. Characteristic for this scenario are high packet loss with recognisable bursts paired with low link utilisation. The high packet loss limits the goodput on individual connections even if the link as such has plenty of spare capacity.
- **Congestion:** The queue may oscillate between overflow and part-empty, but traffic demand is such that young and short TCP flows appear at a high rate. These flows, which have not responded to packet loss yet, provide sufficient traffic to the queue to prevent it from draining completely. Characteristic for this scenario are high packet loss with recognisable bursts paired with high link utilisation, a high number of concurrent flows, a significant number of old stalled flows, and low per-flow goodput.

The next section discusses how we can replicate these scenarios in a network simulator.

3   Modelling an "Island" Link

In the typical "island" scenario, the bulk of data flow is from the Internet to the island, with typical inbound-to-outbound bandwidth ratios of 4:1. Our simulations consider inbound payload data flow only. Our "satellite link" assumes a one-way delay of 240 ms for GEO and of 125 ms for MEO. On the Internet side, the router at the satellite gateway forms the root of a "binary" tree of depth 2 whose edges are links of $\geq$ 1 Gbps and whose leaf nodes are TCP senders with one-way latencies to the gateway of 11, 20, 60 and 80 ms, respectively. This arrangement lets us simulate a variety of latencies of similar orders of magnitude as one would expect in the real world. At the "island end", we use a single TCP receiver connected via a 1 Gbps link to the router.

The next challenge is to simulate the traffic. On real links, tools such as ntop/nprobe [11] and various others can measure the number of active hosts and/or the number of active flows seen during a certain time period. The number of active hosts does not yield the number of active flows, however, even if it permits estimates of the right order of magnitude. Real-world flows also tend to be short in terms of the size in bytes, and those active at one particular time usually have different start times and congestion windows.

This poses the challenge of getting the "flow mix" at least approximately right, as consideration of the extremes shows. Flows that are too small consist of only a few packets – too little to partake in oscillation over a whole cycle. Too few flows are bound to leave the link underutilised and will never fill the queue, so there will be no packet loss and no oscillation – our pre-oscillation scenario. At the other end of the spectrum, in the congestion scenario, one creates new flows at a rate and/or of sizes that are too large. In this scenario, old flows will slow to a crawl because of the congestion and will not complete, while the new flows that have yet to experience packet loss ensure that the queue never empties: Here, we expect plenty of packet loss all the time, a permanently overflowing queue, 100 % link utilisation, and a very high combined goodput. However, because of the large and growing number of flows, the average goodput per flow tends to zero. It is between these poles that one expects queue oscillation.

To obtain a realistic flow size distribution, we assume that the island's Internet traffic is predominantly web traffic. For each new flow that we generate, we choose the byte volume of a flow randomly from the first 100,000 flows in a trace taken at the University of Auckland's border gateway in 2005, which consists of around 85 % web traffic. Figure 1 shows the flow size distribution thus obtained.

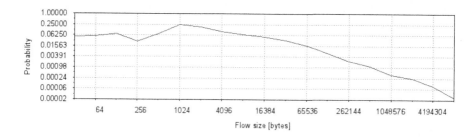

Fig. 1. Flow size distribution for our simulations.

In our simulation, we perform a "sweep" using an estimated flow creation rate. The number of concurrent flows at the beginning of each simulation is zero. It then either grows throughout the simulation or reaches a plateau. Simulations whose number of concurrent flows continues to grow inevitably head for congestion: Older flows stall, and the goodput rate per flow declines. If the number of concurrent flows stabilises, the link can be in either a pre-oscillation or oscillation scenario. We can distinguish between these two scenarios based on the absence or presence of burst packet losses. If the flow creation rate is very low, it is also possible that both the number of concurrent flows and the goodput rate per flow keep growing significantly throughout the experiment, a clear indicator of pre-oscillation.

Another parameter that is difficult to determine is the queue capacity at the satellite gateways, as this information resides with the satellite operators and does not necessarily get documented to their customers. Industry-standard routers typically provision queues for a few to up to several hundred packets, and we tried a number of values in this range. As a general rule, a low capacity queue encourages oscillation as it is easy to fill and easy to clear, whereas a high capacity queue tends to shift the scenario towards congestion.

4   Simulation Results

Our simulation concentrates on the three real-world Pacific Island locations we investigated: Rarotonga in the Cook Islands, Niue, and Funafuti Atoll in Tuvalu. At the time of deployment, Rarotonga had an inbound MEO band-width of 160 Mbps, which has since been upgraded to 332 Mbps, and Niue had 8 Mbps from GEO. The exact GEO bandwidth for Funafuti does not seem to be officially known, however our observations lead us to an estimate of 16 Mbps. Initial daytime observations showed that none of the links was in pre-oscillation mode: Rarotonga and Funafuti oscillated, whereas Niue seemed to have reached permanent congestion.

For the "Niue model", Fig. 2 shows that link utilisation reaches levels close to the maximum around 50 s into the simulation for a queue with a capacity of 50 packets and flows created once every 13 ms. In this scenario, almost all data on the link is goodput. An inspection of packet drops over the experiment time in Fig. 3 shows the high packet loss characteristic of the congestion scenario. Note however that lossless time intervals interleave regularly with those featuring high packet loss – a classic sign of queue oscillation. An event-based inspection of the queue length shows that the queue also frequently drains completely. Figure 4 shows that in this scenario, the number of concurrent flows and the instantaneous average goodput per flow reach a noisy equilibrium towards the end of the sweep.

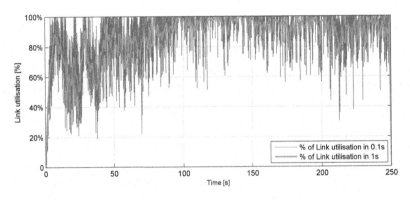

Fig. 2. Simulated link utilisation of the "Niue model". Note that the link is almost fully used after the initial ramp-up.

For the Rarotonga scenario, our initial deployment took place with an inbound satellite link capacity of 160 Mbps. A simulated queue with a capacity of 50 pack-ets also produces oscillation for 160 Mbps if we create a new flow every millisec-ond. Figure 5 shows that this results in a link utilisation of around 60 %, while still producing significant packet loss. Figure 6 shows the bursty nature of the packet losses: During some time periods, the queue drops dozens of arriving pack-ets, during other time periods it drops none. This effect is present during the entire duration of the experiment. The low link utilisation is evidence of complete queue drain, which we have also been able to observe directly on the queue.

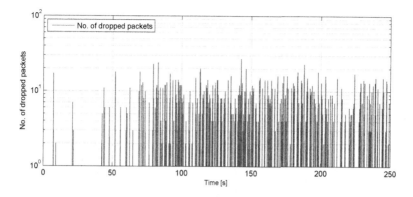

Fig. 3. Bursty packet drops in the simulated "Niue model".

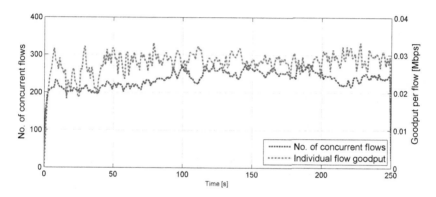

Fig. 4. Simulated number of concurrent flows and average per-flow goodput of the "Niue model": As new flows appear, older flows stall. Higher flow creation rates than the one used in this model lead to a steady rise in concurrent connections and a steady drop in goodput per flow.

Figure 7 shows that number of concurrent flows plateaus, showing that the link copes with demand. The effect of the queue oscillation here is simply a lengthening of flow durations caused by the packet losses.

If one "upgrades" the simulated Rarotonga link to the later value of 332 Mbps and leaves all other parameters (queue capacity, flow creation rate and flow distribution) identical, packet losses *increase* slightly. This is somewhat counterintuitive, but a possible explanation is that the larger queue drain rate causes TCP flows to have larger congestion windows when the queue eventually overflows. This means that there are more packets "in flight" at overflow time that will subsequently be dropped. Goodput in this scenario remains almost unchanged, and the link utilisation predictably drops to around 30 %.

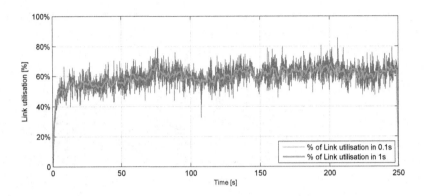

Fig. 5. Simulated link utilisation of the 160 Mbps "Rarotonga model".

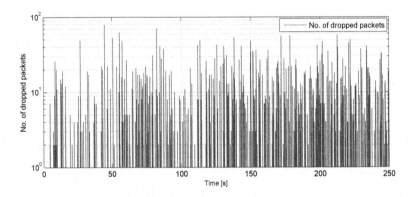

Fig. 6. Bursty packet drops in the simulated 160 Mbps "Rarotonga model".

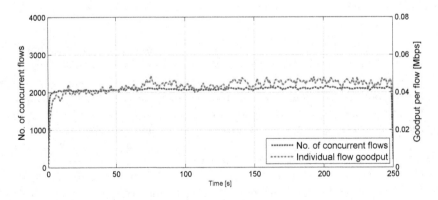

Fig. 7. Simulated number of concurrent flows and average per-flow goodput of the 160 Mbps "Rarotonga model".

5 Tunneling TCP over UDP with Network Coding

Network coding (NC) lets us mask a moderate amount of burst packet loss from the TCP sender and receiver. The type of network coding used in our study is known as *random linear network coding* (RLNC) [8,9].

For simplicity, we discuss the world-to-island traffic direction here only, the opposite direction works analogously. First, we identify an IP network X on the island side that is to benefit from FEC by NC. We then ensure that the off-island NC encoder G_w (which does not need to be placed at or even near the satellite gateway) is the IP gateway to X, i.e., all IP packets to X from anywhere on the Internet are routed to G_w. These packets carry TCP and any other protocols from the TCP/IP suite.

G_w intercepts the packets and groups them into *generations*. A generation is a set of n IP packets $p_1, p_2, \ldots, p_n$ which G_w then turns into $n + \omega$ linear equations with random one-byte coefficients $c_{i,j}$ such that the i-th equation in the set is:

$$\sum_{j=1}^{n} c_{i,j} p_j = r_i,$$

where $c_{i,j} p_j$ is the product of $c_{i,j}$ and each byte in p_j. G_w stores the respective results in each byte of r_i and transmits each such linear equation as a "combination" UDP packet to the on-island decoder G_i. The combination packet contains the $c_{i,j}$, r_i, and other information such as packet length. Without packet loss, G_i thus receives a system of linear equations overdetermined by up to ω equations, whose solution is the generation $p_1, p_2, \ldots, p_n$. G_i solves this system and forwards $p_1, p_2, \ldots, p_n$ to their original destinations in X. Note that more than ω packets need to be lost between G_w and G_i for the generation to become unrecoverable, thus providing FEC for loss bursts of up to length ω.

The UDP combination packets use G_w's IP address as origin and G_i's IP address as destination. Note that the latter, while on the other side of the satellite link, is *not* part of X, but part of a separate IP network Y whose traffic is routed via the satellite gateway. G_i thus needs two interfaces, one with a IP address in X and one with an address in Y. G_w also needs two interfaces: One that receives traffic from the Internet for X, and one that sends UDP combination packets to G_i.

In the reverse direction, G_w and G_i switch roles: G_i acts as the gateway for hosts in X, encoding traffic from X to the Internet into UDP combination packets heading for G_w. Similarly, G_w now decodes these combinations and forwards the packets from X to their destinations on the Internet. This thus establishes a UDP tunnel from the Internet to X and vice versa.

6 Experimental Observations

Our experimental results do not give an entirely uniform picture for the various links:

Niue: The actual scenario on the Niue link has been difficult to measure as the link is also the only electronic means of getting experimental data out. During our visit in December 2014, link utilisation was in excess of 90 % during the day, and packet losses were common. Our simulation indicates that around 200 concurrent flows may be realistic and that typical loss bursts do not exceed a few dozen packets. Distributing these losses across concurrent flows suggests that most flows are unlikely to lose more than one or two packets. This corresponds well to the fact that a generation size of $n = 10$ with $\omega = 2$ was able to mask most of the losses on the real link.

The link into Niue sees sustained peak data rates of around 7.5 Mbps with >7 Mbps recorded for much of the day. Individual conventional TCP connections with 5 MB downloads achieve around 0.3 Mbps. Closer inspection reveals that the actual link transports goodput without redundant retransmissions arriving at the Niue end, which again agrees well with the goodput observations in the simulation above. At the utilisation and data rates observed, the link can thus handle around 25 such parallel connections, note however that most flows are much shorter and take up less bandwidth due to TCP slow start, i.e., the link can actually accommodate a much higher number of connections.

Single TCP connections downloading 5 MB across a TCP/NC tunnel into Niue achieved around 2–2.4 Mbps goodput with very low overhead, i.e., the entire link capacity would be exhausted by 3–4 such connections. Given the high existing link utilisation, this additional performance of even a single connection comes at the expense of conventional TCP goodput. However, if these connections are downloads, the higher goodput rate also shortens the flow. This poses the question as to whether short wideband flows with TCP/NC are better from a user perspective than long thin ones, given that the bulk of bandwidth is taken up by flows that download something.

In Niue, we also investigated the potential of H-TCP and Hybla compared to the standard Cubic TCP used in the Linux kernel. While there were considerable differences between them and Cubic at certain times, neither of the two presented a convincingly strong alternative on this narrowband path.

Funafuti: Conventional TCP on the link into Funafuti arrives at a SilverPeak NX-3700 WAN Optimiser [10], whose exact configuration could not be determined. The NX-3700 supports, among others, two functions that directly impact on our measurements: network memory and parity packets. Network memory is essentially a data compression technique that prevents previously seen data from transiting across the link again. Parity packets protect small sets of subsequent packets by parity bits, such that packets lost on the link can be recovered island-side (in principle, this is a very crude form of network coding). Our observations indicated that both features were active.

Our NC encoder/decoder in Funafuti sits in parallel to the NX-3700, i.e., the standard TCP in our measurements has the benefit of the NX-3700 whereas the TCP/NC traffic does not pass through the WAN optimiser. Figure 8 shows the goodput achieved by individual TCP and TCP/NC connections with 5 MB downloads into Funafuti and the packet loss experienced at the time. The TCP/NC tunnel used $n = 30, \omega = 15$ throughout. Breaks in the curves indicate that the respective download failed. The packet losses into Funafuti are quite significant with up to 10 % and more seen on some weekdays – levels at which conventional TCP simply does not work. Closer inspection of log data reveals burst losses of many hundreds of packets. At some times, packet losses in generations exceed 15 and thus render entire generations undecodable, with a resulting irrecoverable TCP loss rate. The NX-3700 manages to persist in some of these cases, albeit at extremely low goodput rates. At times of moderate packet loss, e.g., in the mornings and evenings, TCP/NC provides significantly better goodput than TCP via the NX-3700. During the night, packet loss drops to negligible levels, and the lower overhead in the TCP connections via the NX-3700 results in better goodput here.

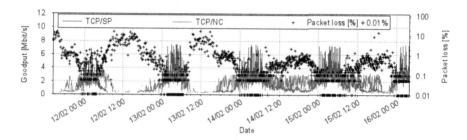

Fig. 8. Measured goodput of individual TCP and TCP/NC connections into Funafuti (Tuvalu) vs. packet loss.

Direct monitoring of the feed from the satellite link to the NX-3700 showed typical daytime link utilisation between 2 and 3 Mbps, far below the estimated 16 Mbps bandwidth. Again, this is a strong indicator for the presence of queue oscillation during these times. Our simulations show neither the low link utilisation observed here, nor the extreme burst error losses. A conceivable explanation for this is that rudimentary error correction such as that performed by the NX-3700 delays detection of packet loss onset by the TCP senders without ultimately being able to mask the packet loss, resulting in more radical TCP back-off.

Rarotonga: The first comparative multi-day measurement of traffic into Rarotonga at the end of January 2015 showed that both TCP and TCP/NC downloads of 20 MB could achieve high goodput of 20–25 Mbps. TCP/NC performed at this level almost continuously, whereas conventional TCP dropped to 5 Mbps and below during daytime peaks with high packet loss. Peak-time link utilisation was typically around 60 % of the available 160 Mbps. This corresponds well with the simulation above.

At the end of May 2015, inbound bandwidth had been increased to 332 Mbps. Telecom Cook Islands was now also able to provide us with hourly averages of the link utilisation as well as of the number of new connections per second. Typical daytime values for the latter were around 800–1000 per second, which greatly assisted in modelling the link for the simulations above. Figure 9 shows that the goodput of our downloads did not benefit from the bandwidth increase at all: Peak-time goodput deteriorated significantly, despite hourly link utilisation peaking at only 150 Mbps – just 45 % of the available capacity. As predicted by the simulation, daytime packet loss (not shown) had also increased and frequently exceeded 1 % – more than sufficient to slow down TCP. Note that the periods of low goodput are once again characterised by high packet loss. They occur during times when link utilisation is comparatively high but still well below 50 %. The queue thus clearly oscillates here, too. Previously common goodput rates of 20 Mbps and more are now the exception and occur mostly during times of low and rising link utilisation. The rate at which new TCP connections commence on the link also follows a diurnal pattern, albeit with maxima both at the time of worst TCP goodput and of good TCP and TCP/NC goodput. Actual TCP goodput observed for our successful downloads was never below 0.6 Mbps, i.e., higher than the average of around 0.045 Mbps predicted by the simulation. A likely cause for this is TCP slow start: Our downloads were much longer than the average TCP flow and therefore had more opportunities to increase their congestion window over time.

TCP/NC gave consistently better goodput (on average 80 % more) during times of low conventional TCP goodput on the link. However, compared to the earlier data series, even this was still a fraction of the goodput at the time. This suggests that burst losses were large enough to damage entire generations on occasion.

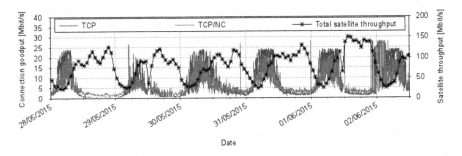

Fig. 9. Measured goodput of individual TCP and TCP/NC connections into Rarotonga and average hourly total satellite link utilisation on 332 Mbps inbound capacity.

7   Conclusions

Signs of TCP queue oscillation were present in all three links investigated. Observed link utilisation on the Niue link was high and short burst errors were observed, however our simulation of the link shows that the high link utilisation in this scenario does not prevent complete queue drain. Rather, the queue seems to refill again rapidly in this case. TCP/NC can achieve significantly higher goodput than conventional TCP. However, since the link is already transporting almost exclusively goodput close to link capacity, the better performance of TCP/NC comes at the expense of the goodput of conventional TCP on this link.

The Tuvalu scenario is not as clear-cut. Observation of the actual link shows significant amounts of burst packet loss and very low link utilisation, as well as very low per-flow goodput for both TCP (via a SilverPeak NX-3700 WAN accelerator) and TCP/NC. All of these are symptoms of severe queue oscillation. However, while our attempts to simulate the link result in oscillation under many scenarios, we have not been able to replicate either the very low link utilisation or the extremely long burst packet losses. Further investigation will be required to establish whether extreme oscillation of this kind can be attributed to different traffic patterns or whether the WAN accelerator plays a role here. TCP/NC is able to provide significantly better goodput on the Tuvalu link at times of moderate packet loss, a condition met for at least several daytime hours on most days.

In the Rarotonga case, simulations of the 160 Mbps link agree well with the empirical data collected from the end of January 2015: Actual link utilisation and packet losses closely match those of a simulated queue with capacity for 50 packets. At the time, TCP/NC was able to provide consistently high goodput close to the best off-peak goodput from conventional TCP. During times of high daytime packet loss, TCP/NC sustains this goodput while conventional TCP goodput often shrinks to a fraction thereof.

Both our simulations and our empirical data from end of May/early June 2015 shows that additional bandwidth need not necessarily result in higher goodput for individual flows, and that the total number of concurrent flows and total goodput may remain virtually the same. The result in this case may simply be a less well utilised link. At the end of May/early June 2015, TCP/NC was again able to provide better goodput than conventional TCP during peak times, albeit at a fraction of the rates observed earlier in the year.

Obtaining a complete picture of the traffic on a particular link is inherently difficult. Information such as flow size distribution, flow creation rates, satellite gateway queue capacity, number of concurrent flows, or off-island latency distributions all have an influence on link behaviour, as do measures such as traffic shaping, rate limiting, or WAN optimisation. Operators in the islands are often not in a position to supply much of this information and data. Moreover, remote longitudinal data collection on site can seem like the communications equivalent of keyhole surgery: Not only does it load the link during experimentation, but also (in the reverse direction) during the retrieval of experimental data. This precludes in particular the retrieval of substantial packet trace files from island sites.

Another challenge is the fact that our TCP/NC experiments to date have not had exclusive access to a link for longer time periods: Our TCP/NC flows always had to share the link with a majority of concurrent conventional TCP flows. This exposes the TCP/NC flows to the burst packet losses experienced by conventional TCP. It is therefore difficult to assess how much of the TCP/NC overhead is required only to cope with the presence of conventional TCP on the link. Access to an exclusive satellite transponder would therefore be desirable in any future investigations.

Acknowledgements. This research was supported by the Information Society Innovation Fund Asia through the Pacific Island Chapter of the Internet Society (PICISOC) and by Internet New Zealand. We would also like to thank the many Internet users and staff of Telecom Cook Islands, Internet Niue, and the Tuvalu Telecommunication Corporation for their patience during this study and for sharing their precious bandwidth with us. We would also like to thank Nevil Brownlee for letting us use his flow traces, which assisted us in modelling our flow size distribution.

References

1. O3b Networks, home page. http://www.o3bnetworks.com/
2. Postel, J.: Transmission Control Protocol, Internet RFC 793
3. Jacobson, V.: TCP Extensions for Long-Delay Paths, Internet RFC 1072
4. Jouanigot, J.M., Altaber, J., Barreira, G., Cannon, S., Carpenter, B. and others: CHEOPS Dataset Protocol: An efficient protcol for large disk based dataset transfer on the Olympus Satellite. CERN, Computing and Networks Division, CERN-CN-93-06 (1993)
5. Kim, J.H., Yeom, I.: Reducing queue oscillation at a congested link. IEEE Trans. Parallel Distrib. Syst. **19**(3), 394–407 (2008)
6. Caini, C., Firrincieli, R.: TCP hybla: a TCP enhancement for heterogeneous networks. Int. J. Satell. Commun. Netw. **22**(5), 547–566 (2004)
7. Leith, D.: H-TCP: TCP Congestion Control for High Bandwidth-Delay Product Paths. Internet Draft, IETF (2008). http://tools.ietf.org/html/draft-leith-tcp-htcp-06
8. Sundararajan, J.K., Shah, D., Médard, M., Jakubczak, S., Mitzenmacher, M., Barros, J.O.: Network coding meets TCP: theory and implementation. Proc. IEEE **99**(3), 490–512 (2011)
9. Hansen, J., Krigslund, J., Lucani, D.E., Fitzek, F.H.: Sub-transport layer coding: a simple network coding shim for IP traffic. In: 2014 IEEE 80th Vehicular Technology Conference (VTC Fall), pp. 1–5 (2014)
10. Silver Peak WAN Optimization Appliances, Appliance Manager Operators Guide, VXOA 6.2, December 2014. http://www.silver-peak.com/sites/default/files/userdocs/appliancemgr_operators_guide_r6-2-5_revn_december2014_0.pdf
11. ntop home page. http://www.ntop.org/

Network Coding over Satellite: From Theory to Design and Performance

M.A. Vazquez-Castro[1]([⊠]) and Paresh Saxena[2]

[1] School of Engineering, Autonomous University of Barcelona,
Campus de Bellaterra, 08193 Barcelona, Spain
angeles.vazquez@uab.es
[2] AnsuR Solutions Barcelona, Advanced Industry Park, c/Marie Curie 8-14,
08042 Barcelona, Spain
paresh.saxena@ansur.es

Abstract. The concept of network coding has greatly evolved since its inception. Theoretical and achievable performance have been obtained for a wide variety of networking assumptions and performance objectives. Even if powerful, such a broad applicability poses a challenge to a unified design approach over different communication networks and systems.

In this work, we propose a (non-reductionist) unified network coding design architectural framework where an ontology of abstraction domains is introduced rather than layer/system/network-specific assumptions and designs. The framework brings together network and system design and seems compatible with upcoming (more general) design frameworks such as software-defined networking, cognitive networking or network virtualization. We illustrate its applicability showing the case of network coding design over DVB-S2X/RCS.

Keywords: Network coding · Satellite communication system

1 Introduction and Contributions

1.1 Evolution of the Concept of Network Coding

The concept of network coding could be to some extent dated back to Yeung and Zhang in [1]. The work was inspired by low earth satellite (LEO) communication networks as illustrated in Fig. 1. The general problem studied was the admissible code rate region of a given network with an arbitrary set of connections. The problem was tackled as a general multiple-source coding problem and only non explicit results were obtained, for which reason subsequent work has prominently been focused on assuming linearity. In the following, we provide a brief account of the evolution of the concept, leaving aside security enforcement that the concept can also provide.

Error-Free Network Information Flow. Alswhede et al. in [2] fully developed the concept with focus on the single-source coding problem. The inspiration in this case was computer network applications in general. The work revealed the fundamental fact

© Institute for Computer Sciences, Social Informatics and Telecommunications Engineering 2015
P. Pillai et al. (Eds.): WiSATS 2015, LNICST 154, pp. 315–327, 2015.
DOI: 10.1007/978-3-319-25479-1_24

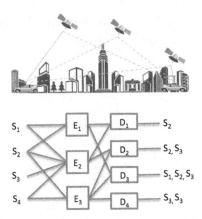

Fig. 1. Illustration of a multisource satellite system along with its logical abstraction as a set of encoders and decoders with arbitrary connections as in [1].

that if coding (rather than mere routing) is applied at the in-network nodes, the source node can multicast the network information flow at the theoretically maximum rate. Such maximum rate, the multicast network capacity, is the smallest minimum cut capacity between the source node and any multicast receiver, what the authors termed as the Max-flow Min-cut for the network information flow. Li et al. [3] further showed that linear network coding and finite alphabet size is sufficient for achievability while Jaggi et al. [4] proposed a deterministic polynomial-time algorithm for network code constructions.

A different thread of work (specifically for networks with non-ergodic error processes, only link failures) was initiated by Koetter and Médard [5]. The authors developed a characterization of linear network coding introducing instrumental tools not only from graph theory but also from control theory and algebra. Specifically, the chosen algebraic framework over finite fields shows how network coding coefficients lead to a matricial network channel transfer whose invertibility drives the solvability of the network (which is equivalent to finding points on algebraic varieties). Alternatively, the same assumption lead to a distributed purely randomized way to find the network coding coefficients in [6], which is capacity achieving.

Given the error-free network assumption in the early work just commented, it can be said that the network coding concept was first discovered aiming at increasing the amount of information transferred across a network (possibly with link-failures). Furthermore, since such error-free networks admit purely logical abstractions, analytically tractable formulations can be obtained as well as analytical closed and/or explicit solutions.

After extensive research assuming error-free (linear) networks, it became apparent that the concept of network coding could be extended to communication networks with links/channels that introduce different types of errors. A number of different proposals exist to tackle this problem each of them illuminating different aspects of it.

Non Error-Free Network Information Flow. Cai and Yeung proposed in [7] to jointly tackle rate maximization and reliability by introducing network error-correcting codes with coherent transmission. The concept of network error correction codes was then to be regarded as a generalization of classical link-by-link error-correcting codes. Subsequent works [8, 9] refined the concept and theoretical coding bounds. Dana et al. in [10] rely on graph-theoretical formulations for the study of a special class of wireless networks, called wireless erasure networks. The multicast capacity under certain assumptions is obtained for such networks revealing a Max-flow Min-cut interpretation. A complementary view is given in Lun et al. in [11], where an (asymptotically) capacity-achieving coding scheme is proposed revealing a fluid-like interpretation of the network flow of "innovative" information.

The algebraic approach to tackle the problem was initiated in [12] by Koetter and Kschischang. They proposed two separated problems (resembling Shannon separation between source and channel coding): the network coding problem for maximizing the network flow (which can be solved assuming error-free links) and the problem of coding for error correction (which can rely on classic coding results and techniques). Accordingly, the proposal is a non coherent transmission model, i.e. neither the source node nor the receiver nodes are assumed to have knowledge of the network coding coefficients. Moreover, it is assumed that the underlying network-coded network implements random linear network coding. Their propose codes as collections of subspaces (for which distance metric driven constructions are developed). The novelty with respect to classic block codes is that while in classical coding theory codewords are vectors, in subspace codes each codework is itself an entire vector space. This can be so due to the fact that the underlying linear network is vector-space preserving. This property ensures that if the source information is represented by a subspace of a fixed vector space and a basis of this subspace is injected into the network, the information can be recovered at the receivers. A worst-case (or adversarial) error model is considered obtaining the maximum achievable rate for a wide range of conditions. Constructions of subspace codes based on rank-metric codes are proposed in [13].

From this brief account of research directions and results when considering network with errors, it is apparent that the concept of network coding become more general with respect to its original purpose. The amount of information transferred across a network needs to be optimized not only in robustness and throughput to achieve capacity but also in reliability.

Physical-Layer Network Information Flow. The concept of network coding further evolved beyond abstracted network flows. Network coding at the physical layer was originally proposed in [14] following the observation that network coding operations are naturally taking place among bearing electromagnetic waves. The well-known additive property of electromagnetic waves can be captured as network coding combinations over the field of complex numbers. For this reason transmission should take place over structured codes as shown in [15–17]. Topologies of interest at the physical layer usually present a higher level of structure than the purely logical ones, the interested reader can check [18] for a survey.

Network Coding for SATCOM. Several works have studied the application of network coding over satellite systems. There are specifically two directions of work on the use of NC in satellite systems.

The first direction focuses on throughput improvement. In [19, 20], network coding is shown to enhance throughput by load balancing and allocating coded packets across different beams in multi-beam satellite systems. Further, works in [21, 22] take advantage of orthogonal transmission available using multi-link reception in multi-beam satellite systems. In this work, network coding is shown to provide benefits in terms of higher throughputs subject to the location of user terminals. Throughput improvement in satcom is then found to depend on enabling extra degrees of freedom in the network, without which network coding alone cannot make much difference.

A different direction looks at the application of network coding to counteract packet losses and guarantee higher reliability and additional performance requirements. It has been shown that network coding together with congestion control algorithms can provide many-fold improvements than existing transport layer protocols [23–25]. In addition, it has been shown in [26], that unequal-protection aware overlapping network coding together with congestion control algorithms can provide improvements in quality-of-experience (QoE) of video streaming. Further, network coding implementation at the link layer has shown to provide several advantages in terms of reliability, complexity, delay, etc. for unicasting and multicasting over satellite networks with re-encoding at intermediate nodes [27, 29]. In [30], it is also shown that it can be used to counteracting prediction failures of the handover procedures in smart gateway diversity satellite systems.

1.2 Discussion and Contributions

From the above brief account, it is apparent the evolution of the concept of network coding for which a rich body of theoretical results exist. Accordingly, it opens up a wide variety of potential applications in general and in satellite communications in particular. Even if powerful, such a broad applicability poses a challenge to a unified design approach over different communication networks and systems.

In this work we propose a network coding design architectural framework, which can be mapped to whatever software and/or hardware resources are provided by the underlying abstracted systems and networks. In particular, our contributions are:

- Proposal of a unified network coding design architectural framework with the following distinguishing features:
  - *Non-reductionist*: Differently to current standard practice, rather than layer/system/network-specific assumptions and designs, an ontology of abstraction domains is introduced. Specific network codes or performance targets can be freely chosen by the designer.
  - *Backward-Forward Compatible:* The proposed framework is backward compatible with existing networks and system. Most importantly, it seems compatible with upcoming more general design frameworks such as: software defined networking, cognitive networking or network virtualization.

– *Inter-disciplinary*: Network, system and coding experts/designers can easily collaborate in the design based on a common design framework with well defined ontological domains.
• We illustrate its applicability for a use case of network coding over Digital Video Broadcasting by Satellite – Second Generation (Extension)/Return Channel Satellite (DVB-S2X/RCS) [31].

2 Proposal of Network Coding Design Framework

2.1 Ontology of Abstraction Domains

Theoretical results in network coding literature usually make use of networking terms such as packets to refer to information units or flows to refer to sequences of them. However, such terms actually refer to mathematical objects, such as vectors or symbols living in some algebraic object. Moreover, while for a networking designer a packet usually refers to the network layer, for a satellite system designer a packet may refer to the link layer data unit of some system data plane.

In order to solve this problem of ambiguity, we build upon the fact that the human brain naturally disambiguates linguistic terms by relating to the *references* they apply to [32]. Such references can be physical or not. If they are physical, the brain makes the unambiguous description of the physical object (this is the way humans learn language and start communicating). In case they are not, which is our case, the brain should be given a reference/description of the non-physical (abstract) object (this is the way e.g. programmers or system designers work) together with the identification of the domain the term belongs to. The challenge in network coding is that we need a terminology for the basic data unit that can also serve for system-level design of a networking technology, which moreover need clear maps to certain mathematical objects.

We propose a clear distinction between the following two abstraction domains for network coding design[1]:

– *Mathematical-Abstract Domain*: This is the domain of network coding theoretical design and the *reference* of the terms to be used for this domain can be taken from the original sources of the corresponding network coding concepts, e.g. network flow in [2].
– *Logical-Abstract Domain*: This is the domain of our proposed network coding design architectural framework. It includes the necessary mapping between mathematical and logical objects so that theoretical network codes can be incorporated into this design domain. We propose as *reference* of the terms for this domain already existing networking terms.

Consequently, in the rest of this paper we will not use mathematical-abstract domain terminology. Note that our proposed logical-abstract domain abstracts away both system-level domain (functionalities of physical system elements) and networking-level

---

[1] Note that this distinction is simpler and well established in classic coding theory.

domain. The latter, will only be specified for a specific network coding design, which will have a clear reference in the standard layered communication model.

2.2   Logical-Abstract Domain: Proposed Design Architectural Framework

This is the domain of our proposed architectural framework. This means the designer (or team of designers) working in this domain do not need to work with mathematical objects. This can be so, because the designer should have access to the necessary mappings between mathematical and logical objects. This way, mathematical models and codes can be incorporated into the abstract-logic framework. Note that the framework is not meant to verify whether or not network coding provides an advantage w.r.t. routing, this should be known in advance (see e.g.) [34]. The framework allows a team of designers to collaborate in order to meet a given set of requirements within a unified abstraction domain jointly considering coding, networking and system level/design considerations. In order to do so, a **reference abstraction level** shall be chosen according to the requirements. The natural choice is to define a sub/intra-layer in the standard IP/TCP protocol stack at which network coding will operate. Such a choice inherently sets the appropriate space/time scales and resolution for the design. The framework then is not reductionist, as the networking domain is abstracted away, but uses the TCP/IP protocol stack for setting the reference level of abstraction and interactions for every specific design. Once settled the abstraction level of the logical-abstract design domain, a **NC functionalities design toolbox** allows the designer to choose functional design blocks *according to the technical requirements and design objectives*. The designer should also identify which external functionalities (w.r.t the reference layer) the architecture should interact with (e.g. other protocols that can make network coding related functionalities more effective and efficient).

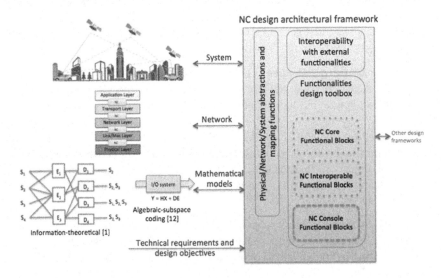

Fig. 2. Proposed NC design architectural framework.

The resulting architectural framework is schematically illustrated in Fig. 2. The output should be a complete network coding functional architecture at the reference communication layer meeting the design requirements and corresponding trade-offs and can be directly implemented as a protocol and/or provided to other more general design frameworks. This way, the framework seems compatible with more general design frameworks such as: software defined networking, cognitive networking or network virtualization. This is so because the design framework can incorporate any underlying software and/or hardware resources and constraints in the same way as any of these other frameworks. In addition, in-network encoding nodes define network coded virtual routes, which can cognitively adapt to network/system dynamics.

2.3 Network Coding Design Toolbox

The proposed architectural framework design provides elementary functionality design blocks (FBs), which can be related to technical requirements (see Fig. 2). The designer may choose existing coding schemes as inputs to the framework (e.g., Greedy and Sparse random codes [35], BATS codes [36], FUN codes [37], DARE [33] and Fulcrum Network Codes [38]) based on the requirements and design target. Such schemes may or may not be compatible with existing forward error correction (FEC) schemes. We propose a preliminary set of NC FBs distributed into three hierarchical levels based on their significance, universality and availability.

NC Core Blocks: These blocks provide the basic core architectural functionalities, elementary coding/re-encoding/decoding operations and logical interpretation for NC usage. In general, they are the main FBs that are foreseen to be present in any NC architecture:

- **NC Logic FB:** It drives the overall network coding solution based on some network coding design, for which a mapping mathematical-logical is given. The scheme can be intra-session/inter-session, coherent/incoherent, large files/streaming, systematic/ non-systematic, etc. Per-node functionalities interact to decide on the use of coding coefficients (random/deterministic), generate and supply them to NC coding FB.
- **NC Coding FB:** It takes care of all the data processing steps including encoding, re-encoding, decoding operations, encapsulation processes, adding/removing headers, etc. It is the central FB of the NC architecture interacting with several others building blocks to allow data processing and flowing.

NC Interoperable Blocks: These blocks provide the interoperability of NC core architecture with external functional blocks. They interact with other external system/protocol level protocols and provide a comprehensive solution for congestion control, cross-layer optimizations, etc. that drive the functionalities of NC core blocks. Examples are:

- **NC Resource-Allocation:** It is responsible for adaptive network coding and allocates optimal coding parameters taking into account different tradeoffs.
- **NC Congestion Control:** It is responsible for managing congestion. It is primarily used when NC is implemented in the upper layers of the protocol stack. It interacts with other cross-layer congestion-control algorithms.

NC Console Blocks: These blocks provide the basic support functionalities (of storage, feedback, etc.) to NC core blocks and NC interoperable blocks, examples are NC storage, NC feedback manager and NC signaling.

3 Design Illustration

In this section, we illustrate the applicability of the proposed framework for the design of network coding for satellite systems over DVB-S2X/RCS. Specifically, two different use cases are presented: kernel-aware network coding unicast [28] and kernel-agnostic (overlay) network coding multicast [26, 29]. The main relevance of these use cases here is to show that the proposed design framework allows the underlying DVB-S2X/RCS standard to be properly abstracted if needed resulting in different operative solutions.

In general, satellite systems suffer from packet losses due to time-variant fading. Hence, we assume encoding at the source to counteract packet losses (erasures) and one in-network encoding node (not necessarily at the satellite) between the source and the sink for the unicast case and between the source and the two sinks for the multicast case. The chosen design objectives of network coding application here is to guarantee quality-of-service (QoS) in terms of reliability and transmission rates for both cases and also QoE for multicast (see [26] for details). Note that given the specific characteristics of the underlying system, the transmission is not assumed interactive.

The architecture design is shown in Fig. 3. Due to the limited space in this article, functions for the different induced time/space scales/resolution for each case are not explained here. The design for the in-kernel unicast has been chosen to be compliant with the existing FEC in the standard at the link-layer (LL) and hence the NC reference layer is identified between the IP layer and the link layer, which we term as (systematic) network coding (LL-SNC). The NC storage FB interacts directly with the physical storage and interacts frequently with NC core blocks to support coding functionalities. The NC feedback manager analyses the feedback and provide support to NC resource allocation FB to evaluate optimal coding parameters. The interactions may or may not be frequent depending on the dynamics coherence time. Finally, the NC signalling FB is responsible to process all the signalling parameters within the NC packets. These parameters (like packet size, frame number, coding scheme ID, systematic/non-systematic packet ID, etc.) are transferred across the network within the NC packets and NC signalling block is responsible for its management. The NC coding scheme (e.g. mapped from some mathematical design) is then responsible for all the coding/re-encoding/decoding operations and NC packetization/de-packetization. Also note that the middle level of FBs or the proposed design serves two main purposes. First, it is interoperable with external functions and performs cross-layer interactions with other protocols. Based on these cross-layer interactions and feedback on the network statistics (packet loss report), optimal coding parameters (code rate, frame length, etc.) can be chosen in order to guarantee QoS and QoE. Such optimal coding parameters are then passed to top-level NC core functions for NC data processing. The bottom level hierarchy consists of NC support blocks that include NC feedback manager FB, NC storage FB and NC Signalling FB.

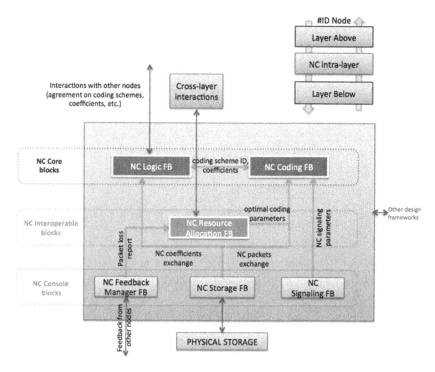

Fig. 3. Illustration of functional architecture network coding design over DVB-S2X/RCS.

In Figs. 4 and 5 we show illustrative performance results for the QoS design objectives. Figure 4 shows the kernel-aware design performance as compared to existing LL-FEC in DVB-S2X/RCS. The simulation assumes realistic satellite setups with light rainfall (20 % packet erasures) at each link and available frame lengths (N = {50,256}) in the standard. The overall reliability is plotted against the code rate. Our results show that up to 53.85 % higher reliability is guaranteed for a code rate of 0.7. This particular code rate is relevant here as LL-SNC achieves 100 % reliability. Similarly with the frame length (N = 50, hence, lower delay), up to 16.28 % higher reliability is guaranteed using LL-SNC. A more detailed description can be found in [28].

For the kernel-agnostic case, the reference layer is identified between the application layer and transport layer. In Fig. 5., we show illustrative performance results with one intermediate *logical* node and two sinks (users) with realistic channel conditions (light rainfall, 20 % for the common link and 20 % and heavy rainfall, 60 % packet erasures for the multicast links) and realistic transmission rates.

Achievable rates are shown compared to the theoretical Max-flow Min-cut. In addition, a region with lower reliability but with a 43.17 % rate increase is also shown in which the sinks with worst channel conditions will recover fewer packets.

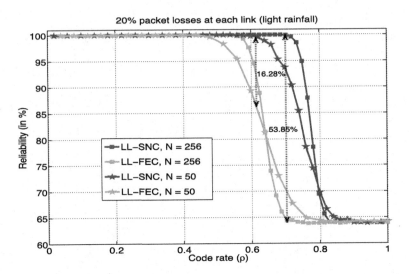

Fig. 4. Illustrative performance results of network coded unicast design between IP and link layer (and comparison with current LL-FEC).

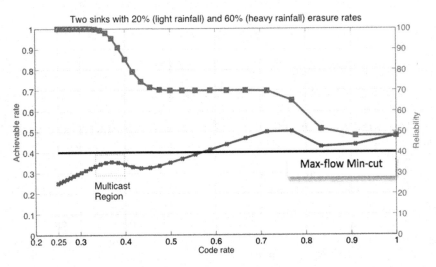

Fig. 5. Illustrative performance results of network coded multicast between application and transport layer.

4   Conclusions and Open Issues

In this work, we propose a (non-reductionistic) unified network coding design architectural framework. The framework brings together coding, network and system design (hence including system-level current standardization practices) within the same ontological level so that interdisciplinary teams can cooperate. The framework is validated by showing its applicability to two different use case scenarios of network coding over satellite. Our proposal contributes to expand satellite communications design beyond purely system-level design over physical elements. Our framework seems compatible with upcoming (more general) design frameworks such as software-defined networking, cognitive networking or network virtualization (see e.g. current activity in Internet research task force, IRTF [39]).

Acknowledgement. The authors acknowledge inter-disciplinary networking support by the COST Action IC 1104.

References

1. Yeung, R.W., Zhang, Z.: Distributed source coding for satellite communications. IEEE Trans. Inf. Theory **45**(4), 1111–1120 (1999)
2. Ahlswede, R., Cai, N., Li, S.-Y.R., Yeung, R.W.: Network information flow. IEEE Trans. Inf. Theory **46**(4), 1204–1216 (2000)
3. Li, S.-Y.R., Yeung, R.W., Cai, N.: Linear network coding. IEEE Trans. Inf. Theory **49**(2), 371–381 (2003)
4. Jaggi, S., Sanders, P., Chou, P.A., Effros, M., Egner, S., Jain, K., Tolhuizen, L.M.G.M.: Polynomial time algorithms for multicast network code construction. IEEE Trans. Inf. Theory **51**(6), 1973–1982 (2005)
5. Koetter, R., Médard, M.: An algebraic approach to network coding. IEEE/ACM Trans. Netw. **11**(5), 782–795 (2003)
6. Ho, T., Koetter, R., Médard M., Karger, D.R., Effros, M.: The benefits of coding over routing in a randomized setting. In: Proceedings of IEEE International Symposium on Information Theory (2003)
7. Cai, N., Yeung, R.W.: Network coding and error correction. In: Proceedings of IEEE Information Theory Workshop 2002, pp. 119–122, Bangalore, India, October 2002
8. Zhang, Z.: Linear network error correction codes in packet networks. IEEE Trans. Inf. Theory **54**(1), 209–218 (2008)
9. Yang, S., Yeung, R.W., Ngai, C.K.: Refined coding bounds and code constructions for coherent network error correction. IEEE Trans. Inf. Theory **57**(3), 1409–1424 (2011)
10. Dana, A.F., Gowaikar, R., Ravi Palanki, R., Hassibi, B., Effros, M.: Capacity of wireless erasure networks. IEEE Trans. Inf. Theory **52**(3), 789–794 (2006)
11. Lun, D.S., Medard, M., Koetter, R., Effros, M.: On coding for reliable communication over packet networks. Phys. Commun. **1**(1), 3–20 (2008)
12. Koetter, R., Kschischang, F.: Coding for errors and erasures in random network coding. IEEE Trans. Inf. Theory **54**(8), 3579–3591 (2008)
13. Silva, D., Kschischang, F., Koetter, R.: A rank-metric approach to error control in random network coding. IEEE Trans. Inf. Theory **54**(9), 3951–3967 (2008)
14. Zhang, S., Liew, S.C., Lam, P.P.: Hot Topic: Physical-layer Network Coding. In: ACM MobiCom, pp. 358–365, September 2006
15. Nazer, B., Gastpar, M.: Compute-and-forward: harnessing interference through structured codes. IEEE Trans. Inf. Theory **57**(10), 6463–6486 (2011)
16. Nazer, B., Gastpar, M.: Reliable Physical Layer Network Coding. Proc. IEEE, Spec. Issue Netw. Coding **99**(3), 438–460 (2011)
17. Vazquez-Castro, M.A.: Arithmetic geometry of compute and forward. In: Proceedings of IEEE Information Theory Workshop (2014)

18. Liew, S.C., Zhang, S., Lu, L.: Physical-layer network coding: tutorial, survey, and beyond. Phys. Commun. **6**(1), 4–42 (2013)
19. Vieira, F., Shintre, S., Barros, J.: How feasible is network coding in current satellite systems ?. In: ASMS Conference and SPSC Workshop, pp. 31–37 (2010)
20. Vieira, F., Lucani, D., Alagha, N.: Load-aware soft-handovers for multibeam satellites: a network coding perspective. In: ASMS Conference and SPSC Workshop, pp. 189–196 (2012)
21. Alegre-Godoy, R., Alagha, N., Vazquez-Castro, M.A.: Offered capacity optimization mechanisms for multi-beam satellite systems In: IEEE ICC, pp. 3180–3184 (2012)
22. Vazquez-Castro, M.A.: Graph model and network coding gain of multibeam satellite communications. In: IEEE ICC, pp. 4293–4297 (2013)
23. Gupta, S., Vazquez-Castro, M.A.: Location-adaptive network-coded video transmission for improved quality-of-experience. In: 31st AIAA International Communications Satellite Systems Conference (ICSSC) (2013)
24. Gupta, S., Pimentel-Niño, M.A., Vazquez-Castro, M.A.: Joint network coded-cross layer optimized video streaming over relay satellite channel. In: 3rd International Conference on Wireless Communications and Mobile Computing (MIC-WCMC) (2013)
25. Cloud J., Leith D., Medard M.: Network Coded TCP (CTCP) Performance over Satellite Networks. In: International Conference on Advances in Satellite and Space Communications (SPACOMM), pp. 53–556 (2014)
26. Pimentel-Niño, M.A., Saxena P., Vazquez-Castro M.A.: QoE driven adaptive video with overlapping network coding for best effort erasure satellite links. In: 31st AIAA International Communications Satellite Systems Conference (ICSSC) (2013)
27. Saxena, P., Vázquez-Castro, M.A.: Network coding advantage over MDS codes for multimedia transmission via erasure satellite channels. In: The 5th International Conference on Personal Satellite Services (PSATS), June 2013
28. Saxena, P., Vázquez-Castro, M.A.: Link Layer Systematic Random Network Coding for DVB-S2X/RCS. In: IEEE Communications Letters, May 2015
29. Saxena, P., Vazquez-Castro, M.A.: Network coded multicast and multi-unicast over satellite. In: The 7th International Conference on Advances in Satellite and Space Communications (SPACOMM), April 2015
30. Muhammad, M., Giambene, G., De Cola, T.: Channel prediction and network coding for smart gateway diversity in terabit satellite networks. In: GLOBECOMM, pp. 3549–3554 (2014)
31. ETSI EN 302 307 V1.2.1, Digital Video Broadcasting (DVB); Second generation framing structure, channel coding and modulation systems for Broadcasting (DVB-S2) (2009)
32. Kripke S.: Naming and Necessity, pp. 193–219. Harvard University Press, Cambridge, Chapter 10 (1979)
33. Saxena, P., Vazquez-Castro, M.A.: DARE: DoF-aided random encoding for network coding over lossy line networks. In: IEEE Communications Letters (2015)
34. Vazquez-Castro, M. A.: Subspace coding over Fq-linear erasure satellite channels. In: 7th International Conference on Wireless and Satellite Systems (2015). (Invited paper)
35. Pakzad, P., Fragouli, C., Shokrollahi, A.: Coding Schemes for line networks. In: IEEE ISIT, pp. 1853–1857 (2005)
36. Yang, S., Yeung, R., Coding for a network coded fountain. In: IEEE ISIT, pp. 2647–2651 (2011)
37. Huang, Q., Sun, K., Li, X., Wu, D.: Just FUN: a joint fountain coding and network coding approach to loss tolerant information spreading. In: ACM MobiHoc, pp. 83–92 (2014)

38. Lucani, D.E., Pedersen, M.V., Heide, J., Fitzek, F.H.P.: Fulcrum network codes: a code for fluid allocation of complexity. In: IEEE Journal on Selected Areas in Communications Submitted for Publication
39. IRTF: Network Coding Research Group (NWCRG). https://irtf.org/nwcrg

WiSATS Session 3

Gateway Selection Optimization in Hybrid MANET-Satellite Network

Riadh Dhaou[1]([✉]), Laurent Franck[2], Alexandra Halchin[1], Emmanuel Dubois[3], and Patrick Gelard[3]

[1] IRIT, TéSA, Université de Toulouse, Toulouse, France
{riadh.dhaou,alexandra.halchin}@enseeiht.fr
[2] Télécom Bretagne - Institut Mines-Telecom, TéSA, Plouzané, France
laurent.franck@telecom-bretagne.eu
[3] CNES, Toulouse, France
{emmanuel.dubois,patrick.gelard}@cnes.fr

Abstract. In this paper, we study the problem of gateway placement in an hybrid mobile ad hoc – satellite network. We propose a genetic algorithm based approach to solve this multi-criteria optimization problem. The analysis of the proposed algorithm is made by means of simulations. Topology dynamics are also taken into account since the node mobility will impact the gateway placement decisions. Our solution shows promising results and displays unmatched flexibility with respect to the optimization criteria.

Keywords: Genetic algorithm · Gateway selection · Mobility

1 Introduction

In the event of forest fires, network services should be provided even though communication infrastructures have been damaged. Mobile ad hoc networks (MANETs) and wireless mesh networks (WMNs) have attracted the attention of researchers. These technologies display multiple advantages such as self-organization and self-healing capabilities, which make them suitable for disaster recovery [1] or search and rescue operations [2]. Services, like video transfer, push-to-talk voice communication, short messaging, and map distribution are topical in that context. One of the most challenging characteristics of MANETS is their mobile nature; nodes are free to move and organize themselves arbitrarily, thus yielding network topologies that may change rapidly and unpredictably. If connectivity to the remote backbones is required, it has to be provided by long haul technologies such as satellite communications. However the integration of MANETs and satellite access raises significant challenges in terms of optimizing network resources, link availability, providing Quality of Service (QoS), minimizing costs and energy consumption [14, 15].

Let's consider a MANET where a subset of nodes called gateways also embed satellite communication capabilities, the problem is to determine a selection of these gateway nodes where satellite access will be actually enabled, thus serving as gateway to remote

© Institute for Computer Sciences, Social Informatics and Telecommunications Engineering 2015
P. Pillai et al. (Eds.): WiSATS 2015, LNICST 154, pp. 331–344, 2015.
DOI: 10.1007/978-3-319-25479-1_25

backbones for the MANET. We refer to this problem as "gateway placement". On the other hand, the selection of gateways will be subject to constraints that reflect operational and communication optimizations: the number of active gateways will be kept minimal and the path from a regular node to its closest gateway must also be short.

In this work we propose and evaluate a Genetic Algorithm (GA) for near optimally solving the gateway placement problem. The GA approach supports optimization with multiple criteria, namely, minimization of the number of hops forming the path between regular node and gateways (the nearest gateway is selected in terms of hops), load minimization of each selected gateway, minimization of the number of paths that share the same link and minimization of the number of gateways. The GA proposed in this paper also takes into account node mobility and regarding this matter the duration of convergence of the algorithm is optimized. Finally, the impact of gateway shadowing by obstacles is also studied.

2   Related Work

In the literature, there is significant work related to the gateway selection optimization problem in wireless mesh networks, mainly focused on improving the performance of WMNs, like gateway allocation, routing, interference management and scheduling. There are some works regarding throughput optimization [1, 3] and also a two-fold optimization, namely, the maximization of the size of the giant component in the network and that of user coverage [4]. However, these solutions assume networks with fixed topologies, and a fixed number of gateways to be placed.

Sharing a similar motivation as ours, exiting research works on wireless mesh networks have seen some focus on gateway selection problem based on the use of genetic algorithms. In [5] an evolutionary algorithm approach is presented for gateway placement, optimizing the throughput. Le et al. consider a uniform topology and don't take into consideration the optimization of gateway load. [7] describes a GA approach to network connectivity and user coverage. However, the authors' focus is on finding a feasible solution using different mutation operators. Xhafa et al. in [8] consider another approach based on simulated annealing algorithm for optimizing the placement of mesh routers in WMNs. A drawback of this approach is the high convergence delay of the algorithm.

A key aspect of our work, which distinguishes it from existing research, is the fact that we study gateway placement in a context of a dynamic topology. The goal in [13] is to find a dynamic placement of mesh routers in a rectangular geographical area to adapt to the network topology changes at different times while maximizing both network connectivity and client coverage. This paper is different from ours because their goal is to maximize the size of the greatest sub-graph component and the client coverage using a partial swarm optimization approach. The main drawback of [13] is the use of a purely mathematical mobility model. The work presented in [6] may be extended for mobility. Hoffman et al. use genetic algorithms to solve the joint problem of Internet gateway allocation, routing and scheduling while minimizing the average packet delay in the network. Although their topology network model is suitable for one with mobility, the simulations are made on one instance of this model, resulting in a static topology being used. While our goal is to minimize the number of gateways, the authors in [6] allocate a fixed number of gateways in order to obtain better network performance. Moreover, load balancing between gateways is not ensured.

To the best of our knowledge, there are no prior applications of GAs to select a various number of MANET nodes called gateways that will provide access to the satellite capacity, taking into account mobility.

3   Evaluation Framework

Simulations of wireless networks employ several components critical to the accuracy of the simulations, one of the most important being the mobility model that mimics the user motion by means of factors such as speed, direction, type of field [10, 11]. The simulation scenario that we use relies on a mobility model called FireMobility describing group motion behavior during forest firefighting operations [12]. The model was designed from interviews with French Civil Protection personnel and field guides. The model describes the deployment of several firefighters' columns, each of them being composed of 3 intervention groups plus a command car. Each group is composed of 1 command car, 4 water tank trucks and 4 firemen pairs. A 7-column model accounts for 196 nodes. These nodes are dispatched on a rectangular playground of 1000 m × 1000 m facing and flanking the fire. As the fire moves, they are re-dispatched according to tactics of firefighting and to ensure safety rules [2].

In this paper, we study the gateway placement problem in a wireless network based on the topology model described above. The term of gateway placement corresponds to the identification of which nodes should be activated from a subset of eligible nodes. In a context of firefighting, only the nodes corresponding to vehicles are eligible (about 60 % of all the nodes). In case of network partitioning, the gateway placement has also to take this situation into account so that no node is left isolated.

4   Genetic Algorithm

It was shown that node placement problems are computationally hard to solve optimally, and therefore heuristic and meta-heuristic approaches are used in practice. Heuristic methods are known for achieving near-optimal solution in reasonable time.

The use of genetic algorithms for optimization problems was inspired from natural evolution. The idea behind GA is to consider a population of individuals. Each individual is a candidate solution to the optimization problem. Selection, mutation and crossover operators are used to foster the improvement of individuals' characteristics as generations pass. It is expected that after a sufficient number of generations, individuals representing a (close to) optimal solution will emerge. A corner stone of the GA is the fitness function that expresses how good an individual (i.e., solution) is with respect to the original problem. The purpose of the GA is therefore to maximize (or minimize) the fitness function and indirectly reaching for solution optimality.

4.1   GA for Gateway Placement Problem

In this section, we present an instantiation of a GA for the problem of gateway placement in a MANET-Satellite network. Each individual contains a chromosome made of n genes where n is the total number of nodes in the network. A gene is encoded using a bit. Setting the ith gene to 1 means that the ith node acts as a gateway. The fitness function, to be minimized, will be detailed in the next section and is based on different strategies for the overall optimization of gateway placement. The initialization of the original population serving as seed for the forthcoming generations had to be chosen with care. A first approach is to randomly generate each individual, potentially yielding a lot of sub-optimal or invalid solutions (e.g., having a partitioned network without any gateway serving a partition). Our approach mixes three improvements. First, only individuals resulting in a valid solution are generated (this approach is proposed in [6]). Second, only individuals displaying less than g gateways are generated. The parameter g is defined as a conservative upper bound on the maximum number of gateways that may be required for the network. Finally, because the GA will be used repeatedly on subsequent topology snapshots, the solution (i.e., the best individual) found from the previous snapshot will be used in the current snapshot to initialize 10 % of the population. These three optimizations speed up convergence to an optimal solution. On a 2008 dual Xeon quad core computer, processing a 10-column topology (i.e., 280 nodes) takes about 30 s.

During the evolutionary phase of the GA, successive rounds of selection, crossover and mutation operators take place. The fitness function is computed for each individual and a selection of the best ones is made. The selection is probabilistic, the better the fitness, the higher the chances to be selected. Doing so does not deterministically rule out weaker individuals. Together with crossovers and mutations, it prevents the GA to get trapped in a local optimum. Evolution stops when the fitness function is stable (i.e., within a 10^{-3} interval) for 20 generations.

4.2   Fiteness Evaluation

The fitness function f has a particular importance in GAs because it determines how good a given solution is. Minimizing the fitness function indirectly contributes in the finding of an optimal gateway placement. Several approaches – called metrics – are used for defining the fitness function.

Metric 1: Our goal is to minimize the number of active gateways (g), maximum link load (MLL) which represents the traffic intensity of the most charged link (serving 1 node represents 1 unit of traffic) and the cumulated distance (CD) between regular nodes and their assigned gateway measured in hops:

$$f_1 = 2 * g + MLL + CD$$

Metric 2: By using shortest path routing, the GA may induce congestion in gateways that are centrally located in the network topology. In order to avoid this hotspot effect, *metric 2* introduces another parameter to the fitness function called the maximum gateway load (*MGL*). *MGL* represents the number of regular nodes that the most loaded gateway serves.

$$f_2 = 2 * g + MLL + CD + MGL$$

Metric 3: Each gateway should be evenly loaded in order to prevent any bottleneck phenomenon. We introduce in the fitness function formula a parameter, the maximum difference (MD) among gateways load. The weight of the gateway factor (*g*) is equal to 4 here because there are many parameters related to load.

$$f_3 = 4 * g + MLL + CD + MGL + MD$$

Metric 4: Reducing interferences among network nodes is a primary concern in the development of wireless networks. Only high quality, least interfering routes to the gateways should be selected. In the previous metrics, shortest path computation is used to assign gateways to the regular nodes. In the current metric, we assign gateways to regular nodes using the shortest path metric and also a gateway dependent metric. For each gateway we will assign a threshold that represents the maximum number of nodes that can be served by the same gateway. The use of the threshold is described in the next section.

$$f_4 = 2 * g + MLL + CD + MD$$

4.3   Relation Between GA Optimization and Routing Optimization

Traditionally, most of the solutions developed for gateway discovery and selection in MANETs are based on a hop count metric to minimize the distance to gateways. With that respect, choosing the shortest path is not always a wise choice as traffic may be routed over poor and highly congested paths. Similarly, the GA should not favor the solutions where gateways are close to each other as they can be exposed to collective shadowing and interferences.

[9] describes dedicated metrics for gateway and route selection in multi-hop backbone networks. The best route to any of the available gateways is computed taking also into account the load at the gateways. In [6], the authors use a genetic algorithm to solve the gateway allocation problem and also the routing with the goal of minimizing the average packet delay in the network. They demonstrate that deviating from the shortest hop routing does not significantly improve performance. If a gateway is overloaded, queuing delays may propagate in the network and it is more effective to uniformly distribute the load over multiple gateways. As a consequence, with **metric 4,** we use a combined metric mixing gateway-dependent and hop-count metric. When calculating the shortest path between a node and every possible serving gateway, if the candidate closest gateway already serves a specific number of regular nodes, the node is assigned to the nearest gateway. When compared with **metric 3, metric 4** consists in shifting the task of gateway load optimization from the GA to the routing algorithm.

4.4 Functional Validation

We validated the GA on simple, regular topologies such as rings and grids. The objective of these tests was to check if the solution given by GA is the optimal one and to highlight the factors that can change the GA convergence. The results obtained helped us to refine the fitness function and to figure out how to take into account the satellite communication aspects. From these tests, we defined a population size as 3 times the network size and a mutation probability of 1 over 100 where only one gene in the chromosome is changed per mutation.

5 Results

The GA was tested over topologies implementing a 5-column FireMobility model (i.e., 140 nodes). A topology is composed of 2 000 snapshots at 1 s intervals. In order to ensure the validity of the results, experiments are repeated 20 times with different random seeds and averaged.

The simulation results consider 4 scenarios which relate to metrics defined before: (a) basic optimization, (b) optimization taking into account the node mobility, (c) minimization and distribution of the gateway load and (d) optimization taking into consideration the optimization of the gateway load at routing level. The impact mobility and gateway shadowing (as a result of tree masking) is also covered.

5.1 Scenarios Without Shadowing

Basic Optimization: This scenario is based on metric f_1.

Figure 1 shows that the genetic algorithm does a good work in placing a low - 4 at most - number of gateways. The price to pay is a potential high load per gateway with uneven distribution (Fig. 2). On the field, it will translate in congestion and interferences among nodes. Other results, not shown here, show that regular nodes are at a maximum distance of 2 hops from their gateway with a favorable side effect of minimizing also the maximum link load (MLL).

Optimization with Node Mobility: The second scenario calls for the inclusion of factors related to node mobility and the link between gateway/satellite. Metric f_2 is used.

Figure 4 shows a decrease in terms of gateway load as expected. Even though the number of selected gateways increases (Fig. 3), the other parameters such as maximum link load are unchanged. A close inspection at Fig. 4 reveals that the gateway load is not homogenously distributed. For a 140-node network, the load of one of the 7 gateways of the network is 60, thrice the load of an equivalent homogenous distribution (20 per gateway). This phenomenon can be explained by the fact that the nodes in the topology are not uniformly distributed.

Optimization of the Gateway Load Distribution: The gateway load should be homogeneously distributed and minimized in order to prevent any bottleneck

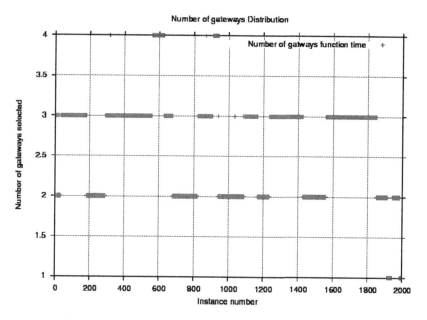

Fig. 1. # of selected gateways at each snapshot of the topology

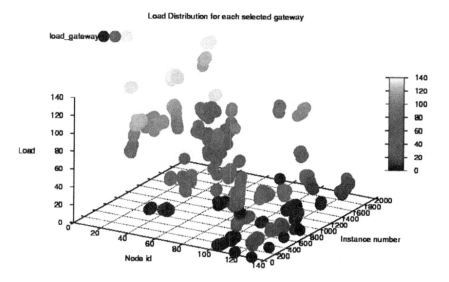

Fig. 2. Load distribution for each selected gateway

phenomena. In order to achieve an even gateway load, the third metric is used:

$$f = 4*g + MLL + CD + MGL + MD$$

Figures 5 and 6 show that less gateway nodes are necessary, the load slightly increases but is homogeneously distributed among gateways. Concerning the optimization of the others metrics like link load, we have to admit that results show as a

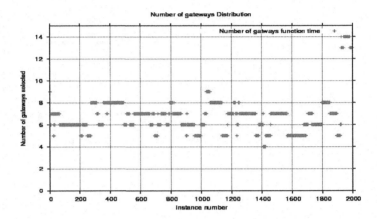

Fig. 3. # of selected gateways at each snapshot of topology

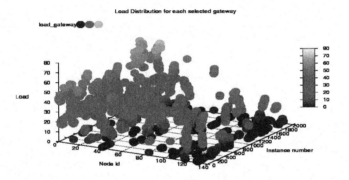

Fig. 4. Load distribution for each selected gateway

significant increase (MLL is tripled, while it was optimal for f1 and f2). It is a direct result of the fitness function where the gateway load factor is more represented than the two other link related factors (MLL and CDD). The choice of gateway nodes is also different from the prior scenarios. These three first scenarios also demonstrate one key advantage of GA over heuristics based approaches: adapting the optimization criteria (i.e., the fitness function) is made easily and does not require a redesign of the core optimization algorithm.

Load Balancing with Optimization of Gateway Load at Routing Level: The simulations are realized with the fourth metric: f = 2*g + MLL + CD + MD. For each regular node the affiliation to a gateway is based on the shortest path metric and also taking into account the gateway load.

The number of selected gateways at each topology snapshot and the load of each gateway are shown in Figs. 7 and 8. The main gain obtained using this method is that the load is well distributed over the number of gateways. The results concerning the number of selected gateways are satisfying regarding the network size. The optimization of the maximum link load is also taken into account (MLL decreases to 2).

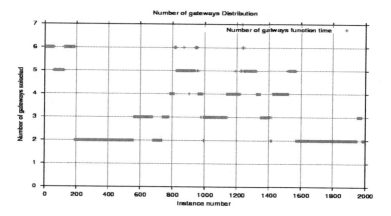

Fig. 5. # of selected gateways at each snapshot of topology

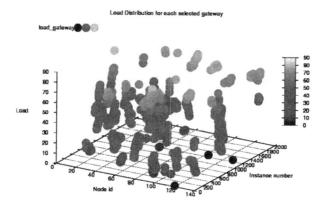

Fig. 6. Load distribution for each selected gateway

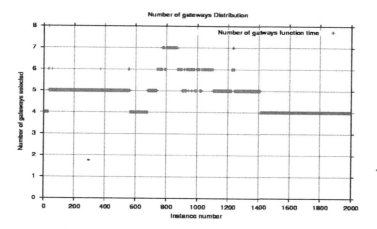

Fig. 7. # of selected gateways at each snapshot of topology

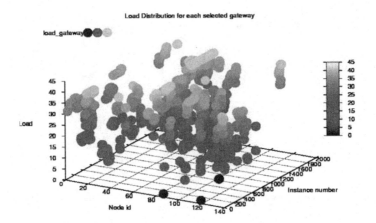

Fig. 8. Load distribution for each selected gateway

Figure 11 (left) shows simulation results concerning the mean and standard deviation of the number of selected gateways obtained using the four metrics. If the most important issue is to optimize the satellite interface usage, then the first metric should be favored since it performs well in terms of number of selected gateways and also of link load. However, this can only be achieved at the cost of high gateway load. If the main concern is to reduce the gateway load and also to optimize the other parameters, the second metric is preferable. In order to have a reasonable number of gateways and an evenly distributed load among gateways the fourth metric should be used. A disadvantage of this last metric is the fact that the link load increased. In our opinion it is more important to have stable solutions and number of gateways reduced despite of a link load increased by 1 (number of paths per link).

5.2 Impact of Mobility

Gateway placement is evaluated at each snapshot, it is therefore important to assess how long a gateway keeps its role. Figures 9 and 10 show the CDF, mean and standard deviation of gateway lifetime. Metric 4 offers the best stability.

5.3 Gateway Shadowing

Obstacles like trees can cause satellite signal impairments for certain gateways. This phenomenon is called shadowing. To assess the impact of shadowing on gateway placement, we performed a systematic shadowing of the nodes selected as gateways. In this context, it translates into marking as "non eligible" a node that was selected as gateway and resuming gateway placement.

Figures 11, 12 and 13 present the results obtained in terms of number of selected nodes, link load and path cumulated distance after systematically shadowing each node selected as gateway. The results are presented averaged for each simulation computed with the four metrics presented earlier.

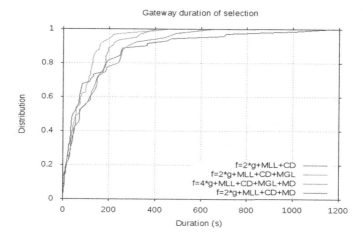

Fig. 9. CDF of duration for one gateway placement

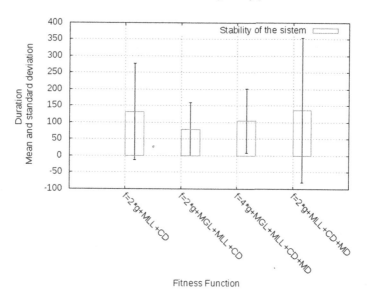

Fig. 10. Mean and standard deviation of gateway duration of selection

Making a synthesis on the results obtained with shadowing, we can say that there is no significant increase of the number of selected gateways. After shadowing, the number of gateways is reduced by 1 at maximum. The link load is optimized and Fig. 12 highlights that the fourth metric improves link load.

Figure 14 shows the percentage of the gateways that are kept established after shadowing. The second metric is the best performer in terms of gateway stability. We noticed that the algorithm has the tendency to choose complementary pairs of nodes with a strong impact on gateway change when shadowing occurs. Figure 15 shows the percentage of regular nodes that are served by the same gateway after shadowing. As expected, the second metric yields the larger stability ratio before/after shadowing. Placements with less gateways are also likely to suffer from the most drastic changes.

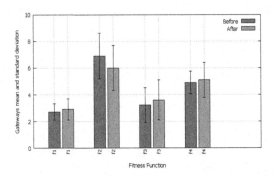

Fig. 11. # selected gateways before/after shadowing

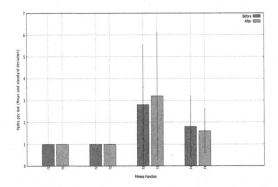

Fig. 12. Link load distribution before/after shadowing

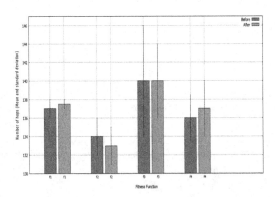

Fig. 13. Path cumulated length distribution before/after shadowing

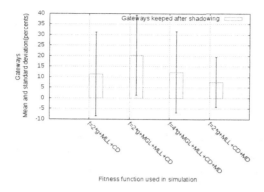

Fig. 14. Percentage of the solution kept after shadowing

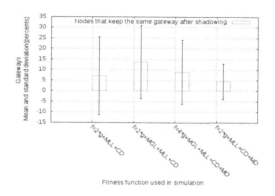

Fig. 15. Percentage of the nodes that are served by the same gateway after shadowing

6 Conclusion

In this work we have presented a Genetic Algorithm approach for the problem of gateway placement in hybrid MANET- Satellite network with the goal to minimize the number of gateways and to optimize various parameters such as: gateway load, link load, path cumulated distance and also the convergence time of the algorithm. The problem was studied on a topology model called FireMobility modeling firefighters action during forest fires. The suitability of GA to this problem was demonstrated. We showed by means of simulations the suitability of different metrics that may be used. Specifics of satellite communications such as gateway shadowing were also discussed. Other criteria, such as the network density [16] may also be considered in the future.

Acknowledgements. This work is partly funded by the French Space Agency under contract CNES R&T R-S12/TC-0008-018 "SatManet".

References

1. Liu, W., Nishiyama, H., Kato, N., Shimizu, Y., Kumagai, T.: A novel gateway selection method to maximize the system throughput of wireless mesh network deployed in disaster areas. In: IEEE International Symposium on Personal, Indoor and Mobile Radio Communications 2012, pp. 771–776 (2012)
2. Hamdi, M., Franck, L., Lagrange, X.: Gateway placement in hybrid MANET-satellite networks. In: IEEE Vehicular Technology Conference, pp. 1–5, September 2012
3. Li, F., Wang, Y., Li, X., Nusairat, A., Wu, Y.: Gateway placement for throughput optimization in wireless mesh networks. Mob. Netw. Appl. **13**(1–2), 198–211 (2008)
4. Xhafa, F., Sanchez, C., Barolli, L.: Ad hoc and neighborhood search methods for placement of mesh routers in wireless mesh networks. In: IEEE International Conference on Distributed Computing Systems Workshops 2009, pp. 400–405 (2009)
5. Le, D., Nguyen, N.: A new evolutionary approach for gateway placement in wireless mesh networks. Int. J. Comput. Netw. Wirel. Commun. **2**, 550–555 (2012)
6. Hoffman, F., Medina, D., Wolisz, A.: Optimization of routing and gateway allocation in aeronautical ad hoc networks using genetic algorithms. In: International Wireless Communications and Mobile Computing Conference, pp. 1391–1396 (2011)
7. Xhafa, F., Sanchez, C., Barolli, L.: Genetic algorithms for efficient placement of router nodes in wireless mesh networks. In: IEEE International Conference on Advanced Information Networking and Applications 2010, pp. 465–472 (2010)
8. Xhafa, F., Barolli, A., Sanchez, C., Barolli, L.: A simulated annealing algorithm for router nodes placement problem in wireless mesh networks. Simul. Model. Pract. Theory **19**(10), 2276–2284 (2011)
9. Ashraf, U., Abdellatif, S., Juanole, G.: Gateway selection in backbone wireless mesh networks. In: IEEE Wireless Communications and Networking Conference 2009, pp. 2548–2553 (2009)
10. Pazand, B., McDonald, C.: A critique of mobility models for wireless networks simulation. In: Computer and Information Science (2007)
11. Sankar, R., Prabhakaran, P.: Impact of realistic mobility models on wireless networks performance. In: IEEE International Conference (WiMoB 2006), June 2006
12. Franck, L., Hamdi, M., Rodriguez, C.G.: Topology modeling of emergency communication networks: caveats and pitfalls. In: TIEMS (The International Emergency Management Society) Workshop (2011)
13. Lin, C.: Dynamic router node placement in wireless mesh networks: a pso approach with constriction coefficient and its convergence analysis. Inf. Sci. **232**, 294–304 (2013)
14. Dhaou, R., Gauthier, V., Tiado, M., Becker, M., Beylot, A.-L.: Cross layer simulation: application to performance modelling of networks composed of MANETs and satellites. In: Kouvatsos, D.D. (ed.) Next Generation Internet: Performance Evaluation and Applications. LNCS, vol. 5233, pp. 477–508. Springer, Heidelberg (2011)
15. Salhani, M., Dhaou, R., Beylot, A.: Terrestrial wireless networks and satellite systems convergence. In: The 25th AIAA International Communications Satellite Systems Conference. The American Institute of Aeronautics and Astronautics (2007)
16. Paillassa, B., Yawut, C., Dhaou, R.: Network awareness and dynamic routing: the ad hoc network case. Comput. Netw. **55**(9), 2315–2328 (2011)

A Pragmatic Evaluation of Distance Vector Proactive Routing in MANETs via Open Space Real-World Experiments

Thomas D. Lagkas[✉], Arbnor Imeri, and George Eleftherakis

The Univeristy of Sheffield International Faculty, CITY College Leontos Sofou Str. 3,
54626 Thessaloniki, Greece
{tlagkas,aimeri,eleftherakis}@city.academic.gr

Abstract. Mobile Ad hoc Networks constitute a promising and fast developing technology that could significantly enhance user freedom. The flexibility provided by such networks is accompanied by unreliability due to notably dynamic conditions that render routing quite problematic. For that reason, the research community has proposed multiple protocols claimed to address this issue, however, only few have been tested via real experiments, while even fewer have reached maturity to become readily available to end users. The main purpose of this paper is to pragmatically evaluate a promising, complete, and finalized MANET protocol via real-world experimentation in open space environment. The considered protocol, with the acronym B.A.T.M.A.N, which is based on distance vector proactive routing, was tested in different networking scenarios that revealed its ability to satisfactorily handle traffic under different conditions.

Keywords: MANET · Proactive routing · Distance vector · B.A.T.M.A.N

1 Introduction

The rapid technological development during the last years offered significantly better communication opportunities between people worldwide. Extensive use of Internet by millions of people enhances collaboration and information sharing. As networking activities are expanded, the demand to enable fast and reliable exchange of information between users increases and its fulfilment becomes more challenging. To meet these requirements, new types of networks have emerged and have been combined with traditional networks.

Taking into consideration the dynamic features of modern users' behaviour, the need for mobility support in regards to communications is inevitable. Apparently, wireless networks are playing a crucial role in providing this type of support. Meanwhile, technology enhancements via the development of more powerful devices allow the adoption of advanced and complex software, which leads to increased demands for network capacity. In this context, multi-hop ad hoc networks where introduced to address networking issues in infrastructureless environments [1]. The most popular type of such networks draws significant interest from the related industry and research community;

© Institute for Computer Sciences, Social Informatics and Telecommunications Engineering 2015
P. Pillai et al. (Eds.): WiSATS 2015, LNICST 154, pp. 345–358, 2015.
DOI: 10.1007/978-3-319-25479-1_26

Mobile Ad Hoc Networks (MANETs) are multi-hop networks with the ability to support user mobility [2].

MANETs were initially intended for military environments and disaster or emergency operations, attributed to their ability of networking without being depended on a fixed infrastructure. Nowadays, this type of networking is considered promising for everyday tasks as well and is expected to contribute directly to the enhancement of existing wireless and cellular systems. In order to understand the basic concept of multi-hop ad hoc networks, it is initially clarified that the simplest ad hoc network, or peer to peer network, is the direct connection of two stations or mobile devices which lie inside each other's range. This type of networks, where entities always communicate directly in pairs, is known as single hop, since data are sent using only one hop, from a specific device to another, therefore there is no need for routing decisions. Bluetooth piconet (Master – Slave) is a typical example of single hop network [3].

The main limitation of single hop networks is the requirement for nodes to be mutually in range in order to communicate. To overcome this restriction, the multi-hop ad hoc model was introduced. In general, a multi-hop ad hoc network can be considered as the union of three or more wireless devices that form an autonomous system connected via wireless links, which do not rely on a fixed base station or predefined network architecture and they are free to dynamically and unpredictably enter or leave the network. The basic prerequisite for the realization of such a system is the responsibility of nodes in range to dynamically discover each other [4]. Multi-hop networking allows packet forwarding in an ad hoc fashion, where the intermediate nodes enable end-to-end packet delivery between out of range nodes.

The possible applications and potential uses of MANETs are practically endless; new application fields keep rising leading to the certainty that this type of networking can find wide acceptance in the near future. In fact, some of the related individual application fields have now matured enough to constitute new areas of research. General purpose MANETs refer to infrastructureless scenarios, where there is no central authority in charge. Hence, in such cases, network behaviour totally depends on the participating devices; as a result there are significant complexities and design concerns due to unpredictable topology changes and battery constraints. A really challenging environment for the deployment of MANETs is military. One of the first needs for infrastructureless networking was originated by military services for the interconnection of soldiers and vehicles in the battlefield. The harsh and highly dynamic conditions of such an environment place significant limitations in realizing reliable communications. For that reason, MANETs are introduced as promising approach. The use of ad hoc networking by emergency services is also a leading application field. The inability to rely on existing infrastructure in cases of disaster increases the demand for dynamic connections.

Another related architectural concept that has attracted significant attention from both the research community and industry is the combination of mobile nodes with fixed networks, also known as hybrid MANETs. The flexibility and scalability of this type of networks allow easy extension of the services provided by the existing infrastructure over a large area, while allowing direct communication between the mobile entities. A promising example is VANET (Vehicular Ad hoc Network), which consists of communicating vehicles as well as fixed devices along the transportation infrastructure (signs, traffic lights, road sensors). The possible individual applications can take advantage of the following three communication types: inter-vehicle, vehicle-to-roadside, and inter-roadside [5].

The main feature that distinguishes MANET from any other type of network is its ability to effectively route information over unreliable and dynamic links in a changing topology. For that purpose, numerous protocols have been proposed, which most of the times are evaluated through theoretical models, simulators or custom prototypes, raising this way concerns about immediate practical applicability. The main motivation of this paper was to explore the actual network behaviour when applying a widely available and ready to use MANET routing protocol via experimentation in real-world scenarios. The rest of the paper is organized as follows: Sect. 2 studies background issues in multi-hop ad hoc routing and related state of the art work, the next section presents the followed network evaluation methodology, Sect. 4 provides and discusses the experimental results, and the paper is concluded in Sect. 5.

2 Background

2.1 MANET Routing Protocols

The most important characteristic of MANETs is efficient data routing and forwarding. Routing is responsible to identify a path toward a destination, and forwarding is in charge of delivering packets through this path. Even though MANETs are quite promising for the future of networking, several challenges must be considered, such as scalability, quality of service, energy efficiency, bandwidth constraints, device heterogeneity, and security. In combination with the unreliable nature of wireless networks makes clear why traditional routing protocols for wired networks are not sufficient for MANETs, where the routing process should take into account the topology dynamism and unpredictability. For that reason, a number of MANET routing protocols have been recently proposed in literature, which can be classified as proactive, reactive, and hybrid [1, 4].

Proactive routing protocols dictate the exchange of routing control information periodically and on topological changes. Typical examples of proactive routing protocols for multi-hop ad hoc networks are: Destination Sequenced Distance Vector (DSDV) [6], Global State Routing (GSR) [7], Hierarchical State Routing (HSR) [8], Optimized Link State Routing Protocol (OLSR) [9], and Better Approach to Mobile Ad Hoc Networks (B.A.T.M.A.N) [10]. Reactive routing protocols create forwarding paths on-demand. Typical examples of reactive (on-demand) routing protocols are: DSR (Dynamic Source Routing) [11], AODV (Ad-Hoc on-Demand Distance Vector) [12], TORA (Temporally Ordered Routing Algorithm) [13], and ABR (Associativity Based Routing) [14]. Hybrid routing protocols are actually the combination of proactive and reactive routing protocols, which means that routes within node's zone are kept up-to-date proactively, whereas distant routes or routes in node's neighbouring zones are set up via reactive routing protocols.

OLSR is one of the most popular protocols for MANETs. It is a proactive, link state routing protocol which employs periodic message exchange to update the topological information in each node for neighbourhood discovery and topology information dissemination, making the routes always available when required. This protocol is optimized for multi-hop ad hoc networks, since it compacts message size and reduces the number of retransmissions needed to flood these messages. Specifically, OLSR includes three generic mechanisms [9]: neighbour sensing, efficient flooding of control traffic, and sufficient diffusing of topological information for optimal routes provision.

Even though OLSR is currently one of the most widely adopted routing protocols in MANETs, it has significant drawbacks. Inefficient bandwidth usage is considered as one of the main weaknesses of OLSR, since each node periodically sends updated information regarding network topology throughout the entire network. Moreover, in order to reduce network flooding, MultiPoint Relays (MPRs) are used to forward topological messages. Thus, in a highly dynamic network environment with rapidly moving nodes, the efficiency of OLSR in supporting data forwarding heavily depends on the network's ability to fulfil frequent exchanges of control messages [15]; a process which is quite unreliable. Despite the fact that latest versions of the protocol are enhanced with new features, the existing limitations remain challenging, due to the rapid growth of mesh networks and the protocol behaviour when calculating the whole topology. For instance, calculating a network topology consisting of 450 nodes takes several seconds for a small CPU [16, 10]. For these reasons, the development of alternative approaches became imperative.

2.2   B.A.T.M.A.N. Routing Protocol

A new solution known as B.A.T.M.A.N algorithm offers a decentralized fashion of spreading topology information by dividing the knowledge of best end-to-end path to all network nodes. The intention is to maintain the knowledge only for the best next hop to all other nodes in the network, thus, there is no need to keep information about the entire network. Moreover, B.A.T.M.A.N offers a flooding mechanism which is event-based and timeless, in order to prevent the increase of opposing topology information and also to restrict the quantity of flooding mesh topology messages. This mechanism contributes in the network performance by limiting control-traffic overhead, making the protocol suitable for networks composed of unreliable links.

According to the algorithm implemented in the B.A.T.M.A.N protocol, nodes announce their presence to their neighbours by transmitting broadcast messages known as originator messages or OGMs. Moreover, the neighbours re-broadcast the OGMs to inform their neighbouring nodes about the presence of the initiator of the OGM message in the network. This process continues until the initiator's OGM is delivered to all nodes, hence, the network is flooded with originator messages. The OGM packet size is 52 bytes including IP and UDP headers. It contains the originator address, the address of the node transmitting the packet, a TTL value and a sequence number. If the mesh includes poor quality wireless links, the OGMs that follow unreliable paths suffer high packet loss or delay, so OGMs that travel over high quality links propagate faster and more reliably. Given that an OGM may be received numerous times by a node, it can be distinguished by the included sequence number. Moreover, "each node re-broadcasts each received OGM at most once and only those messages received from the neighbour which has been identified as the currently best next hop (best ranking neighbour) towards the original initiator of the OGM are used". This is known as selective flooding of OGMs, used to announce the presence of a node in a mesh network. In a nutshell, the working principle is that each node maintains only information about the next link through which the node can find the best route, unlike OLSR, where nodes broadcast "Hello" messages to maintain topological information about the entire network.

B.A.T.M.A.N. advanced (often referenced as B.A.T.M.A.N-adv) is the latest version of the related proactive distance vector routing protocol and is under continuous improvement. It is actually an implementation of the B.A.T.M.A.N protocol at layer 2 of the ISO/OSI model, in the form of a Linux kernel module. In fact, the terms "B.A.T.M.A.N" and "B.A.T.M.A.N-adv" are now used interchangeably, since the latest version of the protocol is the only real option today. It is noted that most of the routing protocols for wireless networks, including the previous implementation called B.A.T.M.A.Nd, transmit and receive routing information and make relevant decisions at layer 3 by manipulating the kernel routing tables. Over the years, with the intention to improve routing performance, B.A.T.M.A.N has evolved from layer 3 to layer 2, without alternating the principles of the underlying routing algorithm. Layer 2 implementation of B.A.T.M.A.N (i.e. B.A.T.M.A.N-adv) transports data traffic as well as routing information using raw Ethernet frames. This is achieved by emulating a virtual network switch of all participating nodes, until the encapsulated traffic is forwarded and delivered to the destination node. In this manner, network topology changes do not affect the participating nodes, since they appear to be link local and unaware of the network topology.

B.A.T.M.A.N-adv is implemented as a kernel driver, in order to provide minor packet processing overhead under heavy load. The objective is to utilize a minimum number of CPU cycles for packet processing, considering that when in user space each packet had to go through the "read()" and "write()" functions to the kernel and back, which procedure was limiting the available bandwidth especially in low-end devices. B.A.T.M.A.N-adv resolves this problem, since it is implemented in Linux kernel.

This work adopts the B.A.T.M.A.N-adv protocol to evaluate the network behaviour of distance vector proactive routing in MANETs with ready-to-use solutions under realistic conditions. Toward this direction, we deployed open space scenarios and employed suitable network evaluation tools, described in the next section.

3 Evaluation Methodology

3.1 Evaluation Tools and Metrics

In order to setup and reveal diagnostic information for the testing network, the Batctl tool was employed [17]. It can be used to configure the B.A.T.M.A.N-adv kernel module and also for presenting information regarding originator tables, translation tables, and debug log. Batctl also includes commands such as ping, traceroute, and tcpdump which are modified to layer 2 functionality. For instance, we used the command "batctl tcpdump interface" to sniff traffic in the forwarder (middle) node.

Furthermore, indicators about the quality of the wireless links were evaluated using the JPerf (Java Performance and Scalability Testing) measurement tool [18]. JPerf is the graphical frontend for Iperf [19], written in Java. Therefore, all the features of Iperf are also supported by JPerf, with the difference that the latter provides a graphical interface which enables easy setup and output visualization. Iperf is a client-server application able to measure bandwidth, latency, jitter, and loss over a network link.

The last evaluation tool that was used in our experiments was a socket-based application we developed for the specific purpose. Our goal was to create controlled conditions, where individual parameters could be configured and tested. The developed software focuses on measuring data loss over TCP and UDP communications. The application operates in client-server mode, it was developed in Java, and offers a simple and effective user interface.

Regarding the network metrics that were considered for the network evaluation, the following were measured during the experiments: Bandwidth (maximum achievable data rate in bits per second), Loss (data sent but not successfully received), RTT (Round Trip Time), and Jitter (variation in delay of received packets). The measurements were taken individually for each one of the different scenarios and for various packet sizes.

3.2   Experimental Setup and Scenarios

The experiments were conducted in open outdoor space allowing adequately long distances between nodes, which enforces routing as stations get out of range. Moreover, the experimental environment makes possible the formation of clear topologies, where there is enough space for nodes to move, hence, to evaluate network behaviour under mobility conditions.

For the purposes of our experiments, four laptops where setup and used as ad hoc nodes. In each laptop, B.A.T.M.A.N-adv was installed, along with the necessary evaluation tools. The ad hoc network was formed using the laptops' Wi-Fi Network Interface Cards. The nodes were elevated approximately 40 cm from the ground. The main specifications of each laptop are the following:

- 1 Dell Inspiron N5110 – NIC: Qualcomm Atheros Dell Wireless 1702 (802.11b/g/n)
- 2 IBM ThinkPad X.41- NIC: Qualcomm Atheros AR5212 (802.11a/g/n)
- 1 Dell Latitude E6400 – NIC: Intel Wireless Wi-Fi Link 4965AGN (802.11a/g/n)

The first testing scenario is illustrated in Fig. 1 and is considered as the base (control) scenario. It is noted that for clarity reasons in the following four figures representing the testing scenarios the circles do not denote ranges, but illustrate connectivity between the corresponding nodes. The first scenario is actually a single hop network, since its topology consists of only two nodes running the B.A.T.M.A.N-adv protocol and operating in a client/server mode. There is no routing in this scenario, due to direct connectivity. To establish the connection, the nodes must be in range; the distance between the nodes is 85 m. This scenario is used to compare results against the other scenarios where routing actually takes place.

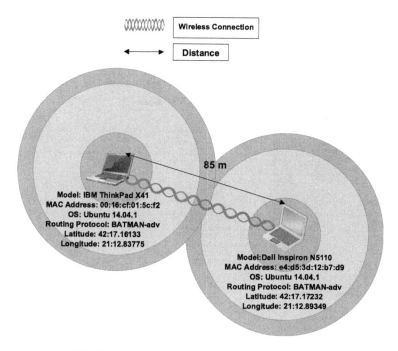

Fig. 1. First experimental scenario network topology

The second scenario consists of three nodes forming a multihop network, as shown in Fig. 2. The node with MAC address "00:16:cf:01:62:56" (source) is placed out of range of the node with MAC address "e4:d5:3d:12:b7:d9" (destination). The node with MAC address "00:16:cf:01:5c:f2" (forwarder) is placed between the two nodes in order to allow the creation of a routing path. So, the source node actually uses the forwarder node in order to transmit packets to the destination node. It is noted that the ground between the source and the forwarder is flat, so the line of sight is good, whereas there is some curvature between the forwarder and the destination. The main intention of this scenario is to reveal the behaviour of the routing protocol in a dual-hop network without mobility.

In the third scenario, the nodes are placed exactly at the same positions as in the second scenario, as shown in Fig. 3. The only difference is that the middle (forwarder) node is in a moving state, so it is mobile (not static). Specifically, it moves with human walking speed in a square area of 30-by-30 m during all experimental measurements. It is important to note that the height of the forwarder is around 1.5 m above the ground, since it is kept in hand while moving, which provides a better line of sight. The intention here is to explore the performance of B.A.T.M.A.N-adv, when there is relative mobility in the routing path.

The fourth scenario is the most complex one; it involves four nodes deployed at different locations. Three nodes are static and one is moving with human walking speed, as illustrated in Fig. 4. The nodes with MAC addresses "e4:d5:3d:12:b7:d9", "00:16:cf: 01:5c:f2" and "00:16:cf:01:62:56" are static, whereas the node with MAC address

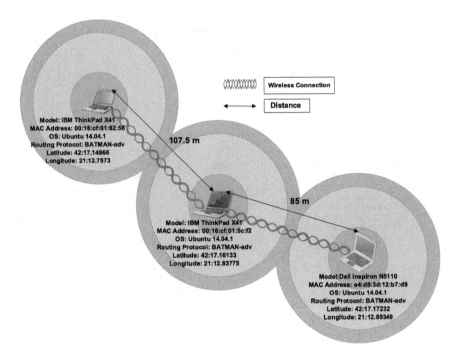

Fig. 2. Second experimental scenario network topology

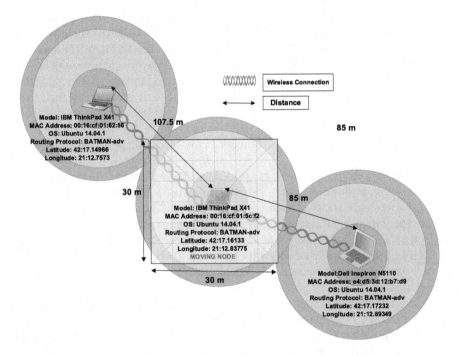

Fig. 3. Third experimental scenario network topology

"00:24:d8:a3:1b:b4" is mobile. The static node at the bottom acts as server, while the mobile node acts as client. The two middle nodes perform data forwarding. Our intention here is to evaluate the ability of the routing protocol to dynamically switch forwarders, hence, alternating routing paths.

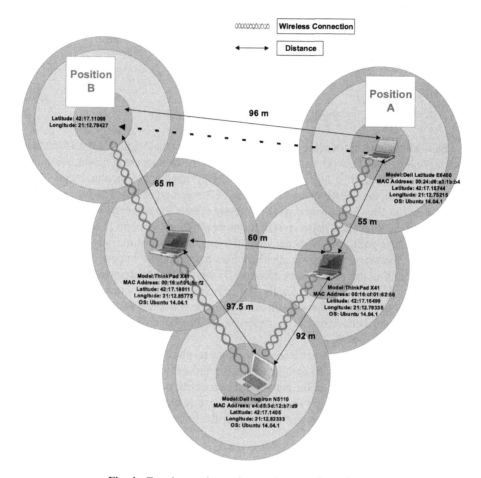

Fig. 4. Fourth experimental scenario network topology

4   Experimental Results and Discussion

In this section, we provide and discuss the results of the experiments conducted based on the aforementioned methodology, employing the described tools and implementing the presented scenarios. Our goal is the evaluation of the performance of B.A.T.M.A.N-adv, as a representative ready-to-use distance vector proactive routing protocol for MANETs, via comparative experimental results under realistic open-space conditions.

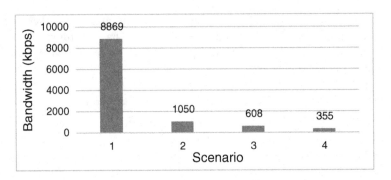

Fig. 5. Bandwidth achieved using JPerf

Bandwidth, in terms of achieved data rate, is one of the most significant performance metrics and reveals network capacity. It is defined as the supported transmission rate from source to destination. In our experiments, bandwidth measurements were performed using the JPERF tool and refers to TCP communication. The results, which are depicted in Fig. 5, show that routing greatly affects the achieved bandwidth. It is evident that the direct link between source and destination (scenario 1) allows successfully delivering significantly higher amount of traffic in the same time interval, compared to the other scenarios. Moreover, mobility also has a notable impact on the specific metric and this is the reason why scenario 3, which dictates forwarder movement, performs worse than scenario 2. Lastly, the complex conditions present in scenario 4, where mobility is combined with path alternation, lead to the worst performance. It is noted that similar behaviour can be observed for the same reasons in the following presented results, as well.

Figure 6 presents the average Round Trip Time results collected for all four scenarios using the batctl-ping tool. The specific metric is representative of the experienced delay when data is transmitted over the network. As expected, the single hop topology of the first scenario induces the lowest delay. On the other end, mobility and forwarder switching delays lead to worst performance for the fourth scenario considering RTT.

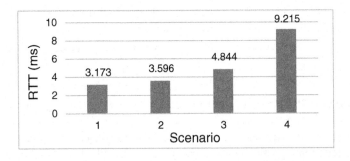

Fig. 6. Average Round Trip Time (RTT) using batctl-ping

In order to have a better view of the resulted latency, we have also conducted related measurements using our custom socket-based tool. The specific experiment involves the

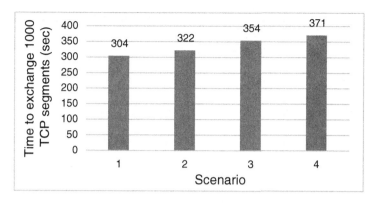

Fig. 7. Time required to exchange 1000 TCP segments using custom socket-based tool

establishment of a bidirectional TCP communication, where a 1400-byte segment is created every 100 ms and transmitted over the network. As soon as it is received by the destination node, the same segment is sent back to the source. Figure 7 shows the time needed for the successful completion of 1000 segments exchange (i.e. 1000 segments sent back and forth). It can be seen that the more the hops and the less stable the topology is, the more the time required for the exchange.

Figure 8 depicts packet loss as percentage of UDP datagrams not received over the total datagrams sent. JPerf was employed to generate 2 MB/s UDP traffic and transmit it over the network towards the destination node. It is evident that the highly dynamic conditions present in the fourth scenario lead to unreliable data paths, which cause significantly increased datagram losses.

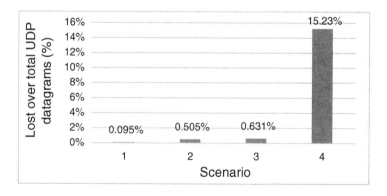

Fig. 8. Percentage of lost UDP datagrams over total sent using JPerf

The last metric that is considered throughout our experiments is jitter. This is a significant indication of the network's ability to efficiently support traffic in a consistent manner causing minimum variations. These delay variations have a major impact on the Quality of Service (QoS) provided especially to multimedia network traffic. This is definitely a challenge for unstable networks, such as dynamic MANETs, as it becomes evident from the results depicted in Fig. 9. Apparently, the highly unreliable conditions present in the fourth scenario lead to so much increased jitter, which actually prohibits serving good quality multimedia streams.

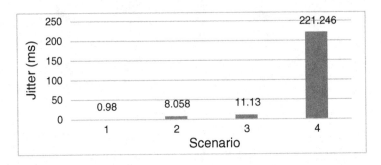

Fig. 9. Jitter in UDP communication using JPerf

Summing up the presented results, it is clear that the B.A.T.M.A.N routing protocol can definitely support communications over a MANET, however, the particular network characteristics affect performance to a significant degree. Specifically, the existence of multiple hops notably limits the available network capacity, meaning that the supported data rate is quite decreased. Mobility also has a notable effect on network behaviour, however, when it does not lead to route changes, the impact is not major. Considering real-world network applications, we could deduce based on the experimental results that B.A.T.M.A.N can satisfactorily support data communications over MANETs when they are not time sensitive, however, in cases where the highly dynamic conditions cause excessive path alternations, reliability is significantly affected and the quality of the provided service is marginal.

5   Conclusion

One of the main challenges of modern networking is meeting the rapidly growing requirements while facilitating participants' autonomy. Working towards that direction, the routing protocols developed for MANETs try to handle the highly dynamic conditions and enhance connectivity. This paper provided an evaluation of a promising MANET protocol which is readily available to end users. B.A.T.M.A.N was installed and configured in different mobile nodes, while four networking scenarios were designed for deployment in open space. The performed real world experiments managed to reveal network behavior under different conditions via studying the collected metrics. Specifically, the results made evident that the protocol is able to satisfactorily serve traffic under most considered conditions, however, there is a great impact on performance when the number of hops or the degree of mobility increase. As expected, the type of network applications which are affected the most are the ones that are quite sensitive to extensive variations, such as real-time streams. In the future, we plan to apply more MANET protocols and perform multi-node real-world experiments to evaluate the performance of specific multimedia streams, emulating the actual usage of the corresponding applications. In that manner, conclusions on optimal protocol configuration can be drawn, as well as directions can be provided for improving existing routing techniques and possibly introducing new more efficient ones.

References

1. Conti, M., Giordano, S.: Multi-hop ad hoc networking: the theory. IEEE Commun. Mag. **45**(4), 78–86 (2007)
2. Mueller, S., Tsang, R.P., Ghosal, D.: Multipath routing in mobile ad hoc networks: issues and challenges. In: Calzarossa, M.C., Gelenbe, E. (eds.) MASCOTS 2003. LNCS, vol. 2965, pp. 209–234. Springer, Heidelberg (2004)
3. Ghosekar, P., Katkar, G., Ghorpade, P.: Mobile ad hoc networking: imperatives and challenges. IJCA Spec. Issue MANETs (3), 153–158 (2010)
4. Hoebeke, J., Moerman, I., Dhoedt, B., Demeester, P.: An overview of mobile ad hoc networks: applications and challenges, 60–66 (2007)
5. Abedi, A., Ghaderi, M., Williamson, C.: Tradeoff, distributed routing for vehicular ad hoc networks: throughput-delay. In: IEEE International Symposium on Modeling, Analysis & Simulation of Computer and Telecommunication Systems (MASCOTS), Miami Beach, FL, pp. 47–56 (2010)
6. Perkins, C., Bhagwat, P.: Highly dynamic destination-sequenced distance-vector routing (DSDV) for mobile computers. ACM SIGCOMM Comput. Commun. Rev. **24**(4), 234–244 (1994)
7. Chen, T.-W., Gerla, M.: Global state routing: a new routing scheme for ad-hoc wireless networks. In: IEEE International Conference on Communications, Atlanta, GA, pp. 171–175 (1998)
8. Joa-Ng, M., Lu, I.-T.: A peer-to-peer zone-based two-level link state routing for mobile ad hoc networks. IEEE J. Sel. Areas Commun. **17**(8), 1415–1425 (1999)
9. Clausen, T., Jacquet, P., Adjih, C., Laouiti, A., Minet, P., et al.: Optimized Link State Routing Protocol (OLSR). INRIA Network Working Group (2003)
10. Neumann, A., Aichele, C., Lindner, M., Wunderlich, S.: Better approach to mobile ad-hoc networking (BATMAN). IETF Draft (2008)
11. Johnson, D., Hu, Y., Maltz, D.: The dynamic source routing protocol (DSR) for mobile ad hoc networks for IPv4 (2007)
12. Perkins, C.E., Park, M., Royer, E.M.: Ad-hoc on-demand distance vector routing. In: Second IEEE Workshop on Mobile Computing Systems and Applications (WMCSA 1999), New Orleans, LA, pp. 90–100 (1999)
13. Park, V.D., Corson, M.S.: A highly adaptive distributed routing algorithm for mobile wireless networks. In: Sixteenth Annual Joint Conference of the IEEE Computer and Communications Societies. Driving the Information Revolution (INFOCOM 1997), Kobe, vol. 3, pp. 1405–1413 (1997)
14. Toh, C.-K.: Associativity-based routing for ad hoc mobile networks. Springer Wirel. Pers. Commun. **4**(2), 103–139 (1997)
15. Singh, V., Kumar, M., Jaiswal, A.K., Saxena, R.: Performance comparison of AODV, OLSR and ZRP protocol in MANET using grid topology through QualNet simulator. Int. J. Comput. Technol. **8**(3), 862–867 (2013)
16. Open-mesh: BATMANConcept. In: Open-mesh. http://www.open-mesh.org/projects/open-mesh/wiki/BATMANConcept

17. Langer, A.: Batctl. In: Open Mesh. http://downloads.open-mesh.org/batman/manpages/batctl.8.html
18. Grove, A.: JPerf. In: Sourceforge. http://sourceforge.net/projects/jperf/
19. Iperf: Iperf - The TCP/UDP Bandwidth Measurement Tool. https://iperf.fr

Elastic Call Admission Control Using Fuzzy Logic in Virtualized Cloud Radio Base Stations

Tshiamo Sigwele[(✉)], Prashant Pillai, and Yim Fun Hu

Future Ubiquitous Networks Lab, Faculty of Engineering and Informatics,
University of Bradford United Kingdom, Bradford, UK
{t.sigwele,p.pillai,y.f.hu}@bradford.ac.uk

Abstract. Conventional Call Admission Control (CAC) schemes are based on stand-alone Radio Access Networks (RAN) Base Station (BS) architectures which have their independent and fixed spectral and computing resources, which are not shared with other BSs to address their varied traffic needs, causing poor resource utilization, and high call blocking and dropping probabilities. It is envisaged that in future communication systems like 5G, Cloud RAN (C-RAN) will be adopted in order to share this spectrum and computing resources between BSs in order to further improve the Quality of Service (QoS) and network utilization. In this paper, an intelligent Elastic CAC scheme using Fuzzy Logic in C-RAN is proposed. In the proposed scheme, the BS resources are consolidated to the cloud using virtualization technology and dynamically provisioned using the elasticity concept of cloud computing in accordance to traffic demands. Simulations shows that the proposed CAC algorithm has high call acceptance rate compared to conventional CAC.

Keywords: Qos · CAC · C-RAN · Cloud computing · Fuzzy Logic · Virtualization

1 Introduction

A plethora of mobile phones and multimedia services in recent years has resulted in gigantic demands for larger system capacities, higher data rates over large coverage areas in high mobility environments. Hence, Radio Access Networks (RAN) have tremendously grown so complex and are becoming so difficult to manage and control. Maintaining Quality of Service (QoS) for real time and non-real time services while optimizing resource utilization is a major challenge due to poor and ineffective Radio Resource Management (RRM) schemes. Call Admission Control (CAC) is a RRM scheme that offers an effective way of avoiding network congestion and plays a key role in the provision of guaranteed QoS in the RAN. The basic function of a CAC algorithm is to accurately decide whether a new or handoff call can be accepted into a resource-constrained network without violating the service commitments made to the already admitted calls. A good CAC scheme aims to optimize Call Blocking Probability (CBP), Call Dropping Probability (CDP) and system utilization. There are many drawbacks

© Institute for Computer Sciences, Social Informatics and Telecommunications Engineering 2015
P. Pillai et al. (Eds.): WiSATS 2015, LNICST 154, pp. 359–372, 2015.
DOI: 10.1007/978-3-319-25479-1_27

facing conventional CAC schemes which make them unsuitable for future mobile communication systems. First, conventional CAC approaches in cellular networks suffer uncertainties due to real time processing of radio signals and the time varying nature of parameters such as speed, location, direction, channel conditions, available power, etc. Many of these traditional CAC schemes are ineffective leading to incorrect request admission when the network is actually incapable of servicing the request or incorrect rejection when there are actually enough resources to service the request. Some of these CAC schemes tend to assume network state information is static. However, in practice the network is dynamic and values measured keep changing.

Second, as stated in previous work [1], traditional CAC schemes are based on stand-alone RAN Base Station (BS) architectures. These BSs are preconfigured for peak loads and have unshared processing and computation resources located in the BS cell areas. These BS resources cannot be shared to address varied traffic needs on other cell areas, causing poor BS utilization, high CBP and CDP. Intelligent CAC schemes based on intelligent decision making techniques like Fuzzy Logic are a promising solution and solve the problem of imprecision and uncertainties cellular networks. The schemes mimic the cognitive behaviour of human mind without the need for complex mathematical modelling making them adaptive, less complex, flexible and suitable to cope with the rapidly changing network conditions cellular networks.

This paper presents an intelligent elastic CAC scheme using Fuzzy Logic in a centralized Cloud RAN (C-RAN), herein termed as eFCC. C-RAN was introduced as a way of solving the drawbacks of conventional RAN [2] by pooling BS resources to a centralized cloud as shown in Fig. 1. Virtualization concept is used on general purpose CPUs (e.g. x86 or ARM processors) to dynamically allocate BS processing resources to different Virtual BS (vBS) and air interface standards. Several air interface standards can be supported and the so called 'tide effects', where traffic of each BS changes rapidly, can also be minimized. The rest of this paper is organized as follows. Section 2 presents the C-RAN architecture that is adopted in this paper. Related work on CAC schemes will be presented and analysed in Sect. 3. The proposed eFCC scheme is described in Sect. 4. Section 5 presents the simulation model and the obtained performance results. Finally, conclusion and further works are presented in Sect. 4.

2   C-RAN Architecture

C-RAN is a paradigm shift for next-generation RANs. C-RAN is described using four C's which stand for; clean, centralized processing, collaborative radio and real-time cloud computing. The C-RAN architecture adopted in this paper is shown in Fig. 1. The C-RAN concept separates the radio and antenna parts from the digital baseband parts and pools multiple Baseband Units (BBUs) in a central office. These digital only BSs, called Virtual BS (vBS), are linked via fiber or the Common Public Radio Interface (CPRI) to remote radio heads (RRHs). Using the concept of Virtualization which separate software from hardware, the digital functionality in the BS hotel is shifted to the cloud and the BSs are abstracted as vBSs. General purpose processors (GPPs) like X86 and ARM processors are used to house physical baseband resources and using cloud

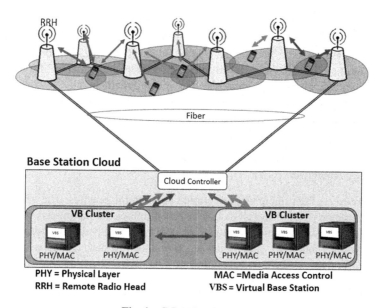

Fig. 1. C-RAN architecture.

computing virtualization concept, multiple vBS virtual machines (VMs) are dynamically provisioned in accordance to traffic demands. The aim is to dynamically scale up and down vBSs in accordance with the traffic fluctuations and the tide effect [2].

The key advantages of C-RAN based network architecture are summarised as follows:

1. Simpler and faster system roll-out and network expansion. Only RRHs need to be installed to expand coverage. Increased network load is combated with cloud hardware (HW) upgrade.
2. Support multi-cell collaborative signal processing. This reduces interference and handover latencies.
3. Improved CAPEX and OPEX.
4. Sharing processing power effectively utilizes the BS.
5. C-RAN is based on open platform, can support multiple standards, and smooth evolution and easy upgrading.
6. Maximum resource sharing and great flexibility to allocate processing resources with traffic demands.

3 Related Work

There are many ways of categorizing CAC schemes like Parameter based, measurement based, utility based, centralized/distributed, static/dynamic etc. Comprehensive surveys can be found here [3–6]. Conventional CAC approaches suffer uncertainties due to time varying nature of parameters (speed, location, direction, channel conditions, available power, etc.) Many of these CAC schemes make erroneous decisions as they tend to assume network state information is static. However, in practice the network is dynamic and values measured keep fluctuating. Intelligent CAC schemes based on intelligent decision making techniques solve the problem of error and uncertainties in conventional schemes. They are adaptive and flexible, thus making them suitable to cope with the rapidly changing network conditions and bursty traffic that can occur in cellular wireless networks to give an efficient network management scheme. Some of the commonly used intelligent decision making mechanisms used for CAC are described below:

3.1 Genetic Algorithm (GA)

GA's are based on the principles of evolution and natural genetics. The genetic algorithms are directed random search techniques used to look for parameters that provide optimal solution to the problem. The notion of a GA is the survival of the fittest in nature, which implies that the 'fitter' individuals are more likely to survive and have chance of passing their features to the next generation. In [7], a CAC scheme using GA has been proposed for roaming mobile users with low handoff latency in Next Generation Wireless Systems. The scheme provides high network utilization, minimum cost, minimum cost but it is not suitable for real time applications since GA is very slow and cannot be used for real time decision making.

3.2 Neural Networks (NN)

NN's are algorithms that are patterned after the structure of the human brain. NN's have learning and adaptive capabilities that can be used to construct intelligent computational algorithms for traffic control. A neural approach for CAC with QoS guarantee in multimedia high-speed networks is proposed in [8]. It is an integrated method that combines linguistic control capabilities and the learning abilities of a neural network. Even though the scheme provides higher system utilization, it requires large computational resources working in parallel. A novel learning approach to solve the CAC in multimedia cellular networks with multiple classes of traffic is presented in [9]. The near optimal CAC policy is obtained through a form of Neuro-Evolution algorithm. This method guarantees that the specified Call Dropping Rate (CDR) remain under a predefined upper bound while retaining acceptable Call Blocking Rate (CBR). This scheme is black box learning approach since the knowledge of its internal working is never known. The scheme also requires high processing time.

3.3 Fuzzy Logic (FL)

In this paper, FL scheme is used because of its simplicity and robustness [5]. FL techniques resembles the human decision making with an ability to generate precise solutions from certain or approximate information. FL avoids uncertainties and computational complexities brought by many CAC schemes. It does not require precise inputs, and can process any number of inputs. FL incorporates a simple, rule based approach based on natural language to solve control problem rather than attempting to model a system mathematically. The main point of interest in eFCC scheme is that it make use of FL to handle uncertainties in the network. A Fuzzy CAC Scheme for High-speed Networks was proposed in [10]. Even though the author used Fuzzy to better estimate equivalent capacity, he does not show how the schemes performs in terms of acceptance rate. In [11], the author propose a fuzzy CAC approach scheme for long Term Evolution (LTE). Even though the proposed scheme shows better call rejection than the Quality index based approach, the CAC scheme is based on standalone BS architecture with low BS utilisation. A method of fuzzy admission control for multimedia applications (MFAC) scheme is proposed in [12]. In this method, for multimedia applications, QoS and Congestion Control (CC) have more parameters and thus two additional fuzzy based controllers: Fuzzy QoS Controller (FQC) and Fuzzy Congestion Controller (FCC) have been introduced to the fuzzy based admission controller, allowing better estimation of QoS parameters. The drawbacks of this scheme is that it has many fuzzy controllers that can magnify CAC complexity and computation latency.

4 Proposed Scheme: Elastic Fuzzy C-RAN CAC

In our proposed scheme, baseband signals from multiple cells are no longer processed on their stand-alone Baseband Units (BBU) but processed in the cloud using the concept of cloud computing. The BS traffic from various cells are processed on a General Purpose Processor (GPP). The GPP is software defined enabling multiple radio signal from different cells to be processed in one computer platform. This is made possible through virtualization technology where hardware components are abstracted from software components. The advantages of virtualization in our scheme are:

- **Green:** With virtualization, fewer servers are used and less power and less cooling is required. Also, during low traffic, *server consolidation* and *VM migration* can help reduce power by switching off some of the less loaded servers.
- **No Vendor Lock-in:** The virtual machines do not really care what hardware they run on, hence no more tied down to a single vendor
- **Reduced Cost**: Reduced amount of hardware used coupled with less downtime, easier maintenance, less electricity used.
- **Disaster Recovery:** If disaster strike the data center itself, virtual machines can always be moved elsewhere.

Virtual Machines (VMs) are dynamically provisioned to service traffic requests from cells. The VMs are defined as vBS in this paper. The vBS performs baseband signal processing of specific cell traffic. The traffic demand from cells is mapped into baseband processing resource utilization such that every cell traffic is serviced by its own vBS. The proposed model considers a C-RAN cellular network with BS resources being pooled together to the cloud to serve two cell coverage areas as shown in Fig. 2. One of the main components is the Cloud Controller which is comprised of the CAC processor and the Resource Estimator (RE). RE continuously measures and dynamically provisions resources to the different vBSs. The eFCC scheme consists of two key elements: an elastic scheme for dynamically allocating resources to vBS and Fuzzy based CAC.

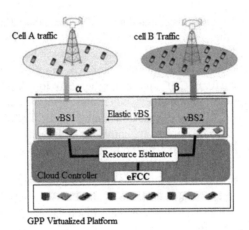

Fig. 2. Elastic C-RAN CAC architecture

The aim is to fairly service requests from cell A and cell B by dynamically provisioning resources from vBS1 and vBS2 respectively while sharing the same processing infrastructure. It is imperative that QoS in the network is not affected hence call blocking probability is used as a performance matrix. If cell A high load compared to cell B, resources are scaled up in an elastic manner. If there is high traffic load on both cells such that the GPP might overload, some of the task for incoming traffic are forwarded to other GPP which can accommodate them and there will be some communication cost inquired. Alternatively, One of the VBS have to be migrated to a less loaded GPP of a less loaded GPP to be able to accommodate the migrated vBS, but migration in real time processing might cause delays and jitters and interrupt service connection.

4.1   Elastic BS Resource Provisioning Model

The resources being shared are processing and computational resources. The processing resources are dimensioned in proportion to correspond with radio resources. Users in coverage areas generate traffic which is aggregated in the BS. This generated traffic from BS is mapped into $\alpha \leq 1$ denoting BS traffic demand as follows:

$$\alpha = \frac{cell^A_{traffic\_load}}{cell^A_{max\_traffic\_load}} \tag{1}$$

where $cell^A_{traffic\_load}$, $cell^A_{max\_traffic\_load}$ is the current load in cell A and peak traffic load for cell A respectively. As such, α also denote the utilization of a BS hence the amount of resources required to service the traffic from cell A is calculated as follows:

$$vBS1_{capacity} = \alpha * Cloud^{total}_{capacity} \tag{2}$$

where $vBS1_{capacity}$ is a virtual BS provisioned in the cloud to services total traffic from cell A and $Cloud^{total}_{capacity}$ is the total amount of processing resources in the cloud. Similarly for cell area B, $\beta \leq 1$ denote the utilization in cell area B:

$$\beta = \frac{cell^B_{traffic\_load}}{cell^B_{max\_traffic\_load}} \tag{3}$$

where $cell^B_{traffic\_load}$, $cell^B_{max\_traffic\_load}$ is the current load in cell B and peak traffic load for cell B respectively. $vBS2$ which is the amount of resources required to service traffic from cell B is calculated as follows:

$$vBS2_{capacity} = \beta * Cloud^{total}_{capacity} \tag{4}$$

The proposed scheme of GPP sharing is suitable during low traffic times, but when traffic surges, some vBS are forced to be migrated to other less loaded GPP's.

Using the proposed method, two vBSs can be dynamically and elastically provisioned based on the traffic demands from the two coverage areas or RRHs. It is assumed that there are 3 classes (VoIP, video streaming and file download). The traffic classes have different QoS requirements. The arrival rates of new calls of class i, $i = 1, 2...K$, are assumed to form a Poisson process with a mean arrival rate λ_i. The service times of new calls of class i are assumed to follow an exponential distribution with mean service time $1/\mu_i$.

4.2   Structure of Fuzzy Logic Controller (FLC)

In order to make a more accurate decision for call acceptance, this paper proposes a fuzzy based CAC scheme, called eFCC scheme. The Fuzzy Logic Controller (FLC) called eFCC is the main part of the FL and its basic elements are shown in Fig. 3. It takes in four inputs and outputs the decision of call acceptance or rejection. The four inputs are: Available Capacity Ac, effective capacity of a new call request Ec, network congestion parameter Nc via Congestion Indicator (CI). The output of the eFCC is admittance decision Ad. The RE measures the total capacity on ongoing connections and dynamically scale out resources in the cloud based on the traffic demand.

4.3   Membership Functions

The term sets of Nc, Ac, Ec and Ad are defined as follows:

- $T(Nc) = \{Negative, Positive\} = \{N, P\}$
- $T(Ac) = \{NotEnough, Enough\} = \{NE, E)$
- $T(Ec) = \{small, medium, big\} = \{sm, me, bi\}$
- $T(Ad) = \{R, WR, AR, WA, A\}$

For $T(Ad)$, $R = Reject$, $WR = Weak\ Reject$, $AR = Average\ Reject$, $WA = Weak\ Accept$ and $A = Accept$. Triangular functions as MF are chosen for simplicity. The membership functions for input and output linguistic parameters are shown in Figs. 4, 5, 6 and 7. The values of MF function have been chosen based on commonly used values of MFs in various literature.

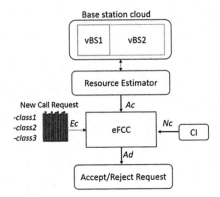

Fig. 3. eFCC CAC model

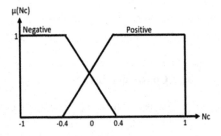

Fig. 4. Membership function for network congestion.

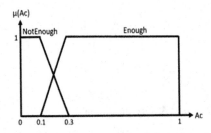

Fig. 5. Membership function for available capacity.

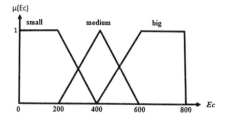

Fig. 6. Membership function for effective capacity.

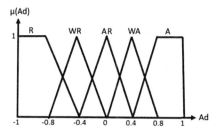

Fig. 7. Membership function for admittance decision.

4.4 Fuzzy Rule Base

The Fuzzy Rule Base consists of a series of 24 fuzzy rules, shown in Table 1. These control rules are of the following form: IF "condition", THEN "action". Example: if Nc is *Negative* and Ac is *NotEnough* and Ec is *Medium*, then *Weak Reject* the request.

5 Simulation and Results

5.1 Simulation Parameters

The posed scheme aims to provide elastic provisioning of capacities for the vBSs from the cloud. MATLAB Simulink was used to simulate the proposed scheme which is compared with a baseline CAC scheme in [1]. The CAC scheme in [1] is applied in C-RAN but without Fuzzy. For simulation and performance evaluation the following three traffic classes shall be considered: VoIP (class1), Video streaming (class2) and file download (class3). Call requests of the three traffic classes were generated with Poison process with parameters shown in Table 2. The value of λ was varied from 0.01 to 1 with every simulation. 100 calls were generated for each traffic class. The simulation time was kept at 500 s. It is assumed that the QoS of a traffic class is equivalent to Ec. Membership Functions for congestion factor and available capacity are shown in Figs. 4 and 5. The MF for Ec in Fig. 6 ranges from non-normalized limits of 0 to 800 Kbps to accommodate all traffic classes from Table 1. The output of the FL is the variable Ad which is the decision factor with five possible outcomes. Triangular and trapezoidal MF are used since they are simple and are suitable for real time traffic. The total capacity was kept constant for the baseline scheme which is based on stand-alone BS and then varied elastically for the eFCC scheme which is based on C-RAN.

Table 1. Fuzzy rule base

Rule	N_c	A_c	E_c	A_d
1	N	NE	Sm	AR
2	N	NE	Me	WR
3	N	NE	Bi	WR
4	N	E	Sm	WA
5	N	E	Me	AR
6	N	E	Bi	WR
7	P	NE	Sm	WA
8	P	NE	Me	AR
9	P	NE	Bi	WR
10	P	E	Sm	A
11	P	E	Me	A
12	P	E	Bi	A
13	N	NE	Sm	R
14	N	NE	Me	R
15	N	NE	Bi	R
16	N	E	Sm	AR
17	N	E	Me	AR
18	N	E	Bi	R
19	P	NE	Sm	WR
20	P	NE	Me	R
21	P	NE	Bi	R
22	P	E	Sm	AR
23	P	E	Me	AR
24	P	E	Bi	WR

Table 2. Simulation parameters

Traffic type	Parameters			
	λ (calls/s)	$1/\mu$ (s)	Ec (kb/s)	Priority
Voip	varied	300	12.2	1
Video streaming	varied	300	768	2
File download	varied	300	144	3

5.2 Simulation Results

The CBP is measured for every service class to evaluate the performance of the proposed CAC scheme. 100 calls are simulated while changing call arrival from 0.01 calls per second (36 calls/h) to 1 call per second (3600calls/h) for each scheme. The CBP for the two schemes are shown in Figs. 8, 9 and 10. In Fig. 8 it can be seen that the CBP for VoIP for eFCC is lower than for baseline scheme by 34 %. This is because the baseline resources get exhausted quickly, while for eFCC, resources are elastically pooled from the cloud to serve incoming requests. Both schemes completely flattens for $\lambda > 0.8$ since calls are rejected to maintain QoS. Figure 9 shows the CBP for video streaming application. In this scenario also the eFCC scheme outperforms the baseline scheme by a factor of approximately 75 %. Figure 10 shows the CBP for file download application. Here eFCC also outperforms the baseline scheme at a much larger average factor of 90 %. File downloads in baseline follows a constant behavior of CBP while with eFCC CBP increases constantly.

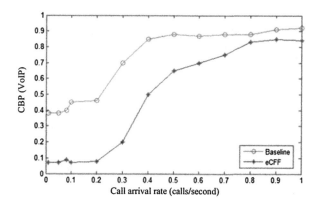

Fig. 8. Call blocking probability (VoIP).

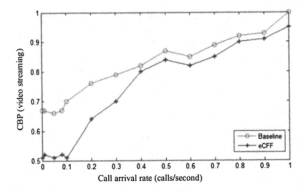

Fig. 9. Call blocking probability (video streaming).

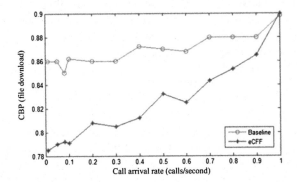

Fig. 10. Call blocking probability (file download).

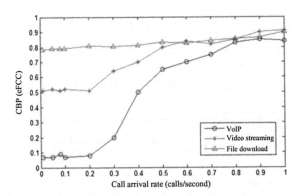

Fig. 11. Call blocking probability (eFCC).

From the results, it can be seen that in general the CBP increases as traffic load increases for both the schemes since the network resources get exhausted as more requests are serviced. It can also be seen that the CBP of the proposed eFCC scheme is lowest for all traffic classes in all the application scenarios. This is so because eFCC elastically provision resources that are pooled from the cloud hence increasing call acceptance rate. Figure 11 shows the CBP for all traffic classes simulated under eFCC scheme. It can be observed that high priority traffic have lower CBP that low priority traffic. This is necessary to maintain QoS of the overall network.

6 Conclusion

The paper presents an intelligent elastic fuzzy logic based call admission control scheme in cloud radio access networks. A combination of fuzzy and C-RAN improve uncertainties and imprecations in conventional CAC schemes and combat the problem of constrained resources by sharing BS resources in the cloud. The scheme take the advantage of cloud computing's elasticity and resource sharing to dynamically provision resources based on traffic demands. The simulation results shows that for heterogeneous traffic classes, the proposed scheme has lower CBP than conventional CAC. The scheme proposed has better robust performance and outperforms the conventional schemes. In future the system will be extended to consider handoff calls and additional performance measures like system utilization and call dropping. The energy saving gained by virtualization and vBS migration will be the main target in future.

References

1. Sigwele, T., Pillai, P., Hu, Y.F.: Call admission control in cloud radio access networks. In: The 2nd International Conference on Future Internet of Things, Barcelona, Spain (2014)
2. CMRI: C-RAN: the road towards green RAN. In: C-RAN International Workshop, Beijing, China (2010)
3. Ahmed, M.H.: Call admission control in wireless networks: a comprehensive survey. IEEE Commun. Surv. Tutorials **7**, 49–68 (2005)
4. Niyato, D., Hossain, E.: Call admission control for QoS provisioning in 4G wireless networks: issues and approaches. IEEE Netw. **19**, 5–11 (2005)
5. Qilian, L., Karnik, N.N., Mendel, J.M.: Connection admission control in ATM networks using survey-based type-2 fuzzy logic systems. IEEE Trans. Syst. Cybern. **30**, 329–339 (2000)
6. Yanbing, L., Man, M.: Survey of admission control algorithms in IEEE 802.11e Wireless LANs. In: 9th Conference on Future Computer and Communication, pp. 230–233 (2009)
7. Karabuda, D., Hung, C., Bing, B.: A CAC scheme for multimedia applications based on fuzzy logic. In: Proceedings of ACM Symposium on Applied Computing, Nicosia, Cyprus, pp. 1151–1158 (2004)
8. Cheng, R.G., Chang, C.J., Lin, L.-F.: A QoS-provisioning neural fuzzy connection admission controller for multimedia high-speed networks. IEEE/ACM Trans. Netw. **7**, 111–121 (1999)
9. Yang, X., Bigham, J.: A call admission control scheme using neuroevolution algorithm in cellular networks. In: 20th IJCAI, Hyderabad, India (2007)
10. Barolli, L., Koyama, A., Yamada, T.S., Yokoyama, Suganuma, T., Shiratori, N.: A fuzzy admission control scheme for high-speed networks. In: 12th International Workshop on Database and Expert Systems Applications, pp. 157–161 (2001)

11. Ovengalt, C.B.T., Djouani, K., Kurien, A.: A fuzzy approach for call admission control in LTE networks. Procedia Comput. Sci. **32**, 237–244 (2014)
12. Barolli, L., Durresi, M., Sugita, K., Durresi, A., Koyama, A.: A CAC scheme for multimedia applications based on fuzzy logic. In: 19th International Conference on Advanced Information Networking and Applications, pp. 473–478 (2005)

Implementation of DVB-S2X Super-Frame Format 4 for Wideband Transmission

Christian Rohde[1]([✉]), Holger Stadali[1], Javier Perez-Trufero[2], Simon Watts[2], Nader Alagha[3], and Riccardo De Gaudenzi[3]

[1] Fraunhofer Institute for Integrated Circuits (IIS), Am Wolfsmantel 33, 91058 Erlangen, Germany
{Christian.Rohde,Holger.Stadali}@iis.fraunhofer.de
http://www.iis.fraunhofer.de/dvbs2x
[2] Avanti Communications, Cobham House, 20 Black Friars Lane, London, UK
{javier.pereztrufero,simon.watts}@avantiplc.com
[3] European Space Agency (ESA), ESA ESTEC, Noordwijk, The Netherlands
{Nader.Alagha,Riccardo.de.Gaudenzi}@esa.int

Abstract. Recently the extension of the digital video broadcasting second generation standard for transmission over satellite (DVB-S2) has been finalized in order to achieve a higher spectral efficiency without introducing fundamental changes to the complexity and structure of the common DVB-S2 standard [1]. Therefore, this extension is termed DVB-S2X. In this paper, we focus on a more powerful physical layer frame structure, known as Super-Frame (SF), which has been adopted as optional waveform container in Annex E of the DVB-S2X specification [2]. The paper provides insights to capabilities of the SF structure in support of emerging system applications. Analytical results of the SF performance are complemented by the performance results obtained from an end-to-end testbed implementing SF format 4, which is optimized for wideband transmission and very low SNR reception conditions. The testbed includes prototype design of modulator and demodulator featuring the SF generation and detection capability. The prototype devices are able to operate at a wide range of signal-to-noise ratios and at high symbol rates. This design represents the basis for planned over-the-air tests using a single wideband satellite transponder to demonstrate the feasibility of transmitting and receiving 1 Gbit/s.

Keywords: Digital video broadcasting (DVB) · Super-frame synchronization · Wideband transmission · Very low SNR · Broadcast satellite systems (BSS)

1 Introduction to DVB-S2X Super-Framing

The recently introduced DVB-S2X standard [2] keeps the physical layer frame (PLFRAME) structure from DVB-S2 [1]. It results in variable frame size depending on chosen modulation and coding (MODCOD) and pilots On/Off signaled

© Institute for Computer Sciences, Social Informatics and Telecommunications Engineering 2015
P. Pillai et al. (Eds.): WiSATS 2015, LNICST 154, pp. 373–387, 2015.
DOI: 10.1007/978-3-319-25479-1_28

by the slightly modified physical layer header (PLH). The key innovations are new MODCODs for finer granularity in spectral efficiency and an extended SNR range, lower roll-offs compared to DVB-S2, a special frame for very low signal-to-noise ratio (VL-SNR) burst mode reception, and channel bonding to allow coordinated data spreading over different signals [4].

The use of variable coding and modulation (VCM) and adaptive coding and modulation (ACM) schemes may introduce additional complexity at the receiver due to the time-variant PLFRAME size. Furthermore, the non-constant PLFRAME size dramatically complicates the implementation of techniques such as multi-user multiple-input multiple-output (MU-MIMO) transmission over satellite [5] or beam-hopping/-switching in multi-spot-beam satellite systems [6]. Therefore, the SF concept has been included to the DVB-S2X specification providing additional regular framing structures in order to enable these and future techniques as well as to support receiver synchronization in severe channel conditions [3].

1.1   Common Components of Super-Frames

The SF concept follows a simple rule to provide a common container that allows hosting different format-specific contents as illustrated in Fig. 1. In essence, each SF consists of exactly 612,540 physical layer symbols that are split among the following fields:

- Start of SF (SOSF) is composed of 270 symbols containing a known sequence, chosen from a set of orthogonal Walsh-Hadamard (WH) sequences, to be used to detect the SF and to mark the start of the SF.
- SF format indicator (SFFI) consists of 450 symbols, embedding 4 bits of signaling information (robustly coded and spread over 450 symbols) that is used to identify 16 distinct SF formats.
- A common data/signaling field of 611,820 symbols can be allocated in several different ways to the actual physical layer frames according to the SFFI.
- The SF aligned scrambling sequence that is applied to the entire length of the SF symbols, with possibility of individual scrambling for signaling and payload elements. The scrambling sequence could also vary per beam in a multi-beam system.

All SF formats share these common elements and a constant SF length, independent of the container content. A periodic repetition of the SF helps the demodulator to perform the detection and the synchronization operation. Furthermore, from the SF format that is decoded from the SFFI, the receivers can identify the content of the SF. This allows the receivers to decode only the desired formats and to discard other SF formats while maintaining the synchronization to the SF structure and removing the carrier and symbol clock uncertainties.

Currently, five different SF formats are specified in DVB-S2X Annex E [2] to be carried in the SF container:

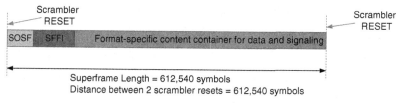

Superframe Length = 612,540 symbols
Distance between 2 scrambler resets = 612,540 symbols

- 720 symbols for SOSF + SFFI
- Format-specific allocation of 612540 − 720 = 611820 symbols

Fig. 1. Common structure of the super-frame of DVB-S2X Annex E [2].

- Format 0: DVB-S2X but with SF-aligned pilots and including the new VL-SNR frame for VL-SNR burst-mode reception
- Format 1: Legacy support of DVB-S2 but with SF-aligned pilots
- Format 2 and 3: Bundled PLFRAME formats suitable for precoding and beam-hopping applications
- Format 4: Flexible multi-purpose format optimized for wideband transmission and large SNR range

1.2 Format-Specific Components of Format 4

SF format 4 provides support for wideband carrier transmission since each PLH carries information about the PLFRAME time slicing number (or stream ID) that could be used at the receiver to apply PLFRAME filtering to decode only a target subset of all PLFRAMEs. This format is also suitable for VL-SNR operation due to the availability of PLFRAME spreading factors 2 and 5.

Format 4 offers four different PLH protection levels which are signaled by the super-frame header (SFH):

- Level 0: Standard protection (size 2 slots) using BPSK modulation and overall code rate 1/10
- Level 1: Robust protection (size 4 slots) using BPSK modulation and spreading 2 leading to an overall code rate 1/20
- Level 2: Very robust protection (size 10 slots) using BPSK modulation and spreading 5 leading to an overall code rate 1/50
- Level 3: High efficiency mode (size 1 slot) using QPSK modulation and overall code rate 1/8.75 due to puncturing

The signaling of the PLH robustness level as well as a pointer to the first PLH in the SF enable the support of VCM/ACM transmission scenarios, which includes VL-SNR applications but also high efficiency signaling for the high SNR and high throughput case. Although the protection level can change on a per SF basis, which is advantageous for traffic shaping, the most optimized PLH protection always assures such that each served receiver (with individual SNR conditions) in a SF can perform successful PLH tracking.

Finally, four different dummy frame types are defined in addition to the conventional DVB-S2 dummy frame. The four types result from the combinations of "arbitrary or deterministic content" and "regular size dummy frame or dummy frame termination at end of SF". The first criterion exploits that the signaled MODCOD of the PLH has to be respected for the dummy frame generation, which shall provide a large variety of (non-linear) channel estimation means suitable for application of predistortion techniques. The second criterion determines whether a dummy frame has regular length according to the signaled MODCOD or terminates with the end of the SF (dynamic SF padding), which is of value for SF-wise processing as required, e.g., for beam-hopping.

2   Super-Frame Synchronization Performance

The common elements of the SF (SF length, SOSF and SFFI structure) allow common processing techniques for SF detection and frame synchronization independently of the actually transmitted SF content format. A common SF detection processor solely relies on the structure as presented in Fig. 1 but needs a priori knowledge of the SOSF Walsh-Hadamard sequence index and the SF scrambler configuration. The SF synchronization can be further enhanced by utilizing the SFFI underlying spreading sequences structure.

A reliable detection of the start of SF is essential for the overall synchronization and operation of receivers adopting SF structure. Accordingly, robust detection techniques are discussed in the following to allow operation at VL-SNR and in the presence of carrier frequency and clock frequency uncertainties.

2.1   Super-Frame Detection Algorithms

The SF detection relies on correlation algorithms, similar to many other frame synchronization solutions. In the following sub-sections, different algorithms for correlation peak detection are introduced and compared in terms of robustness and detection performance taking into account channel imperfections.

Conventional Full Correlation. A conventional approach for SF detection is to carry out cross-correlation of the receiver input samples and the full-length reference sequence and compare to a threshold for correlation peak detection. Let $c[k]$ denote the correlator output at time instant k. We use the Euclidian norm of the correlator function for correlation peak analysis, i.e.

$$b[k] = \text{abs}(c[k])^2 = \text{Re}\{c[k]\}^2 + \text{Im}\{c[k]\}^2. \tag{1}$$

We also assume an additive white Gaussian noise (AWGN) present at the correlator input. Apart from a correlation peak, the correlator output $c[k]$ as a time-discrete random variable will have Gaussian distribution [7]. Thus, the noise component of $b[k]$ has a Chi-square distribution with two degrees of freedom. This noise component is defined as $n_b[k]$ with probability density function

$$f(n_b, \sigma_{n_b}) = \frac{1}{2\sigma_{n_b}^2} \cdot e^{-\frac{n_b}{2\sigma_{n_b}^2}} \tag{2}$$

for values $n_b \geq 0$. Herein, $\sigma_{n_b}^2$ denotes the variance of the noise component. Note that this also holds in case of only noise at the correlator input in absence of a signal, which results in a reduced $\sigma_{n_b}^2$. To calculate the false alarm probability $Pr(FA)$, i.e. the probability that a correlation noise sample is above the threshold instead of the true correlation peak, we use

$$Pr(FA) = Pr(n_b > b_{\text{thresh}}) = 1 - Pr(n_b \leq b_{\text{thresh}})$$

$$= 1 - \int_0^{b_{\text{thresh}}} f(n_b, \sigma_{n_b}) \, \partial n_b = e^{-\frac{b_{\text{thresh}}}{2\sigma_{n_b}^2}}. \tag{3}$$

This represents the probability that a noise sample $n_b[k]$ occurs at any sampling instant k, which is greater than a threshold b_{thresh}. Thus, we can calculate from (3) the appropriate threshold for a target false alarm probability $Pr(FA_{\text{target}})$ by

$$b_{\text{target}} = -\ln\left(Pr(FA_{\text{target}})\right) \cdot 2\sigma_{n_b}^2. \tag{4}$$

Consequently, a scaling factor $S = -\ln(Pr(FA_{\text{target}}))$ results from the target false alarm probability. Knowing that the correlation noise within $c[k]$ is Gaussian, an estimation of the actual value of $2\sigma_{n_b}^2$ is achieved by averaging over $b[k]$. Thus, an adaptive threshold results as

$$b_{\text{thresh}}[k] = S \cdot \text{mean}(b[k]) \approx S \cdot 2\sigma_{n_b}^2, \tag{5}$$

which meets b_{target} on average. This is known as Constant False Alarm Rate (CFAR) correlator peak detector [8].

Note that the operator $\text{mean}(\cdot)$ is averaging the peak detector input samples except for the ones for which the value exceeds the threshold i.e. $b[k] > b_{\text{thresh}}[k - k_{\text{guard}}]$ with e.g. $k_{\text{guard}} = 2$ or 3. This is to avoid misdetection because the correlation peak would already be part of the average for $k_{\text{guard}} = 0$ [9].

Subblock-Based Abs-Square Algorithm. The performance of full-length cross-correlation algorithm degrades considerably due to carrier frequency offset (CFO). Even in the presence of small CFOs, which are very likely to occur during the acquisition phase, the performance degradation is quite high. With increasing CFOs the correlation peak value decreases, which leads to a higher probability of misdetecting valid peaks. The correlation loss gets larger for longer sequences like the 270 symbols of the SOSF or even worse in case of the 720 symbols SOSF+SFFI length.

As a remedy, we divide the correlation window into SB subblocks. In other words, the correlator is applied to SB subblocks of the SOSF, so that each correlation subblock has a length of $L_{SB} = \lfloor \text{SOSF length}/SB \rfloor$. Consequently, $c_1[k]$ is the result of correlating the input signal with the first part of the SOSF and $c_2[k]$ is related to the second part and so on. Note, that this splitting is related to the original full length correlation by

$$c[k] = \sum_{i=1}^{SB} c_i[k - (SB - i) \cdot L_{SB}] = \sum_{i=1}^{SB} c_i'[k], \qquad (6)$$

which represents also a common method to parallelize a correlation. For the sake of easier notation the variable $c_i'[k]$ already includes the necessary delays.

As a subblock combining scheme, we can derive the absolute square of each subblock correlation result by

$$b_{\text{Abs-Square}}[k] = \sum_{i=1}^{SB} \text{abs}\left(c_i'[k]\right)^2. \qquad (7)$$

This represents a straight forward extension of the previous algorithm for the full correlation but with phase removal. Accordingly, the noise component in $b_{\text{Abs-Square}}[k]$ again features a Chi-Square distribution but here with $2 \cdot SB$ degrees of freedom. Thus, the false alarm probability is computed as [7]:

$$Pr(FA) = 1 - Pr(b_{\text{Abs-Square}}[k]) > b_{\text{thresh}}) = e^{-\frac{b_{\text{thresh}}}{2\sigma_{n_b}^2}} \cdot \sum_{k=0}^{SB-1} \frac{1}{k!} \cdot \left(\frac{b_{\text{thresh}}}{2\sigma_{n_b}^2}\right)^k. \qquad (8)$$

In analogy to (5), we calculate the detection threshold by

$$b_{\text{thresh}} = S \cdot \text{mean}(b_{\text{Abs-Square}}[k]) \approx S \cdot SB \cdot 2\sigma_{n_b}^2. \qquad (9)$$

This estimate enables to insert into (8), which yields

$$Pr(FA) = e^{-S \cdot SB} \cdot \sum_{k=0}^{SB-1} \frac{1}{k!} \cdot (S \cdot SB)^k. \qquad (10)$$

Therefore, the scaling factor S can be determined for a given target $Pr(FA)$ and number of subblocks SB with iterative methods.

Subblock-based Cross-Correlation Algorithm (XCorr). An alternative subblock combining scheme is based on conjugate complex multiplication of the subblock correlation results in analogy to the differential post integration method [10]. The scheme is abbreviated as XCorr and defined by

$$b_{\text{XCorr}}[k] = \text{abs}\left(\sum_{i=1}^{SB-1} c_i'[k] \cdot c_{i+1}'^*[k]\right). \qquad (11)$$

Due to the cross-correlation of the subblock correlation results, phase rotations due to CFO is transformed to a common delta phase, which enables a coherent summation of the cross-correlation results. In [10], $Pr(FA)$ is calculated for a given threshold but unfortunately there is no simple closed form expression of the threshold calculation for a given $Pr(FA)$ target. In order to determine a CFA rate, the scaling factor S in our case was determined empirically from the analysis of the noise histogram at a given target $Pr(FA)$.

Simulation Results. In order to analyze the impact of the CFO on the correlation peak detection performance, we first assume perfect symbol timing synchronization. Further below, we will drop this assumption to sketch a cold-start acquisition. We compare the following configurations:

- Full Corr. (270): Full Correlation w.r.t. the 270 symbols of the SOSF is applied, which is independent of the SFFI content by disregarding the SFFI.
- Full Corr. (720): Full Correlation w.r.t. the 720 symbols of SOSF+SFFI is applied, which either assumes a single target SFFI content or is realized with 16 parallel instances according to the 16 different SFFI code-words.
- Abs-Square (48 SBs): In order to exploit the SFFI structure (spreading by a factor of 30) for information removal, a subblock size of 15 symbols is selected for further analysis to ease the comparison with the XCorr algorithm. Here, the abs-operation performs information removal from the SFFI. This yields in total $18 + 30 = 48$ SBs from SOSF+SFFI for evaluation.
- XCorr (33 SBs): The maximum subblock size of the XCorr algorithm is here 15 symbols because of the pairwise conjugate complex multiplication of subblock correlation results for SFFI information removal. Although this yields 48 SBs from SOSF+SFFI, the SFFI information removal is limited to conjugate complex multiplication of disjoint pairs of subblock correlation results, which finally leads to an effective number of 15 SBs from the SFFI. This means in total $18 + 15 = 33$ SBs.

The rather small subblock size of 15 symbols shall enhance the CFO robustness. Note that a general XCorr SB size optimization w.r.t. SNR maximization under CFO is given by the so-called CHILD rule [10]: Choose the SB size close to $3/(8\nu)$ with the relative CFO $\nu = \Delta f \cdot T$. A worst-case CFO of $\nu = 0.02$ yields 18.75, which is already quite close to the chosen SB size.

The two subblock-related algorithms exploit the SFFI structure in a non data-aided (NDA) fashion opposed to the algorithm Full Corr. (720), where up to 16 parallel correlators each of 720 symbols length can be required due to the unknown SFFI sequence.

In Fig. 2, peak detection performance is shown versus E_s/N_0 for the described algorithms and different relative values of CFO. In all cases, a $Pr(FA) = 10^{-5}$ is targeted. This corresponds to 6 false peak detections on average per SF. However, due to the SF regular pattern, these false alarms can be easily discarded since their occurrence distance does not match the length of the SF. In fact, the distance between two consecutive SOSF or SOSF+SFFI detections shall be very close to an integer multiple of the SF length. The uncertainty is mainly due to potential symbol slips of a timing loop that is typically a very small portion compared to the SF length. Following the same logic, the regularity of SF structure also helps to compensate for occasional missing the peak detection since the occurring time of the start of SF can be predicted from previous SFs.

Figure 2 illustrates the sensitivity of the peak detector using full correlation algorithms to CFO values of 0.01 and 0.02. As shown in Fig. 2 the two

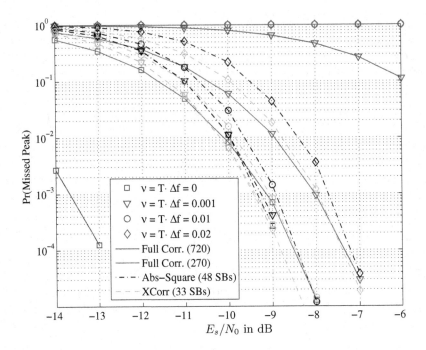

Fig. 2. $Pr(MP)$ versus E_s/N_0 of different algorithms with target $Pr(FA) = 10^{-5}$.

subblock-based algorithms (Abs-Square and XCorr) outperform the full correlation algorithms with a slight advantage for the XCorr one. Nevertheless, even the subblock-based algorithms show some notable degradation at CFO of $\nu = 0.02$.

2.2   Algorithms for SF-aided Timing Synchronization

Especially under VL-SNR conditions and low roll-offs as well as uncompensated sampling frequency offset (SFO) and CFO, a timing loop scheme [11] cannot converge properly. As a remedy, one can first start SF-detection which would suffer from the arbitrary sample timing. Thus, we describe an enhanced detection technique to maintain at 1 sample per symbol processing while tolerating the initial sample timing inaccuracy. The proposed approach implements two parallel SB related peak detection instances, instance A and B. Instance A performs best in case of sampling phase $\tau = 0$ and instance B aims for regenerating a high correlation result under the worst-case sampling phase offset of $\tau = 0.5$. Once a correlation peak grid is found, an estimate for the SFO can be calculated from the correlation peak distance and forwarded to the then activated timing loop for initialization. This provides significant support in VL-SNR configuration of the timing loop [12].

The hybrid or dual processing approach introduces additional evaluation logic after the two instances. Finally, the two hybrid or dual processing schemes are:

- Hybrid Abs-Square algorithm related processing:
  Instance A: $b_{\text{Abs-Square,A}}[k] = \sum_{i=1}^{SB} \text{abs}(c_i'[k])^2$
  Instance B: $b_{\text{Abs-Square,B}}[k] = \sum_{i=1}^{SB} \text{abs}(c_i'[k] + c_i'[k-1])^2$
- Hybrid XCorr algorithm related processing:
  Instance A: $b_{\text{XCorr,A}}[k] = \text{abs}(\sum_{i=1}^{SB-1} c_i'[k] \cdot c_{i+1}'^*[k])$
  Instance B: $b_{\text{XCorr,B}}[k] = \text{abs}(\sum_{i=1}^{SB-1} c_i'[k] \cdot c_{i+1}'^*[k] + c_i'[k-1] \cdot c_{i+1}'^*[k-1])$

Each instance uses its own threshold comparison step including threshold adaptation but the same scaling factor S is applied. A SF-detection is therefore reported if at least one of the two instances leads to a peak value above the threshold. Note that the improved $Pr(MP)$ by hybrid processing doubles the $Pr(FA)$ compared to the original approach.

In Fig. 3, we compare the $Pr(MP)$ performance for a CFO of $\nu = 0.01$, where the random sampling phase τ is assumed to be equally distributed over $\tau \in [-0.5, 0.5]$. Due to this randomness, Fig. 3 shows the mean $Pr(MP)$ which indicates averaging over τ. The worst-case roll-off of 5 % is considered and a configured target $Pr(FA) = 10^{-5}$ per each instance. Figure 3 shows a severe performance degradation compared to Fig. 2, where only $\tau = 0$ is valid. The hybrid processing (dashed curves) can improve the detection performance compared to the non-hybrid algorithms (solid curves) with a modest increase in

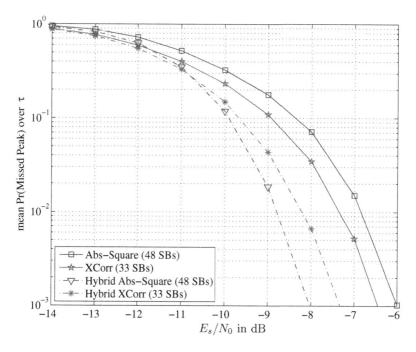

Fig. 3. Mean $Pr(MP)$ averaged over sampling phase offset τ versus E_s/N_0 for a CFO of $\nu = 0.01$ and roll-off 5 %.

the computation load. The Abs-Square approach benefits more from the hybrid processing than the XCorr approach, since the hybrid Abs-Square algorithm provides a gain of 2 dB in E_s/N_0 at mean $Pr(MP) = 10^{-3}$ compared to a hybrid XCorr gain of only 0.8 dB. However, the hybrid XCorr has been selected for implementation because it inherently provides a CFO estimate.

3 Implementation of End-to-End Wideband Transmission

3.1 Testbed in the Laboratory

A hardware testbed has been developed that implements the DVB-S2X super-framing using format 4 [13]. More specifically, the goal is to study, develop and validate the system and key technologies for the next generation of high-speed IP-based broadcasting and broadband access in future Ka band or Q/V bands satellite systems with wideband transponders. The overall structure of the wideband testbed is depicted in Fig. 4, where the involvement of companies is represented by their logos. It provides the following features and testing capabilities:

– End-to-End Multimedia Satellite Broadcasting System
– Demonstration of > 1 Gbit/s over a single wideband transponder
– Proof-of-concept for multiplexing of different service types and QoS per stream
– Symbol Rates up to 400 Mbaud for future wideband transponders
– Support of a wide SNR range from 20 dB down to −10 dB
– Realistic satellite channel models and impairments

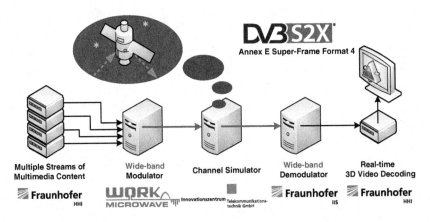

Fig. 4. Visualization of the testbed for end-to-end wideband transmission.

Features of Key Components. The key components of the testbed are the wideband modulator and demodulator. The FPGA-based platform of the modulator relies on a newly developed board. In terms of wideband capabilities, the modulator supports high-speed multi-stream processing including prioritization schemes for different inputs:

- Up to 12 ASI inputs for transport streams with each 213 Mbit/s max data rate, which can be also internally reconfigured to TSoIP-to-TS converters
- Up to 4 integrated GSE Encapsulators with each up to 400 Mbit/s IP throughput

The demodulator platform combines a Xilinx Virtex 7 evaluation board VC707 and an ADC TI ADS5400EVM within the DT4950 industry PC. The demodulator supports time-slicing, where a subset of input streams can be selected at physical layer for simultaneous decoding and decapsulation. Robust synchronization methods are implemented for reliable operation down to $-10\,$dB SNR.

Since both devices are compatible with the DVB-S2X specification implementing super-framing format 4, they support the complete roll-off range (5 %, 10 %, 15 %, 20 %, 25 %, 35 %) and all MODCODs related to QPSK, 8PSK, 16APSK, 32APSK for operation modes CCM, VCM, and ACM.

Furthermore, the channel simulator provides emulation of all effects of transmission over satellite with a signal bandwidth of up to 600 MHz. Apart from modeling satellite payload components like input/output filters or the non-linear amplifier and common implementation related impairments like phase-noise and frequency offset, rain-fading and mobility profiles can be applied to the signal.

Example Testbed Results. The test platform allows for live demonstration of broadcast transmission of advanced multimedia video/IPTV services. The wideband solution enables the transmission of a large number of channels and it supports high throughput applications such as 3DTV (Stereo at HD resolution) or multiple simultaneous HDTV resolutions (such as 720 p and 1080 p) or ultra HDTV. The receive side real-time video decoding directly reflects the impairments when QoS requirements of specific streams are not met due to worse channel conditions, e.g. switching back from 3D to 2D rendering. Moreover, the testbed allows analyzing the impact of all potential channel impairments as well as the gains due to activated counter-measures at receiver or transmitter. For example, increasing the symbol rate is limited w.r.t. the satellite filter characteristics and bandwidth introducing more and more inter-symbol interference. When applying signal equalization, the inter-symbol interference is compensated and the enhanced symbol rate directly yields increased throughput.

Examples of measurement results concerning this optimization of symbol rate R_s over transponder bandwidth B_{sat} and characteristics are shown in Fig. 5, where each point reflects the decoding threshold for a specific combination of symbol rate, roll-off, and MODCOD. Active channel impairments are satellite input/output filters and AWGN, wherefore the testbed achieves an overall implementation loss of only 0.2 to 0.3 dB w.r.t. the theoretical decoding thresholds (dashed vertical lines). One can observe for each MODCOD (colored boxes) that driving the symbol rate and therefore the achieved spectral efficiency (relative to the transponder bandwidth) leads to degrading the decoding thresholds. Therefore, higher SNR values are required corresponding to a right-shift compared to the previous point of lower symbol rate. This is due to remaining inter-symbol interference and less received signal power because of attenuation at the band edges, which reflects the trade-off for symbol rate optimization.

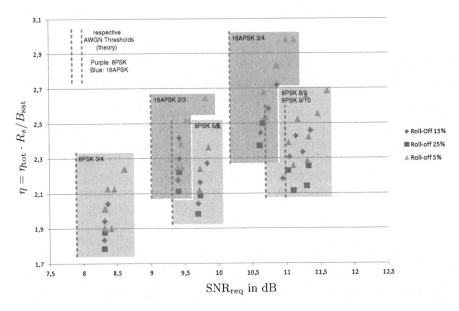

Fig. 5. Achieved spectral efficiency η relative to the satellite input/output filter bandwidth B_{sat} versus required SNR.

3.2   Planned Over-the-Air Tests

As an extension to the laboratory demonstration, an over-the-air live demonstration is also planned [14]. The main objective is to develop, integrate, verify and optimize further enhanced transmission techniques through testing over real-world wideband satellite transponders that allow for data rates over 1 Gbit/s on a single carrier per transponder transmission.

To achieve this goal, the testbed devices are further enhanced to support

- Higher order constellations (64APSK to 256APSK) according to the DVB-S2X specification
- Advanced static and dynamic predistortion techniques as well as enhanced equalization schemes
- Optimization of ACM algorithms, where the feedback loop from the receiver side to the gateway will be at IP level.

In order to qualify the availability and feasibility of a potential 1 Gbit/s link, various throughput analyses are carried out by means of link budget calculations, which depend on link impairments like weather condition statistics, satellite configuration like e.g. OBO optimization, and waveform configuration. Obviously, the final trade-off is between availability (target > 99.5 %) and net throughput on IP layer (target > 1 Gbit/s) for a given carrier.

In Fig. 6, simulation results for clear sky conditions show that the ambitious throughput targets are met in 31 % of the user locations in a beam footprint. More specifically, the 1080.7 Mbit/s physical data rate corresponds to 1019.4 Mbit/s IP data rate due to 2.67 % super-framing overhead and a 3 % worst-case GSE overhead. To achieve this, the HYLAS 2 satellite model is configured to have a carrier bandwidth of 230 MHz and 4 dB output back-off of the amplifier. For the transmit signal, the MODCOD 64APSK 5/6, a roll-off of 5 %, and a symbol rate of 219 Mbaud are used. Although these results hold for a user antenna size of 2.4 m, also tests with 1.2 m size are envisaged.

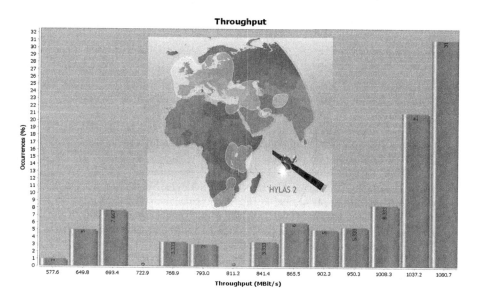

Fig. 6. Rate of throughput occurrence (derived from roll-off 5 % and different MOD-CODs) originating from randomly distributed user locations.

4   Conclusion

An introduction to the new SF format 4 of the DVB-S2X standard suitable for wideband transmission has been presented. This emphasizes that the super-frame concept opens up possibilities to improve the flexibility and resilience to distortions as well as being future proof due to the super-frame content format signaling. Specifically, the configuration features of format 4, e.g., in terms of scalability in SNR range have been shown. And simulation results have demonstrated the robustness of the waveform elements.

A hardware testbed has been established to validate the system and key technologies for the next generation of high-speed IP-based broadcasting and broadband access in future Ka band or Q/V bands satellite systems with wideband transponders, where symbol rates of up to 400 Mbaud are achieved and a wide SNR range from -10 to $20\,\mathrm{dB}$ is covered. As one example of a large variety of test and configuration cases, symbol rate optimization measurement results have been presented. Furthermore, early results for the planned over-the-air demonstration of wideband transmission show that transmission of $1\,\mathrm{Gbit/s}$ net data rate is feasible with the developed equipment.

Acknowledgments. The work on super-frame design, verification, and implementation were carried out in an European Space Agency project [13]. The planned over-the-air tests are also part of an European Space Agency project [14]. Opinions, interpretations, recommendations and conclusions presented in this paper are those of the authors and are not necessarily endorsed by the corresponding company or organization.

References

1. ETSI EN 302 307–1 V1.4.1 (2014–11), Digital Video Broadcasting (DVB); Second generation framing structure, channel coding and modulation systems (...); Part 1: DVB-S2
2. Draft ETSI EN 302 307–2 (2014–10), Digital Video Broadcasting (DVB); Second generation framing structure, channel coding and modulation systems (...); Part 2: DVB-S2 Extension (DVB-S2X)
3. Draft ETSI TR 102 376–2 V1.1.1 (2014-xx), User guidelines for the second generation system for Broadcasting, Interactive Services News Gathering and other broadband satellite applications; Part 2: S2 eXtensions (DVB-S2X)
4. Digital Video Broadcasting (DVB), White Paper on the use of DVB-S2X for DTH applications, DSNG & Professional Services, Broadband Interactive Services and VL-SNR applications. DVB Document A172 (2015)
5. Caire, G., Debbah, M., Cottatellucci, L., De Gaudenzi, R., Rinaldo, R., Mueller, R., Gallinaro, G.: Perspectives of adopting interference mitigation techniques in the context of broadband multimedia satellite systems. In: 23rd AIAA International Communications Satellite Systems Conference (ICSSC), Rome (2005)
6. Angeletti, P., Fernandez Prim, D., Rinaldo, R.: Beam hopping in multi-beam broadband satellite systems: system performance and payload architecture analysis. In: 24th AIAA International Communications Satellite Systems Conference (ICSSC), San Diego (2006)
7. Proakis, J.G.: Digital Communications, 3rd edn. McGraw Hill Inc, New York (1995)
8. Scharf, L.: Statistical Signal Processing: Detection, Estimation, and Time Series Analysis. Addison Wesley, New York (1991)
9. De Gaudenzi, R., Giannetti, F., Luise, M.: Signal recognition and signature code acquisition in CDMA receivers for mobile communications. IEEE Trans. Vehic. Tech. **47**(1), 196–208 (1998)

10. Villanti, M., Salmi, P., Corazza, G.E.: Differential post detection integration techniques for robust code acquisition. IEEE Trans. Com. **55**(11), 2172–2184 (2007)
11. Gardner, F.M.: A BPSK or QPSK timing error detector for sampled receivers. IEEE Trans. Com. **COM–34**(5), 423–429 (1986)
12. Rohde, C., Stadali, H., Lipp, S.: Flexible Synchronization Concept for DVB-S2X Super-Framing in Very Low SNR Reception. Accepted for publication at Ka Conference, Bologna (2015)
13. European Space Agency Project, Wide-band Direct to Home (WiDiHo), Contract No. 4000103596/11/NL/AD. http://artes.esa.int/projects/wide-band-direct-home-widiho-itt6613
14. European Space Agency Project, Ultra-High Throughput Transmission Through Wideband Ka Transponder (TARGETS), Contract No. 4000110170/14/NL/EM

Technology Impact on Agricultural Productivity: A Review of Precision Agriculture Using Unmanned Aerial Vehicles

H.S. Abdullahi[(✉)], F. Mahieddine, and R.E. Sheriff

Faculty of Engineering and Informatics, University of Bradford, Richmond Road, Bradford, UK
{H.S.Abdullahil,F.Mahieddine,R.E.Sheriff}@bradford.ac.uk

Abstract. Technology application to agricultural productivity is thought to be the solution to meet food demand of the growing population. In a rapidly changing world, with the prospect of decreasing arable land due to urbanization and industrialization, agricultural output requires a 70 % increase in production levels and efficient growth in the harvesting, distribution and consumption of the resources, to meet demand. There are innovations in Information and Communications Technology that can be applied to the agricultural sector in areas of precision farming, use of farm management software, wireless sensors, and use of agricultural machinery. Remote sensing technology is playing a key role through precision agriculture. This paper highlights ways in which precision agriculture is impacting on agriculture with the use of unmanned aerial vehicles for image capturing, processing and analysis.

Keywords: Agricultural productivity · Geographical information system · Global Positioning System · Image capturing · Precision agriculture · Unmanned Aerial Vehicle

1 Introduction

The application of technology to agricultural development is fast increasing and widely practiced in many parts of the world, where agriculture serves as a major source of revenue and livelihood. Technology is referred to as the application of scientific knowledge for practical purposes or the use of machinery to better facilitate a process and reduce the intensive manual labour required in agricultural production [1].

Technological applications to the agricultural sector eliminates the stress and tedious manual labour involved in agriculture. It also increases yield and aids proper management of farm input translating into output. The advantages of applying modern technology to agricultural productivity are: more people would eat better, while eradicating hunger and reducing malnutrition from increased production; improved nutrients in food; reduction in the physical pressure on the environment; improved quality of life and living standards as food costs decline, and; increase in savings, as the majority of people spend most of what they earn on food [2]. This serves as an encouragement to young individuals, not interested in the sector due to

© Institute for Computer Sciences, Social Informatics and Telecommunications Engineering 2015
P. Pillai et al. (Eds.): WiSATS 2015, LNICST 154, pp. 388–400, 2015.
DOI: 10.1007/978-3-319-25479-1_29

the subsistence methods previously employed in farming, which were considered tedious and unrewarding. Technology is also playing an important role in marketing in recent times and it is very important to the agro-industry, either at the primary or production, secondary or processing, or tertiary, involving both marketing and packaging stages. Agricultural technology can also play a vital role in solving the problems and concerns relating to the conservation and management of rural resources. Mass production, being aided by new technology and intensive marketing, can only help the agriculturist in exploiting both the domestic market and the international market to all extent, with the volume of production depending on the capital investments, and marketing strategies with the technical capacity used in the production and processing stage [3].

Space technology is playing a key role in its use for agricultural productivity through precision farming and by taking aerial views of farms. Precision farming deals with obtaining the exact readings of some particular events, such as: weather forecasts, to help to prepare lands for farming; and Global Positioning System (GPS) monitoring, and the use of the Unmanned Aerial Vehicle (UAV) systems, which can provide farmers with GPS guidance to position the applicators (Variable rate technology) by applying specific amount of chemicals to specific locations. Precision agriculture is a viable solution to agricultural production due to: (i) the ongoing reduction in the amount of arable land; (ii) projected increase in global population, and; (iii) reduction in cost of agriculture by avoiding wastage in the application of chemicals or water.

Figure 1 shows the classification of agricultural technology.

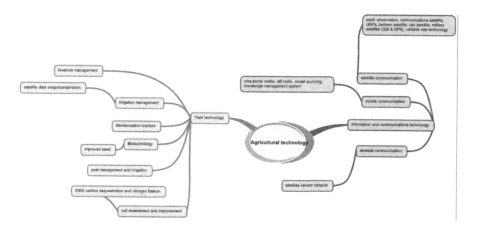

Fig. 1. Classification of agricultural technology

1.1   Precision Agriculture Using UAVs

If appropriately utilised, precision farming technology, which includes yield mapping, automatic steering and variable rate application, can greatly increase the efficiency of farm operations. Precision agriculture is majorly involved with two basic technologies, *viz.*: Geographical Information System (GIS) and GPS technology, which may eventually utilise sensors, monitors and controllers for controlling a farm's equipment, such as shaft monitors, servo motors etc.

Precision farming is a technology that affects the entire production process from extension services to management functions. It is mainly an information technology (IT) based farm management system and it involves a process of data collection, data mapping and analysis, and site specific treatment.

The components of precision agriculture are: spatial referring, crop and soil monitoring, decision support and differential action [4].

1.2 Cycle of Precision Agriculture

The cycle of precision farming involves obtaining images or data to generate map yields, weeds and topography, before application of the herbicides or fertilizer, water (irrigation) and finally to obtain results for implementation.

Before an agriculturist starts the process of precision farming, a good idea and understanding of the soil types, hydrology, micro-climates and aerial photography of the farm are required, as well as an understanding of the variable factors within the fields that effect a yield map. The yield map serves as a confirmation of data of what the farmers have, usually by simply taking an aerial photograph of the farm [4, 5].

Remote sensing used in implementing the technology include platforms such as satellites, aircraft, balloons and helicopters, and a variety of sensors such as optical and near-infrared and RADAR (Radio Detection and Ranging) installed on these platforms for its applications. Diagnostic information derived from images collected from these on-board sensors, such as biomass, Leaf Area Index (LAI), disease, water stress and lodging, can thus assist in crop management, yield forecasting, and environmental protection [6, 7].

The cycle of precision agriculture application is shown in Fig. 2:

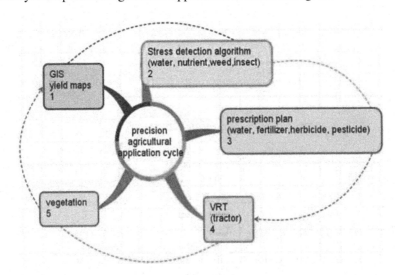

Fig. 2. The cycle of precision agriculture

The process of high resolution imagery collection involves the transmission of low cost multi hi-resolution imagery to a mobile ground station and processing centre, then via the Internet (GPS/WAAS (The Wide Area Augmentation System) based geo referencing) and to GIS (Geographical information system) processing.

Precision agriculture has effectively been in use since 2007, precisely when yield maps and target soil sampling came into use. The yield map and the target maps are used to create prescription take-off maps for phosphate and potash to apply to crops, while the Yara-N sensor is used to apply a variable rate of nitrogen.

The electromagnetic (EM) scans identify different soil types, and this layer of data is used to create variable rate seed maps with the aim to improve crop establishment. Tractors majorly use auto steer with the information provided and the sprayers have auto section control and auto boom height.

All of the different aspects of precision agriculture used have individual gains, but the auto steer used in machineries like tractors probably gives the most advantages due to production efficiency, allowing the operator to concentrate on the job being done rather than on driving the tractor. It also means that if the tramlines are established accurately, all subsequent passes will also be accurate. Variable rate application of fertilizer has allowed some savings on the farm, but more importantly the fertilizer is allocated to the correct areas of the field at the appropriate amount, which makes economic and environmental sense.

One major challenge with precision agriculture is managing and analyzing large data sets but in the future, cloud systems will be used to transfer data from machines or equipment to the office, or mobile devices and *vice-versa*. Another feature of precision agriculture that is being used is weed spotting and site specific in-crop treatments that have been effectively used for weeds, such as blackgrass and other grass weeds [5].

1.3   Benefits of Precision Agriculture/Farming

Precision agriculture helps to monitor the vegetation's physical and chemical parameters by placing electrical conductivity, temperature, nitrates, soil moisture, evapotranspiration and radiation sensors. These ensure that the optimal conditions for plant growth are achieved. This makes the field administration automated by consolidating a Decision Support System (DSS) in the precision agriculture environment in the best conditions for the particular soil and plant species and will be consequently advanced in view of the information acquired by the sensors. The Decision Support System will propose the best period for watering (or whether it is needed), the need to inundate to wash the salt substance because of an abundance in the radicular zone, the need to prepare, and so on. By presenting a precision agriculture framework in the everyday operation of an agrarian misuse, time is spared because of the up to date estimation systems. Information from the sensors is naturally transmitted to a central server and this can be consulted utilizing a Smartphone or Laptop. On the other hand, email or SMS alerts can be modified to tell the field holder when there is a need to flood, prepare or address any issue in their properties. Also, costs regarding water, pesticides and others are improved and can undoubtedly be decreased [9]. In summary, with precision agriculture, the benefits are greater sustainability, higher productivity, increases in economic benefits and environmental protection.

Precision agriculture revolves around data analysis and evaluation with its use in precision soil preparation, precision seeding, Precision crop management and precision harvesting.

1.4 Technologies Employed in Precision Agriculture

The technologies used in precision agriculture are remote sensing using satellites, UAVs, generally involving data capturing process, GIS, GPS and the Variable rate sensors. Geostatistics is an associated technology with precision agriculture, known as a branch of applied statistics that quantifies the spatial dependence and spatial structure of a measured property, and in turn uses that spatial structure to predict values of the property at unsampled locations on the field. Interpolation is a procedure generally used for predicting unknown values of neighboring location. Other technologies employed are spatial modelling, also known as variography, and spatial interpolation, otherwise called kriging [10].

2 UAV's in Precision Agriculture

The use of UAVs in agriculture is fast becoming widespread, while the implementation of aerospace engineering and sensor technology are reducing in cost. UAVs employ cameras to collect images and sensors to compile a set of data to help with monitoring and decision making on the farm.

UAVs collect data at high spatial resolutions enabling differences in crops to be compared by the centimetre rather than the metre, as in the case with satellites. They also provide immediate visual information about large areas of crops, which help farmers with fast decision making.

UAVs are small-sized electronically controlled devices launched from the side of a field or area of interest to the user. The size of UAVs usually affect the size of the payload, respectively. UAVs are guided by either a radio controller and drones generally have been made compatible with Android or Apple Smartphone applications that control the drones. Cameras attached to a UAV take pictures that can be two dimensional, with every pixel linked to a GPS location on the ground.

UAVs can also transmit live videos from flight to the receiving station on the ground. Depending on the type of camera employed, thermal images can be obtained from the UAV [11].

There are basically two types of UAVs available at the moment, as summarized in Table 1:

Table 2, below, highlights the differences between using a satellite and the UAV for obtaining images on the ground.

Figure 3 shows an example of the differences in image quality of an analyzed vegetation spot using both satellite and a UAV. The image on the left is from the satellite with the best resolution of 1 m and on the right is a UAV image highlighting the vegetation clearly with a resolution of 10 cm.

Table 1. Basic UAV Types [11, 12]

UAV	Rotary wings	Fixed wings
Flight duration	Fly duration shorter than fixed wings.	Fly up to an hour(s)
Wind pressure	Can be flown in winds gusting from 20 to 50 mph	Fly in and out of the wind rather than across the wind for satisfactory images
Flexibility in changing direction	Allow new direction during flight for re-direction	Allow new direction upload during flight for re-direction
Price range	$500 to $100,000	$500 to $100,000
Deployable option Resolution	highly deployable ± 1 cm RGB, 5 cm Multispectral	highly deployable ± 8 cm RGB and 15 cm Multispectral

Table 2. Differences between using a Satellite and a UAV for Obtaining Images on the Ground

Observation body	Satellites	UAVs
Area of coverage and usages	Global, covers large areas	Local, covers only a specific location
Air traffic restriction	Not affected by air traffic restrictions	Affected by air traffic
Operation costs, resolution	Costly, slow to task, low resolution (63 cm) and affected by cloud cover	Low operation costs, highly deployable and high resolution (4 to 10 cm), buy in costs are high but gradually falling
Regulations	No regulations	Regulations exist in certain locations like the United states.

UAV's are also useful for farmers, consultants or crop insurance agents for scouting for damages from weather events, harmful insects, diseases and chemical drift on the farm [11].

Images taken during flights are imported to GIS databases, which can be stitched together with special software to generate 3D reconstruction of fields. Normalized Differential Vegetation Index (NDVI) sees the health of crops or highlights stress areas in crops, water nutrients or weeds that they might have missed with eyes, alone. Multispectral cameras used on the UAVs allow farmers to identify crop health, perform risk mitigation and even identify soil health, while the more recent hyperspectral cameras will allow the identification of specific vegetation types in the future. Right after obtaining patterns from images, maps can be obtained from extracted GPS coordinates.

Fig. 3. Satellite image on the left with a UAV image analyzed on the right [13]

A recent approach of using UAVs that improves farmers' decision making is through mapping fields with 3D technology that provides detailed topographical images where soil changes may be present. This can also help in the future to fine tune hybrid detection for higher elevations with more water and higher organic matter [8, 9]. The general term for all flying objects used is unmanned aircraft systems, involving the use of drones, UAVs and aerial survey. It should be noted that using low cost balloons, kites and UAVs for taking aerial photography costs are low compared to the use of conventional satellites with respect to the respective requirements as seen from the table above.

Post processing of data obtained involves observation, crop type, soil moisture, biomass and irrigated land mapping, while imaging involves the use of the following sensors; visual sensor, light detector and ranging sensor, thermal infrared and multi-spectral sensors [15].

Precision technology is found to be certainly most useful in arable farming, where growers can observe fields on a square by square metre basis and apply nutrients to specific locations where required while treating disease problems. Livestock farming also profits in some ways, like the automated feeding systems used in adjusting the quantity and type of feed served to animals according to the need, weight and lactation period of the livestock [14].

The technology can be used on machineries with sensors attached, like a combine harvester spotting a bearing that is running too hot, and a field connect service that gathers information on weather, soil and crops to determine irrigation decisions.

Big data is a word that encompasses any mass of information obtained from the sensors or any recording equipment. The data/information on a farm can range from irrigation, slurry spreading, fungicide applications to dairy rations. Data have always been important in agriculture before precision agriculture, from the amount of nitrogen applied to fuel used and a tractor's last service date. Algorithms are what add real monetary value to data collection by transforming pages of machinery coding into clear, interpreted information that can be used to produce maps or yield prediction models [15].

2.1 Benefits of Using UAVs/Drones

Drones can provide farmers with detailed views making it possible to see crops from the air that can reveal patterns exposing everything from irrigation problems to soil variation and even pest and fungal infestations that are not apparent to the human eye. Airborne cameras can take multispectral images, capturing data from the infrared as well as the visual spectrum. These can be combined to provide details of healthy and distressed plants. Drones can be used anytime as desired by the farmer (weekly, monthly, hourly etc.) [16].

Most successes recorded using UAVs are crop scouting, weed management and livestock monitoring. Data implementation and analysis using broadband mobile communications are expected to provide live streaming and also have possibilities for robust images, such as weed mapping [17].

2.2 Limitations of UAVs

The low costs of UAVs (the cost of purchasing a drone is just a part and not inclusive of the cost of processing, collecting and producing images) limit the sensor payload, both in weight and dimension, and they are also not as stable as the high-end sensors resulting in reduced image quality. Since low cost UAVs are usually equipped with light weight engines, this limits their reachable altitude.

Also, the available commercial software packages applied for photogrammetric data processing are set up to support UAV images, as there are no standardized work flows and sensor models currently being implemented. UAV images do not also benefit from the sensing and intelligent human features that limits them from taking actions based on unexpected situations, like avoiding collisions with an oncoming flying object and also requires full understanding before it can be deployed with maximum interaction with its receiving station, usually a laptop.

The frequency of flying UAVs also has to be carefully selected and, finally, there are insufficient regulations on flying UAVs and they are restricted in certain regions as a security precaution [18]. Some other challenges are their inability to take readings during extreme weather conditions like rain.

2.3 Analyzing UAV Images

The processing of images involves starting with obtaining a small inexpensive UAV that maps the fields at high resolution with a multi spectral camera, then the images are geo-referenced and auto-rectified to provide an accurate representation of the field. Now the location of the weed patches is relayed to a larger helicopter type UAV that spot sprays the patches with herbicides, pesticides or even water for irrigation. An existing model for that purpose is the YAMAHA RMAX, which does not take over the conventional crop spraying method but helps to ease spraying on a spot application basis. Crop scouting can be said at this point to be the best or highest use of UAVs, while the aerial imagery is just to locate problem areas [19, 20].

The GPS/INS data onboard the UAV and the ground receiving stations are required to obtain images or predefined points for georeferencing or mosaicking [21, 22]. Other uses include monitoring crop progress before harvest, which cannot be easily done due to weather instability, but the UAV tracks plant growth with a high level of accuracy. In the near future, it is projected that UAVs will be able to observe, monitor and forecast the spread of fungus-like organisms that endanger crops [15]. The Normalized Difference Vegetation Index (NDVI), the Soil Adjusted Vegetation Index (SAVI), and the Green NDVI, which are all vegetation indices, are being used extensively to study various biological parameters quantifiably [22, 23].

NDVI shows the difference between regular red light reflected from plants and near infrared light. Healthy chlorophyll absorbs the red light and reflects near infra-red, while damaged chlorophyll reflects both [24]. One way to make a cheap consumer 3D camera with two lenses is by modifying one for near infra-red [25]. UAV images are usually processed using open source software like the digital terrain model processing tools. The images have to be rectified and merged to an ortho-mosaic for further analysis, while the best result is achieved by photogrammetric processing. Single image processing is used in place of photogrammetric processing when there is no regular image block alignment. The single image rectification involves correcting the optical distortion using a third degree polynomial approach and then rectifying using a nonparametric rectification approach and then to mosaic processing. Image blending is an effective way of producing high quality mosaics, while noting that radiometric variations of overlapping views occur often with UAV images. An improved scale invariant feature transform (SIFT) algorithm is mostly used for UAV imagery but has a disadvantage of memory consumption

Images can be completely processed by using a close range Vision Measurement Software (VMS) photogrammetry software and GOTCHA image matching. Open source GIS software used in UAV images are the Qgis and the open jump [26]. These images can then be compared to the aerial photograph taken on the ground with set GPS coordinates to map out areas of specific needs of herbicide, fertilizer, water and also identification of high yield producing vegetations on the ground depending on request. Figure 4 shows an analyzed image of an aerial farm using UAV.

The figure shows the result of an analyzed image from a UAV showing details on farm status and highlighting problem areas and areas due for harvest.

3   Discussion

This paper focuses on the stages involved in using UAVs to monitor crop growth/health status during growth seasons ensuring that the crops or plants obtain the required amount of treatment needed to avoid losses, while increasing production growth rate. It is concluded from the above that technological applications are the solution to an increased agricultural production required to feed the growing population with the limited available resources.

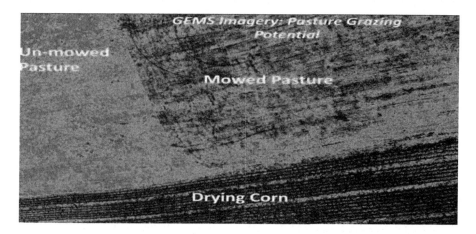

Fig. 4. An analyzed aerial image of an Agricultural Farm showing its status [27]

Ensuring that the specific quantity of required farm input resources like fertilizers, insecticides, herbicides, water to plants and the reduction in farm labour activities are the key areas where the technology has been effective. These technologies also ensure improved soil fertility before planting to post-harvest activities.

From research, it is observed that policies are the key playing factors in the agricultural sector from the land use act, price standardization in farm land inputs, stable market output prices, land ownership acts (laws on small lands and large farms), laws on maintaining the environment where farms are located, provision of credit facilities to farmers for machinery and farm input resource purchase, ensuring the basic standard of living for farmers, etc. All of these form standards making the application of technological services possible. Policy implementation has made remarkable differences in the United Kingdom and Brazil in their production rates, making their economies food sufficient and also exporters of their major food crops. These policies, however, differ from region to region depending on existing laws on agricultural practices.

However, a very clear understanding of these agricultural technologies is usually the first step in its implementation. The costs of applying these technologies are high and not affordable by the majority of farm land owners, especially small scale farmers, but from research, it is observed that the amount required for using the technology is less than losses accounted on harvested products. Also, nutrients in food crops are lower as compared to when technology was applied by ensuring the over usage of chemicals on farmlands and other input resources.

Precision agriculture on the other hand using UAVs is a relatively new technology and strict regulations are still enforced in certain regions about flying objects but the advantages offered surpasses its costs and disadvantages. Data implementation and analysis are areas of great importance, while the software for processing the UAV images are not readily available and difficult to utilise.

Future research work should focus on reducing the cost of UAVs and that of implementing the technology by using low cost components for obtaining high resolution images. Also, continued research is required in the software processing or UAV photogrammetry, which actually generates the result of the true health status of crops at every stage of germination.

With all of these facts, UAVs should be commercialized for agricultural purposes and laws should be enforced to permit its usage following its relevance to the sector.

4   Conclusion

UAV technology in precision agriculture is a robust, timely, cost effective way to obtain viable data on the farm to improve yields and overall profitability in sustainable farming systems. It saves time, increases yields and provides return on investment. The future of agricultural production with less human labour, increased productivity, reduced losses on harvest and planting, all depend largely on these technologies and should be given due consideration in every ramification. The control of the UAVs should be automated for both flying and landing purposes with less human intervention. Also, this technology should be very familiar to farmers and hence extension programmes should be arranged to teach them how to fully utilise and implement the technology.

5   Future Research Work

The future research work here involves the use of multiple sensors to monitor vegetation, ranging from thermal, multispectral, hyperspectral, optoelectronic, crop reflectance to photo electric. This produces results to measure soil nutrient content, detect water stress detection, leaf area index, and chlorophyll index.

These biophysical parameters are then estimated using NDVI, NGRDI (Normalized green red difference index), and ExG (excess green index) vegetation indices before validation. All of which is in an effort to avoid post-harvest loss, pestilence and diseases by monitoring yield, for a uniform soil fertility, which gives maximum production. Also for high profitability by employing low input costs, especially fertilizer N applied to crops, for sustainability and protecting the environment by greatly avoiding ground water contamination and delaying deterioration. The next phase of this research will involve obtaining results from a UAV in a particular field with the necessary simulations, computer model and the appropriate decision support system for real-time applications by variable rate applicators or manual applications, depending on the demand. This phase will be purely technical involving algorithms, processing and analyzing images, while presenting results to the field or field trials. Having performed all field trials for the different requirements such as estimation of nutrient availability, fertilizer N requirement, weed spotting, and irrigation requirements, appropriate seedling and timing, appropriate harvest period, and specific applications will be matched for each exact requirements on a field. This involves matching platforms with specific applications to suit its exact purpose. Also, an interface will be developed where results obtained from the images are input to simulate results as to how they will affect the farm in reality. This will be developed using matlab and special software. Also algorithms will be developed to fully analyze crop health and define specific amount requirements in the plots, which can finally be used in the application maps.

References

1. Esfahani, L., Asadiyeh, Z.: The role of information and communication technology in agriculture. In: The 1st International Conference on Information Science and Engineering (ICISE2009), pp. 3528–3531 (2009)
2. Wiggins, S.: Agriculture, Hunger and Food Security. DFID, United Kingdom (2004)
3. How Collected Data could Benefit your Farm - Farmers Weekly. http://www.fwi.co.uk/machinery/how-collected-data-could-benefit-your-farm.htm
4. Davis, R.J., Ballie, C.P., Schmidt, E.J.: Precision Agriculture Technologies - Relevance and Application to Sugarcane Production. National Centre for Engineering in Agriculture Publication, Toowomba, Queensland (2007)
5. What Is Precision Farming? E. S. Government of Alberta, Alberta Agriculture and Rural Development, Policy and Environment, Taber, Alberta, Canada (1997)
6. Zhang, C., Kovacs, J.M.: the application of small unmanned aerial systems for precision agriculture: a review. Precis. Agric. **13**, 693–712 (2012)
7. Haboudane, D., Miller, J.R., Pattey, E., Zarco-Tejada, P.J., Strachan, I.B.: Hyperspectral vegetation indices and novel algorithms for predicting green LAI of crop canopies: modeling and validation in the context of precision agriculture. Remote Sens. Environ. **90**(3), 337–352 (2004)
8. Farm Managers give the Lowdown on Precision Technology - Farmers Weekly. http://www.fwi.co.uk/business/farm-managers-give-the-lowdown-on-precision-technology.htm
9. Benefits of Precision Agriculture to Increase your Field Productivity – Iris. http://iris.cat/5-benefits-of-precision-agriculture-to-increase-your-field-productivity/
10. Davis, R.J., Baillie, C., Schmidt, E.: Precision Agriculture Technologies-Relevance and Application to Sugarcane Production, Agric. Technol. a Chang. Clim, pp. 114–122 (2009)
11. Precision Agriculture & Hyperspectral Sensors - Surface Optics Corporation. http://surfaceoptics.com/applications/precision-agriculture-hyperspectral-sensors/
12. Technology - URSULA Agriculture. http://www.ursula-agriculture.com/how-we-work/technology/
13. Spying on Europe's Farms with Satellites and Drones - BBC News. http://www.bbc.co.uk/news/world-europe-16545333
14. Grenzdörffer, G., Engel, A., Teichert, B.: The Photogrammetric Potential of Low-Cost UAVs in Forestry and Agriculture. Int. Arch. Photogramm. Remote Sens. Spat Inf. Sci. **1**, 1207–1213 (2008)
15. Technology: Drones Could Change Face of No-Tilling. http://www.precisionfarmingdealer.com/articles/393-technology-drones-could-change-face-of-no-tilling
16. Agriculture. http://www.sgaimaging.com/agriculture/
17. Cheap Drones Give Farmers a New Way to Improve Crop Yields I MIT Technology Review. http://www.technologyreview.com/featuredstory/526491/agricultural-drones/
18. Eisenbeiß, H.: UAV Photogrammetry. University of Technology, Dresden (2009)
19. Autonomous Systems: Opportunities and Challenges for the UK. https://connect.innovateuk.org/documents/3299324/6049173/Autonomous+Systems+Report.pdf/d05a4cd6-fa91-4f4c-99b3-08b672dd967c
20. Specifications I Yamaha RMAX. http://rmax.yamaha-motor.com.au/specifications

21. Laliberte, A., Goforth, M., Steele, C., Rango, A.: Multispectral remote sensing from unmanned aircraft: image processing workflows and applications for rangeland environments. Remote Sens. **3**(11), 2529–2551 (2011)
22. Klingner, S., Becker, M., Schneider, M.: Service engineering in the domain of precision farming. Precis. Agric. **13**, 793–799 (2013)
23. Hunt, E.R., Hively, W.D., Fujikawa, S.J., Linden, D.S., Daughtry, C.S.T., McCarty, G.W.: Acquisition of NIR-green-blue digital photographs from unmanned aircraft for crop monitoring. Remote Sens. **2**, 290–305 (2010)
24. Reflectivity|SSTSoftware. http://www.sstsoftware.com/products-services/imagery-services/reflectivity/
25. Ten Lessons for Farm Drones | Robohub. http://robohub.org/ten-lessons-for-farm-drones/
26. Niethammer, U., Rothmund, S., Schwaderer, U., Zeman, J., Joswig, M.: Open source image-processing tools for low-Cost UAV-based landslide investigations. In: International Archives Photogramm. Remote Sens. Spat. Inf. Sci., XXXVIII–1/, In ISPRS Zurich 2011 Workshop, 161–166 (2011)
27. GEMS: Precision Agriculture Sensor for UAV Multispectral Imaging. https://www.youtube.com/watch?v=icjvpOWijRo

Author Index

Printed in the United States
By Bookmasters